VECTOR ANALYSIS

(For Undergraduate Students of Mathematics, Physics and Engineering)

P. Duraipandian

Kayalal Pachaiyappa

S. CHAND PUBLISHING

S Chand And Company Limited

(ISO 9001 Certified Company)

S Chand And Company Limited
(ISO 9001 Certified Company)
Head Office: D-92, Sector–2, Noida – 201301, U.P. (India), Ph. 91-120-4682700
Registered Office: A-27, 2nd Floor, Mohan Co-operative Industrial Estate, New Delhi – 110 044, Phone: 011-49731800
S. CHAND PUBLISHING
www.**schandpublishing.com;** e-mail: **info@schandpublishing.com**

Marketing Offices:

Chennai	:	Ph: 23632120; chennai@schandpublishing.com
Guwahati	:	Ph: 2738811, 2735640; guwahati@schandpublishing.com
Hyderabad	:	Ph: 40186018; hyderabad@schandpublishing.com
Jalandhar	:	Ph: 4645630; jalandhar@schandpublishing.com
Kolkata	:	Ph: 23357458, 23353914; kolkata@schandpublishing.com
Lucknow	:	Ph: 4003633; lucknow@schandpublishing.com
Mumbai	:	Ph: 25000297; mumbai@schandpublishing.com
Patna	:	Ph: 4011400; patna@schandpublishing.com

First Edition 2014
Reprint 2018
Reprint 2020 (Twice), 2021, 2022, 2023, 2024 (Twice)

Reprint 2026

ISBN: 978-93-843-1969-4 **Product Code:** H6VAN61MATH10ENAA14R

PRINTED IN INDIA

By Shri Shyamlal Printing Press Private Limited, Plot 20/4, Site-IV, Industrial Area Sahibabad, Ghaziabad – 201 010 and Published by S Chand And Company Limited, A-27, 2nd Floor, Mohan Co-operative Industrial Estate, New Delhi – 110 044.

PREFACE

The book aims at a systematic development of the theory of Vector Analysis in scalar fields (Scalar valued functions), vector fields (vector valued functions), pseudoscalar fields and pseudo vector fields, so as to cater the needs of both undergraduate and postgraduate students of Mathematics, Physics, Computer Science, Computer Applications and many other Engineering Sciences. The requirements of line, surface and volume integrals theorems, have been brought out in such a manner that an average student can easily acquire a depth and comprehension in them.

Even a first reading of the book will give the reader a sufficient mastery over the subject. The feature will commend the book to the students.

The text contains many solved examples and there are numerous problems in each chapter all of which are prompted with hints.

Revision has been made on the lines suggested by Professors of Tamil Nadu Universities. The main feature of the revision has been to make the work suitable for self-study. Again some more solved examples have been added. We thank all the Professors for their valuable suggestions. We look forward for further suggestions from them.

We are thankful to the management and the editorial team of S.Chand & Company Pvt. Ltd. for help, support and guidance in publication of this book.

Thanks to all those who have helped us directly or indirectly in preparation of this project.

P. Duraipandian

Kayalal Pachaiyappa

CONTENTS

CHAPTER

1

Differentiation of Vector Functions

A vector $\bar{A} = A_x\hat{i} + A_y\hat{j} + A_z\hat{k}$ can be represented by its three components $\bar{A} = (A_x, A_y, A_z)$.

The magnitude of $\bar{A}$ is $\sqrt{A_x^2 + A_y^2 + A_z^2}$.

Lets consider a vector $\bar{A} = (A_x, A_y, 0)$ and let θ denote the angle that the vector $\bar{A}$ makes in the counterclockwise (anticlockwise) direction with the positive x-axis. Since the z-component is zero, the vector $\bar{A}$ lies in the xOy plane. The x-component and y-component are $A_x = A\cos\theta$, $A_y = A\sin\theta$. In the xOy plane, the vector $\bar{A} = A\cos\theta\,\hat{i} + A\sin\theta\,\hat{j}$.

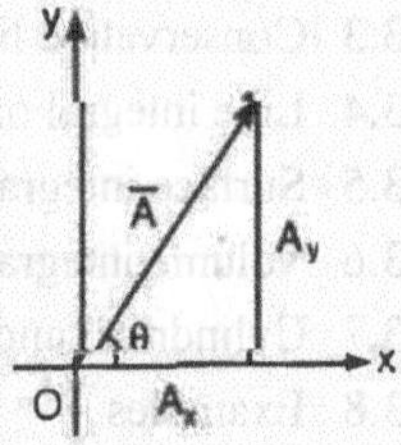

Direction of the vector $\bar{A}$ which makes an angle θ with the positive x-axis, depends on the sign of A_x and A_y in the xOy plane.

When $A_x > 0, A_y > 0, 0 < \theta < \pi/2$, the vector lies in the first quadrant.

When $A_x < 0, A_y > 0, \pi/2 < \theta < \pi$, the vector lies in the second quadrant.

When $A_x < 0, A_y < 0, \pi < \theta < 3\pi/2$, the vector lies in the third quadrant.

When $A_x > 0, A_y < 0, -\pi/2 < \theta < 0$, the vector lies in the fourth quadrant.

1.1. Vector functions

If for each value of a scalar variable u there corresponds a vector **f**, then **f** is said to be a vector function of the scalar variable u. The vector function is written more specifically as **f**(u).

Example. The vector $(a\cos u)\,\mathbf{i} + (a\sin u)\,\mathbf{j} + (bu)\,\mathbf{k}$ is a vector function of the scalar variable u.

It is evident that, in general, both the direction and magnitude of **f**(u) vary as the scalar u varies. Sometimes, in particular, **f**(u) may vary only in direction or only in magnitude.

Constant vector. A vector whose magnitude is a constant and whose direction is in a fixed direction is a constant vector.

If $\mathbf{f}(u)$ is a vector function of u and if $\mathbf{f}(u) = f_1\mathbf{i} + f_2\mathbf{j} + f_3\mathbf{k}$, then f_1, f_2, f_3 are functions of u, i.e., $\mathbf{f}(u) = f_1(u)\mathbf{i} + f_2(u)\mathbf{j} + f_3(u)\mathbf{k}$.

Note. A scalar function has only a magnitude while a vector function has both magnitude and direction. Mass, Temperature are scalar. Position, Displacement, Velocity, Acceleration, Force, Momentum and Torque are vectors.

1.2. Limit of a vector function

A vector $\mathbf{v}_0$ is said to be the limit of the vector function $\mathbf{f}(u)$, as u tends to u_0, if the limit of the scalar function $|\mathbf{f}(u) - \mathbf{v}_0|$, as u tends to u_0, is zero, that is, if $\lim_{u \to u_0} |\mathbf{f}(u) - \mathbf{v}_0| = 0$. This limit is written as

$$\lim_{u \to u_0} \mathbf{f}(u) = \mathbf{v}_0.$$

It can be established from theorems on limits of scalar functions that, if $\mathbf{f}(u) = f_1(u)\mathbf{i} + f_2(u)\mathbf{j} + f_3(u)\mathbf{k}$, then

$$\lim_{u \to u_0} \mathbf{f}(u) = \left\{\lim_{u \to u_0} f_1(u)\right\}\mathbf{i} + \left\{\lim_{u \to u_0} f_2(u)\right\}\mathbf{j} + \left\{\lim_{u \to u_0} f_3(u)\right\}\mathbf{k}$$

and consequently

$$\lim_{u \to u_0} \{\mathbf{A}(u) * \mathbf{B}(u)\} = \left\{\lim_{u \to u_0} \mathbf{A}(u)\right\} * \left\{\lim_{u \to u_0} \mathbf{B}(u)\right\},$$

where * denotes either a plus or a minus or a dot or a cross.

1.3. Derivative of a vector function

A vector function $\mathbf{f}(u)$ is said to be derivable or differentiable with respect to u if

$$\lim_{\Delta u \to 0} \frac{\mathbf{f}(u + \Delta u) - \mathbf{f}(u)}{\Delta u} \text{ exists.}$$

This limit is called the derivative or the differential coefficient of f(u) with respect to u and is denoted by $\frac{d\mathbf{f}}{du}$.

Higher derivatives of $\mathbf{f}(u)$ are defined in the same manner and are written as

$$\frac{d}{du}\left(\frac{d\mathbf{f}}{du}\right) = \frac{d^2\mathbf{f}}{du^2}, \frac{d}{du}\left(\frac{d^2\mathbf{f}}{du^2}\right) = \frac{d^3\mathbf{f}}{du^3}.$$

Remark 1. If $\mathbf{f}(u + \Delta u)$ is written as $\mathbf{f}(u) + \Delta\mathbf{f}$, then $\mathbf{f}(u + \Delta u) - \mathbf{f}(u) = \Delta\mathbf{f}$ and $\frac{d\mathbf{f}}{du} = \lim_{\Delta u \to 0} \frac{\Delta\mathbf{f}}{\Delta u}$.

Remark 2. If $\mathbf{f}(u)$ is a constant vector, then its derivative is a zero vector because

$$\mathbf{f}(u + \Delta u) - \mathbf{f}(u) = 0.$$

Remark 3. The differential coefficient of **f** with respect to u is the the rate of change of **f** with respect to u.

Book work 1.1. (*i*) if ϕ is a scalar function of u and **a**, a constant vector, then $\frac{d(\phi\mathbf{a})}{du} = a\frac{d\phi}{du}$.

(*ii*) If a also is a function of u, $\frac{d(\phi\mathbf{a})}{du} = \frac{d\phi}{du}\mathbf{a} + \phi\frac{d\mathbf{a}}{du}$. ... (1.3.1)

Proof:

$$(i) \quad \frac{d(\phi\mathbf{a})}{du} = \lim_{\Delta u \to 0} \frac{\Delta(\phi\mathbf{a})}{\Delta u} = \lim_{\Delta u \to 0} \frac{(\phi + \Delta\phi)\mathbf{a} - \phi\mathbf{a}}{\Delta u}$$

$$= \lim_{\Delta u \to 0} \frac{\Delta\phi}{\Delta u}\mathbf{a} = \frac{d\phi}{du}\mathbf{a}.$$

$$(ii)\ \frac{d(\phi \mathbf{a})}{du} = \lim_{\Delta u \to 0} \frac{\Delta(\phi \mathbf{a})}{\Delta u} = \lim_{\Delta u \to 0} \frac{(\phi + \Delta\phi)(\mathbf{a} + \Delta \mathbf{a}) - \phi \mathbf{a}}{\Delta u}$$

$$= \lim_{\Delta u \to 0} \left\{ \frac{\phi \Delta \mathbf{a}}{\Delta u} + \frac{\Delta \phi}{\Delta u} \mathbf{a} + \frac{\Delta \phi}{\Delta u} \Delta \mathbf{a} \right\} = \phi \frac{d\mathbf{a}}{du} + \frac{d\phi}{du} \mathbf{a}.$$

Geometric interpretation of the dot product $\bar{A}.\bar{B}$ is the projection of the vector $\bar{A}$ in the direction of the vector $\bar{B}$. Dot product is the product of the projection of $\bar{A}$ in the direction of $\bar{B}$, with the length of $\bar{B}$ *i.e.*, $(A \cos \theta)B$ [or the projection of $\bar{B}$ in the direction of $\bar{A}$, with the length of $\bar{A}$ *i.e.*, $(B \cos \theta)A$].

$$\text{Proj}_B A = \frac{\bar{A} \cdot \bar{B}}{|\bar{B}|} \quad \text{and } \text{proj}_A B = \frac{\bar{A} \cdot \bar{B}}{|\bar{A}|}$$

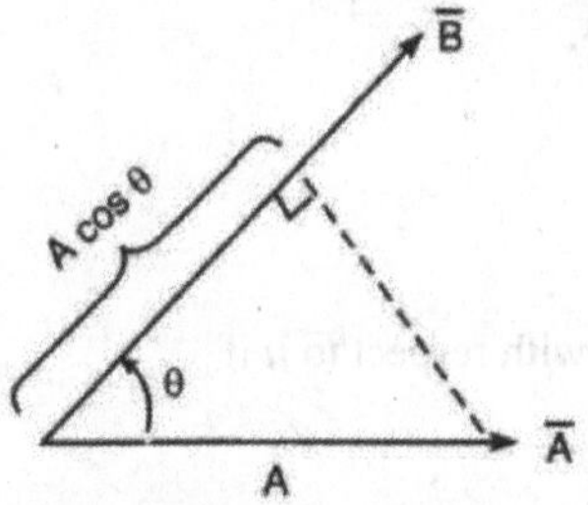

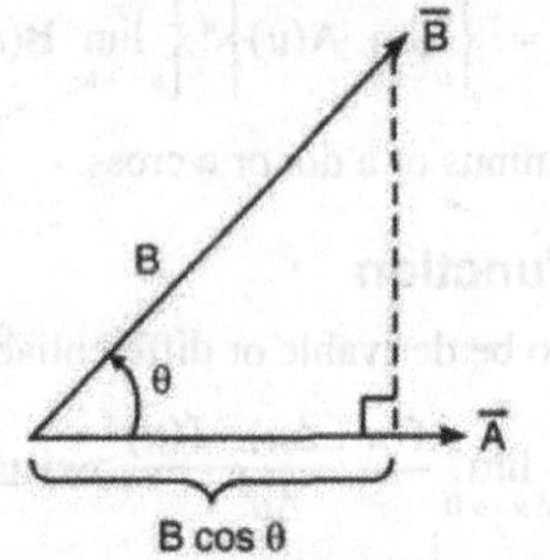

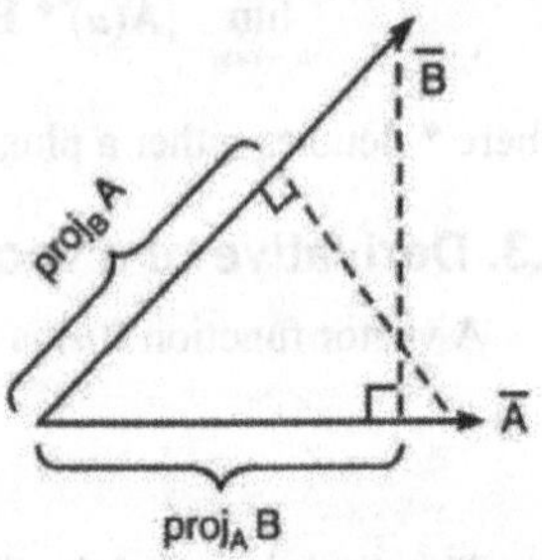

Dot product is positive when $-\pi/2 < \theta < \pi/2$, zero when $\theta = \pm \pi/2$, negative when $\pi/2 < \theta < 3\pi/2$ in case $|\bar{A}|, |\bar{B}| > 0$.

Right-hand Rule for the Direction of Cross Product

Redraw the vectors $\bar{A}$ and $\bar{B}$ so that their tails are touching (at any point in the vector field or at the origin). Draw an arc starting from the vector $\bar{A}$ and finishing on the vector $\bar{B}$. Curl your right hand fingers the same direction as the arc and make a fist. Your right thumb points in the direction of the cross product $\bar{A} \times \bar{B}$. The direction of the cross product $\bar{A} \times \bar{B}$ is perpendicular to the plane formed by $\bar{A}$ and $\bar{B}$.

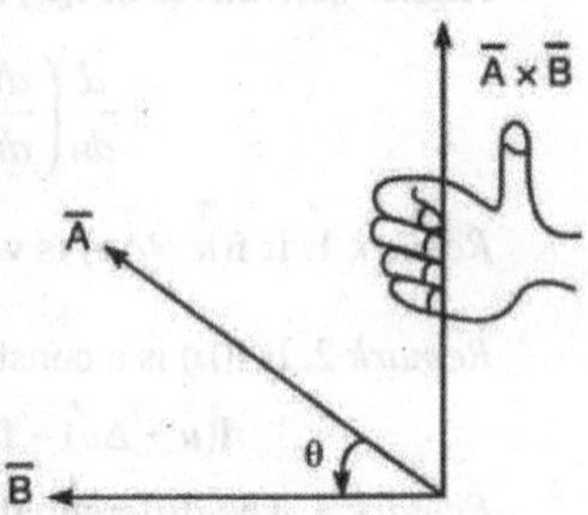

Cross product $\bar{A} \times \bar{B}$ is the product of the projection of the vector $\bar{A}$ in the direction perpendicular to the vector $\bar{B}$, with the length of $\bar{B}$ *i.e.*, $(A \sin \theta)B$ [or the projection of the vector $\bar{B}$ in the direction perpendicular to the vector $\bar{A}$, with the length of $\bar{A}$ *i.e.*, $(B \sin \theta) A$] and the unit vector $\hat{n}$ perpendicular to both $\bar{A}$ and $\bar{B}$.

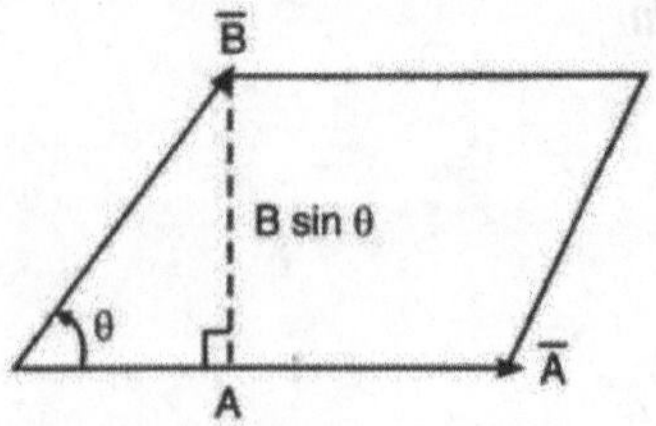

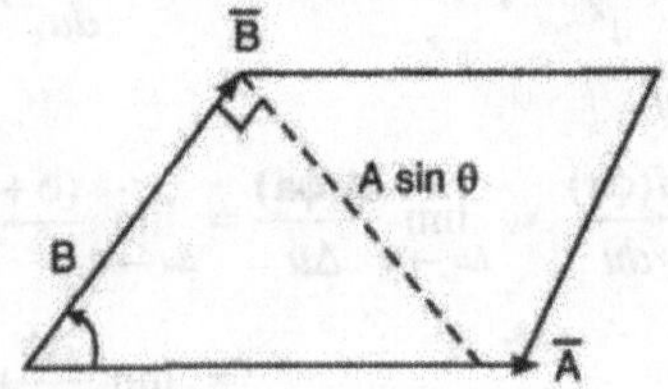

Geometric interpretation to the magnitude of the cross product is the area of the parallelogram.

Two parallel vectors do not have any component perpendicular to their common direction geometrically.

Book work 1.2. If **A**, **B** are functions of scalar variable u,

(i) $$\frac{d(\mathbf{A}+\mathbf{B})}{du}=\frac{d\mathbf{A}}{du}+\frac{d\mathbf{B}}{du}$$

(ii) $$\frac{d(\mathbf{A}\cdot\mathbf{B})}{du}=\frac{d\mathbf{A}}{du}\cdot\mathbf{B}+\mathbf{A}\cdot\frac{d\mathbf{B}}{du} \quad \text{... (1.3.2)}$$

(iii) $$\frac{d(\mathbf{A}\times\mathbf{B})}{du}=\frac{d\mathbf{A}}{du}\times\mathbf{B}+\mathbf{A}\times\frac{d\mathbf{B}}{du} \quad \text{... (1.3.3)}$$

Proof :

(i) $$\frac{d(\mathbf{A}+\mathbf{B})}{du}=\lim_{\Delta u\to 0}\frac{(\mathbf{A}+\Delta\mathbf{A})+(\mathbf{B}+\Delta\mathbf{B})-(\mathbf{A}+\mathbf{B})}{\Delta u}$$

$$=\lim_{\Delta u\to 0}\frac{\Delta\mathbf{A}}{\Delta u}+\lim_{\Delta u\to 0}\frac{\Delta\mathbf{B}}{\Delta u}=\frac{d\mathbf{A}}{du}+\frac{d\mathbf{B}}{du}.$$

(ii) $$\frac{d(\mathbf{A}.\mathbf{B})}{du}=\lim_{\Delta u\to 0}\frac{(\mathbf{A}+\Delta\mathbf{A}).(\mathbf{B}+\Delta\mathbf{B})-\mathbf{A}\,.\,\mathbf{B}}{\Delta u}$$

$$=\lim_{\Delta u\to 0}\left(A\cdot\frac{\Delta\mathbf{B}}{\Delta u}+\frac{\Delta\mathbf{A}}{\Delta u}\cdot\mathbf{B}+\frac{\Delta\mathbf{A}}{\Delta u}\cdot\Delta\mathbf{B}\right)$$

$$=\mathbf{A}\cdot\frac{d\mathbf{B}}{du}+\frac{d\mathbf{A}}{du}\cdot\mathbf{B} \qquad \left(\because \lim_{\Delta u\to 0}\Delta\mathbf{B}=\mathbf{0}\right)$$

(iii) In (ii) replace dots by crosses.

Remark 1. It is important to note here that differentiation of vector functions is similar to differentiation of scalar functions but for the fact that in results pertaining to cross products the order of the vectors is not to be changed during differentiation. This restriction is due to anti-commutative nature of vector multiplication.

Remark 2. The above results can also be established by assuming that $\mathbf{A}=a_1\mathbf{i}+a_2\mathbf{j}+a_3\mathbf{k}$, $\mathbf{B}=b_1\mathbf{i}+b_2\mathbf{j}+b_3\mathbf{k}$. As an illustration we now differentiate **A.B**.

$$\frac{d(\mathbf{A}\cdot\mathbf{B})}{du}=\frac{d(a_1b_1+a_2b_2+a_3b_3)}{du}$$

$$=\frac{da_1}{du}b_1+a_1\frac{db_1}{du}+\frac{da_2}{du}b_2+a_2\frac{db_2}{du}+\frac{da_3}{du}b_3+a_3\frac{db_3}{du}$$

$$=\left(\frac{da_1}{du}b_1+\frac{da_2}{du}b_2+\frac{da_3}{du}b_3\right)+\left(a_1\frac{db_1}{du}+a_2\frac{db_2}{du}+a_3\frac{db_3}{du}\right)$$

$$=\left(\frac{da_1}{du}\mathbf{i}+\frac{da_2}{du}\mathbf{j}+\frac{ja_3}{du}\mathbf{k}\right)\cdot(b_1\mathbf{i}+b_2\mathbf{j}+b_3\mathbf{k})+(a_1\mathbf{i}+a_2\mathbf{j}+a_3\mathbf{k})\,.\left(\frac{db_1}{du}\mathbf{i}+\frac{db_2}{du}\mathbf{j}+\frac{db_3}{du}\mathbf{k}\right)$$

$$=\left\{\frac{d(a_1\mathbf{i}+a_2\mathbf{j}+a_3\mathbf{k})}{du}\right\}\cdot(b_1\mathbf{i}+b_2\mathbf{j}+b_3\mathbf{k})+(a_1\mathbf{i}+a_2\mathbf{j}+a_3\mathbf{k})\,.\left\{\frac{d(b_1\mathbf{i}+b_2\mathbf{j}+b_3\mathbf{k})}{du}\right\}$$

$$=\frac{d\mathbf{A}}{du}\cdot\mathbf{B}+\mathbf{A}\cdot\frac{d\mathbf{B}}{du}.$$

Book work 1.3. If **A, B, C** are functions of the scalar variable u, then

(*i*) $$\frac{d}{du}[\mathbf{ABC}] = \left[\frac{d\mathbf{A}}{du},\mathbf{B},\mathbf{C}\right]+\left[\mathbf{A},\frac{d\mathbf{B}}{du},\mathbf{C}\right]+\left[\mathbf{A},\mathbf{B},\frac{d\mathbf{C}}{du}\right] \quad \text{... (1.3.4)}$$

(*ii*) $$\frac{d}{du}\{\mathbf{A}\times(\mathbf{B}\times\mathbf{C})\} = \frac{d\mathbf{A}}{du}\times(\mathbf{B}\times\mathbf{C})+\mathbf{A}\times\left(\frac{d\mathbf{B}}{du}\times C\right)+\mathbf{A}\times\left(\mathbf{B}\times\frac{d\mathbf{C}}{du}\right) \quad \text{... (1.3.5)}$$

Proof: To reduce the length we shall use 'dash' to denote differentiation.

(*i*) $$\frac{d}{du}[\mathbf{A},\mathbf{B},\mathbf{C}] = \frac{d}{du}[\mathbf{A}.(\mathbf{B}\times\mathbf{C})] = [\mathbf{A}.(\mathbf{B}\times\mathbf{C})]'$$

$$= \mathbf{A}'.(\mathbf{B}\times\mathbf{C})+\mathbf{A}.(\mathbf{B}\times\mathbf{C})' \text{ by (1.3.2)}$$

$$= [\mathbf{A}',\mathbf{B},\mathbf{C}]+\mathbf{A}.\{\mathbf{B}'\times\mathbf{C}+\mathbf{B}\times\mathbf{C}'\} \text{ by (1.3.3)}$$

$$= [\mathbf{A}',\mathbf{B},\mathbf{C}]+[\mathbf{A},\mathbf{B}',\mathbf{C}]+[\mathbf{A},\mathbf{B},\mathbf{C}'].$$

(*ii*) $$\{\mathbf{A}\times(\mathbf{B}\times\mathbf{C})\}' = \mathbf{A}'\times(\mathbf{B}\times\mathbf{C})+\{\mathbf{A}\times(\mathbf{B}\times\mathbf{C})'\} \text{ by (1.3.3)}$$

$$= \mathbf{A}'\times(\mathbf{B}\times\mathbf{C})+\mathbf{A}\times(\mathbf{B}'\times\mathbf{C}+\mathbf{B}\times\mathbf{C}')$$

$$= \mathbf{A}'\times(\mathbf{B}\times\mathbf{C})+\mathbf{A}\times(\mathbf{B}'\times\mathbf{C})+\mathbf{A}\times(\mathbf{B}\times\mathbf{C}').$$

As in the case of scalar functions, if **f**(u) is a vector function of the scalar u, where u is a function of t, then

$$\frac{d\mathbf{f}}{dt} = \frac{d\mathbf{f}}{du}\frac{du}{dt}. \quad \text{... (1.3.6)}$$

1.4 Partial derivatives of vector functions

If for each pair of values of the scalar variables u and v there corresponds a vector **f**, then **f** is said to be a vector function of u and v. Here the vector function is written specifically as **f**(u, v). Similarly vector functions of several variables are defined.

Now we define partial derivatives of a vector function of two variables. Given the vector function **f**(u, v) of the variables u and v.

$$\lim_{\Delta u\to 0}\frac{\mathbf{f}(u+\Delta u,v)-\mathbf{f}(u,v)}{\Delta u} \text{ and } \lim_{\Delta v\to 0}\frac{\mathbf{f}(u,v+\Delta v)-\mathbf{f}(u,v)}{\Delta v}$$

are called the partial derivatives of **f**(u, v) with respect to u and v respectively. These partial derivatives are denoted, as in ordinary Calculus, by the symbols $\frac{\partial\mathbf{f}}{\partial u},\frac{\partial\mathbf{f}}{\partial v}$ and the higher partial derivatives, by the symbols

$$\frac{\partial}{\partial u}\left(\frac{\partial\mathbf{f}}{\partial u}\right)=\frac{\partial^2\mathbf{f}}{\partial u^2},\ \frac{\partial}{\partial v}\left(\frac{\partial\mathbf{f}}{\partial v}\right)=\frac{\partial^2\mathbf{f}}{\partial u\partial v}, etc.$$

Similary the partial derivatives of vectors of more than two variables are defined.

1.5 Velocity of a Particle

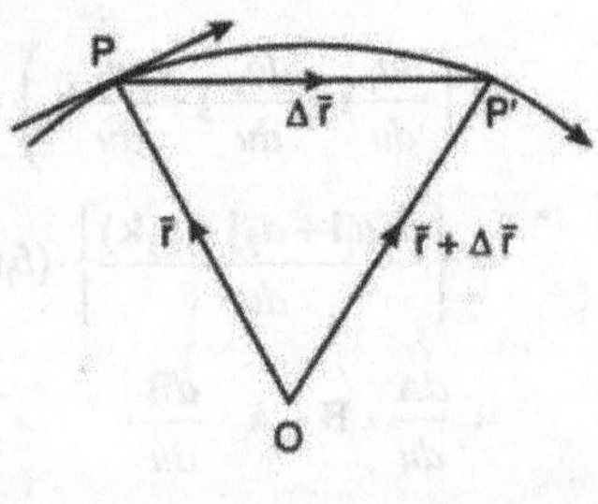

Suppose a particle moves along a curve whose vector equation is $\bar{r} = \bar{r}(t)$, where $\bar{r}$ is a function of t. Let P be its position at time t and P′ be its position at time $t + \Delta t$. Then

$$\overline{OP} = \bar{r},\ \overline{OP'} = \bar{r}+\Delta\bar{r},$$

where $\Delta\bar{r}$ is the increment in $\bar{r}$ corresponding to the increment Δt in t. Now

$$\overline{PP'} = \overline{OP'} - \overline{OP} = (\bar{r} + \Delta\bar{r}) - \bar{r} = \Delta\bar{r}.$$

This is the displacement of the particle in the time interval Δt. So the rate of displacement of the particle at P is

$$\lim_{\Delta t \to 0} \frac{\Delta\bar{r}}{\Delta t} \text{ or } \frac{d\bar{r}}{dt}.$$

But the rate of displacement is the velocity of the particle, normally denoted by $\bar{v}$. Thus

$$\bar{v} = \frac{d\bar{r}}{dt}.$$

Note. The velocity of **P** is a vector along the tangent to the curve at P.

1.6 Examples

1. Show that

$$\frac{d}{du}\left(\mathbf{A}\times\frac{d\mathbf{B}}{du} - \frac{d\mathbf{A}}{du}\times\mathbf{B}\right) = \mathbf{A}\times\frac{d^2\mathbf{B}}{du^2} - \frac{d^2\mathbf{A}}{du^2}\times\mathbf{B}.$$

2. *Solved sum.* **Find the derivatives of A . B and A × B with respect to u if**

$$\mathbf{A} = u^2\mathbf{i} + u\mathbf{j} + 2u\mathbf{k} \text{ and } \mathbf{B} = \mathbf{j} - u\mathbf{k}.$$

(*i*)
$$\frac{d}{du}(\mathbf{A}\cdot\mathbf{B}) = \frac{d\mathbf{A}}{du}\cdot\mathbf{B} + \mathbf{A}\cdot\frac{d\mathbf{B}}{du}$$
$$= (2u\mathbf{i} + \mathbf{j} + 2\mathbf{k}) \cdot (\mathbf{j} - u\mathbf{k}) + (u^2\mathbf{i} + u\mathbf{j} + 2u\mathbf{k}) \cdot (-\mathbf{k})$$
$$= (1 - 2u) + (-2u) = 1 - 4u.$$

Alternatively,
$$\frac{d}{du}(\mathbf{A}\cdot\mathbf{B}) = \frac{d}{du}\{(u^2\mathbf{i} + u\mathbf{j} + 2u\mathbf{k})\cdot(\mathbf{j} - u\mathbf{k})\}$$
$$= \frac{d}{du}(u - 2u^2) = 1 - 4u.$$

(*ii*)
$$\frac{d}{du}(\mathbf{A}\times\mathbf{B}) = \frac{d\mathbf{A}}{du}\times\mathbf{B} + \mathbf{A}\times\frac{d\mathbf{B}}{du}$$
$$= \begin{vmatrix} \mathbf{i} & \mathbf{j} & \mathbf{k} \\ 2u & 1 & 2 \\ 0 & 1 & -u \end{vmatrix} + \begin{vmatrix} \mathbf{i} & \mathbf{j} & \mathbf{k} \\ u^2 & u & 2u \\ 0 & 0 & -1 \end{vmatrix}$$
$$= (-2u - 2)\mathbf{i} + (3u^2)\mathbf{j} + 2u\mathbf{k}.$$

Alternatively,
$$\frac{d}{du}(\mathbf{A}\times\mathbf{B}) = \frac{d}{du}\begin{vmatrix} \mathbf{i} & \mathbf{j} & \mathbf{k} \\ u^2 & u & 2u \\ 0 & 1 & -u \end{vmatrix} = \frac{d}{du}\{(-u^2 - 2u)\mathbf{i} + u^3\mathbf{j} + u^2\mathbf{k}\}$$
$$= (-2u - 2)\mathbf{i} + 3u^2\mathbf{j} + 2u\mathbf{k}.$$

3. *Solved Sum* If $\bar{A} = u\bar{i} + u^2\bar{j} + u^3\bar{k}$, $\bar{B} = u^3\bar{i} + u^2\bar{j} + u\bar{k}$, find

(*i*) $\dfrac{d(\bar{A}\cdot\bar{B})}{du}$ (*ii*) $\dfrac{d(\bar{A}\times\bar{B})}{du}$.

(*i*) **Method 1.** Differentiating $\overline{A}\cdot\overline{B}$ first, we get

$$\frac{d(\overline{A}\cdot\overline{B})}{du} = \frac{d\overline{A}}{du}\cdot\overline{B}+\overline{A}\cdot\frac{d\overline{B}}{du}$$

$$=(\overline{i}+2u\overline{j}+3u^2\overline{k}).(u^3\overline{i}+u^2\overline{j}+u\overline{k})+(u\overline{i}+u^2\overline{j}+u^3\overline{k}).(3u^2\overline{i}+2u\overline{j}+\overline{k})$$

$$=[u^3+2u^3+3u^3]+[3u^3+2u^3+u^3]=12u^3.$$

Method 2. Finding $\overline{A}\cdot\overline{B}$ first, we get

$$\overline{A}\cdot\overline{B} = (u\overline{i}+u^2\overline{j}+u^3\overline{k}).(u^3\overline{i}+u^2\overline{j}+u\overline{k})$$

$$= u^4+u^4+u^4=3u^4.$$

$$\therefore \quad \frac{d(\overline{A}\cdot\overline{B})}{du} = \frac{d(3u^4)}{du}=12u^3.$$

(*ii*) Following the second method,

$$\overline{A}\times\overline{B} = \begin{vmatrix} \overline{i} & \overline{j} & \overline{k} \\ u & u^2 & u^3 \\ u^3 & u^2 & u \end{vmatrix}$$

$$= (u^3-u^5)\overline{i}+(u^2-u^6)(-\overline{j})+(u^3-u^5)\overline{k}.$$

$$\therefore \quad \frac{d}{du}(\overline{A}\times\overline{B}) = (3u^2-5u^4)\overline{i}-(2u-6u^5)\overline{j}+(3u^2-5u^4)\overline{k}.$$

4. Find $\frac{d}{du}(\mathbf{A}\cdot\mathbf{B}), \frac{d}{du}(\mathbf{A}\times\mathbf{B})$, if $\mathbf{A}=\mathbf{i}+u\mathbf{j}+u^2\mathbf{k}, \mathbf{B}=u^2\mathbf{i}-u\mathbf{j}+\mathbf{k}$.

[Ans. $2u, (1+3u^2)(\mathbf{i}-\mathbf{k})+4u^3\mathbf{j}$.**]**

5. Find $\frac{d}{du}(\mathbf{A}\times\mathbf{B})$ in the following cases :-

(*i*) $\mathbf{A}=2u\mathbf{i}+u^2\mathbf{j}, \mathbf{B}=-u\mathbf{j}+\mathbf{k}$. **(1980)**

(*ii*) $\mathbf{A}=5u^2\mathbf{i}+u\mathbf{j}-u^3\mathbf{k}, \mathbf{B}=\sin u\mathbf{i}-\cos u\mathbf{j}$. **(1981)**

[Ans. (*i*) $2u\mathbf{i}-2\mathbf{j}-4u\mathbf{k}$

(*ii*) $\mathbf{i}(-3u^2\cos u+u^3\sin u)-\mathbf{j}(3u^2\sin u+u^3\cos u)+\mathbf{k}(-11u\cos u-\sin u+5u^2\sin u)$.**]**

6. *Solved sum.* **Show that the necessary and sufficient condition for the non-zero vector function $\mathbf{f}(u)$ to be of constant magnitude is $\mathbf{f}\cdot\frac{d\mathbf{f}}{du}=0$, (*i.e.*, $\mathbf{f}$ and $\frac{d\mathbf{f}}{du}$ are perpendicular to each other).** **(1974, 80)**

Let $\mathbf{f}=f\,\hat{\mathbf{f}}$, where f is the magnitude of $\mathbf{f}$. Then $\mathbf{f}\cdot\mathbf{f}=f^2$ which on differentiation gives

$$\frac{d\mathbf{f}}{du}\cdot\mathbf{f}+\mathbf{f}\cdot\frac{d\mathbf{f}}{du} = 2f\frac{df}{du} \text{ or } \mathbf{f}\cdot\frac{d\mathbf{f}}{du}=f\frac{df}{du} \quad (1)$$

Necessity part. If f is constant, then $\frac{df}{du}=0$ and so by (1) $\mathbf{f}\cdot\frac{d\mathbf{f}}{du}=0$.

Sufficiency part. If $\mathbf{f}\cdot\frac{d\mathbf{f}}{du}=0$, then by (1) $f\frac{df}{du}=0$. But $f\neq 0$. So $\frac{df}{du}=0$ and consequently f is a constant.

7. *Solved sum.* Show that the necessary and sufficient condition for the vector function $\mathbf{f}(u)$ to have a constant direction is $\mathbf{f} \times \frac{d\mathbf{f}}{du} = \mathbf{0}$ (*i.e.*, $\mathbf{f}$ and $\frac{d\mathbf{f}}{du}$ are parallel to each other). **(1980)**

Let $\mathbf{f} = f\hat{\mathbf{f}}$. Then $\frac{d\mathbf{f}}{du} = \frac{df}{du}\hat{\mathbf{f}} + f\frac{d\hat{\mathbf{f}}}{du}$.

$$\therefore \quad \mathbf{f} \times \frac{d\mathbf{f}}{du} = \mathbf{f} \times \left\{\frac{df}{du}\hat{\mathbf{f}} + f\frac{d\hat{\mathbf{f}}}{du}\right\}$$

$$= \frac{df}{du}(f\hat{\mathbf{f}} \times \hat{\mathbf{f}}) + (f\hat{\mathbf{f}}) \times \left(f\frac{d\hat{\mathbf{f}}}{du}\right) = f^2\left(\hat{\mathbf{f}} \times \frac{d\hat{\mathbf{f}}}{du}\right). \quad \text{... (1)}$$

Necessity part. If $\mathbf{f}$ has a constant direction, then $\hat{\mathbf{f}}$ is a constant vector and

$\frac{d\hat{\mathbf{f}}}{du} = \mathbf{0}$. So $\mathbf{f} \times \frac{d\mathbf{f}}{du} = \mathbf{0}$ by (1).

Sufficiency part. If $\mathbf{f} \times \frac{d\mathbf{f}}{du} = \mathbf{0}$, then $\hat{\mathbf{f}} \times \frac{d\hat{\mathbf{f}}}{du} = \mathbf{0}$ by (1).

So, either $\hat{\mathbf{f}}$ and $\frac{d\hat{\mathbf{f}}}{du}$ are parallel or $\frac{d\hat{\mathbf{f}}}{du}$ is a zero-vector. They are not parallel but they are perpendicular. Therefore $\frac{d\hat{\mathbf{f}}}{du} = \mathbf{0}$. Hence the direction of $\hat{\mathbf{f}}$ is constant, i.e., the direction of $\mathbf{f}$ is constant.

8. *Solved sum.* Show that, if $\mathbf{f}$ is not of constant direction, then $\left|\frac{d\mathbf{f}}{du}\right| \neq \frac{d}{du}|\mathbf{f}|$.

Differentiating $f^2 = \mathbf{f}.\mathbf{f}$, we get

$$2f\frac{df}{du} = 2\mathbf{f}.\frac{d\mathbf{f}}{du} = 2|\mathbf{f}|\left|\frac{d\mathbf{f}}{du}\right|\cos\theta = 2f\left|\frac{d\mathbf{f}}{du}\right|\cos\theta,$$

where θ is the angle between $\mathbf{f}$ and $\frac{d\mathbf{f}}{du}$. So $\frac{df}{du} = \left|\frac{d\mathbf{f}}{du}\right|\cos\theta$.

But $\mathbf{f}$ is not of constant direction and so $\mathbf{f}$ and $\frac{d\mathbf{f}}{du}$ are not parallel. Therefore θ is not zero.

Consequently,

$$\frac{df}{du} \neq \left|\frac{d\mathbf{f}}{du}\right| \text{ or } \frac{d}{du}|f| \neq \left|\frac{d\mathbf{f}}{du}\right|.$$

9. *Solved sum.* Show that the necessary and sufficient condition for a vector function $\mathbf{f}(u)$ may be constant is $\frac{d\mathbf{f}}{du} = \mathbf{0}$. **(1977, 81)**

The necessity part follows from the definition of derivative. To prove the sufficiency part let

$$\mathbf{f} = f_1\mathbf{i} + f_2\mathbf{j} + f_3\mathbf{k}.$$

If $\frac{d\mathbf{f}}{du} = \mathbf{0}$, then $\left(\frac{df_1}{du}\right)\mathbf{i} + \left(\frac{df_2}{du}\right)\mathbf{j} + \left(\frac{df_3}{du}\right)\mathbf{k} = \mathbf{0}.$

So $\frac{df_1}{du}=0,\ \frac{df_2}{du}=0,\ \frac{df_3}{du}=0$. Thus f_1, f_2, f_3 are constants and consequently **f** is a constant vector.

10. *Solved Sum.* If **f** and **g** are vector functions of u such that $\mathbf{f}\times\frac{d\mathbf{g}}{du}=\mathbf{g}\times\frac{d\mathbf{f}}{du}$ for all values of *u*, show that **f** and **g** are always perpendicular to a fixed direction. **(1976, 82)**

The given condtion is

$$\bar{f}\times\frac{d\bar{g}}{du}-\bar{g}\times\frac{d\bar{f}}{du}=\bar{0} \text{ or } \bar{f}\times\frac{d\bar{g}}{du}+\frac{d\bar{f}}{du}\times\bar{g}=\bar{0}.$$

i.e., $\frac{d(\bar{f}\times\bar{g})}{du}=\bar{0}.$

Therefore $\bar{f}\times\bar{g}$ is a constant vector. So its direction is constant. By the definition of cross product, now $\bar{f}$ and $\bar{g}$ are perpendicular to this constant direction.

11. *Solved sum.* If **e** is a variable unit vector depending on *u*, show that

$$\left|\mathbf{e}\times\frac{d\mathbf{e}}{du}\right| = \left|\frac{d\mathbf{e}}{du}\right|$$

Now **e.e** = 1. Differentiating this with respect to *u*,

$$\frac{d\mathbf{e}}{du}\cdot\mathbf{e}+\mathbf{e}\cdot\frac{d\mathbf{e}}{du}=0 \text{ or } \mathbf{e}\cdot\frac{d\mathbf{e}}{du}=0.$$

∴ The angle between *e* and $\frac{de}{du}$ is 90°. Consequently

$$\mathbf{e}\times\frac{d\mathbf{e}}{du}=|\mathbf{e}|\left|\frac{d\mathbf{e}}{du}\right|\sin 90°\,\hat{\mathbf{n}}=\left|\frac{d\mathbf{e}}{du}\right|\hat{\mathbf{n}}.$$

$$\therefore\quad \left|\mathbf{e}\times\frac{d\mathbf{e}}{du}\right| = \left|\frac{d\mathbf{e}}{du}\right||\hat{\mathbf{n}}| = \left|\frac{d\mathbf{e}}{du}\right|. \qquad (\because |\hat{\mathbf{n}}|=1).$$

12. Show that, if **f** is a vector function of u, then

(*i*) $\frac{d}{du}[f, f', f'']=[\mathbf{f}, \mathbf{f}', \mathbf{f}''']$,

(*ii*) $\frac{d^2}{du^2}[\mathbf{f}, \mathbf{f}', \mathbf{f}'']=[\mathbf{f}, \mathbf{f}'', \mathbf{f}''']+[\mathbf{f}, \mathbf{f}', \mathbf{f}^{iv}]$,

where the dash denotes differentiation with respect to *u*.

13. If $[\mathbf{f}, \mathbf{f}', \mathbf{f}'']=0$, show that $\mathbf{f}\times\mathbf{f}'$ has a fixed direction and that **f** is parallel to a fixed plane.

[Hint : $(\mathbf{f}\times\mathbf{f}')]\times\frac{d}{du}(\mathbf{f}\times\mathbf{f}')=[\mathbf{f}, \mathbf{f}', \mathbf{f}'']\mathbf{f}=0$. So $\frac{d(\mathbf{f}\times\mathbf{f}')}{du}=\mathbf{0}$ and $\mathbf{f}\times\mathbf{f}'$ has a constant direction and **f** is a perpendicular to it.]

14. If $\mathbf{r}=\mathbf{a}\cos\omega t+\mathbf{b}\sin\omega t$, where **a**, **b** are constant vectors and ω, a constant scalar, show that

$$\mathbf{r}\times\frac{d\mathbf{r}}{dt}=\omega\mathbf{a}\times\mathbf{b},\ \frac{d^2\mathbf{r}}{dt^2}=-\omega^2\mathbf{r}. \qquad \textbf{(1981, 83, 85)}$$

[Hint: LHS = $\cos nt\,\bar{a}+\sin nt\,\bar{b})\times[n(-\sin nt\,\bar{a}+\cos nt\,\bar{b})]$]

15. *Solved sum.* **If *a, b, w* are vector functions of a scalar variable *u* and if $\frac{da}{du} = w\times a, \frac{db}{du} = w\times b$,**

then show that $\frac{d}{du}(a \times b) = w \times (a \times b)$. **(1975)**

$$\frac{d}{du}(a \times b) = \frac{da}{du}\times b + a\times\frac{db}{du}$$
$$= (w \times a) \times b + a \times (w \times b)$$
$$= (w.b)\, a - (a.b)w + (a.b)w - (a.w)b$$
$$= w \times (a \times b).$$

16. *Solved sum.* **t, n, b are three mutually perpendicular unit vectors whose directions vary with a scalar variable *u*. Show that t′, n′, b′ are coplanar, where the dash denotes differentiation with respect to u.**

We have to show that [t′, n′, b′] = 0.

Differentiating $t = n \times b,\ t' = n' \times b + n \times b'$.

So $[t', n', b'] = [n' \times b + n \times b', n', b']$
$$= (n'.n')(b.b') + (n.n')(b'.b') - (n'.b')(n'.b + n.b')$$
$$= (n'.n')(b.b') + (n.n')(b'.b') - (n'.b')\frac{d}{du}(n.b)$$
$$= 0 \text{ since } n.b = 0 \text{ and } n.n' = b.b' = 0 \ [\text{from } n.n = b.b = 1].$$

$\therefore$ **t′, n′, b′ are coplanar.**

17. The position vectors of points on a curve are given by $r = u^2 i - uj + (2u + 1)k$, where u is the parameter. Find the following at the point $u = 0$:

$$\frac{dr}{du}, \frac{d^2r}{du^2}, \left|\frac{dr}{du}\right|, \left|\frac{d^2r}{du^2}\right|, \frac{dr}{ds}\left(=\frac{dr}{du}\Big/\left|\frac{dr}{du}\right|\right)$$ **[Ans. $2k - j, 2i, \sqrt{5}, 2, (2k - j)/\sqrt{5}$.]**

18. Prove the following results if A and B are vector functions and ϕ, a scalar function of the scalar parameters *u* and *v* :-

(i) $\frac{\partial}{\partial u}(A+B)=\frac{\partial A}{\partial u}+\frac{\partial B}{\partial u}$ *(ii)* $\frac{\partial}{\partial u}(\phi A) = \left(\frac{\partial \phi}{\partial u}\right)A+\phi\left(\frac{\partial A}{\partial u}\right)$

(iii) $\frac{\partial}{\partial u}(A\cdot B) = \left(\frac{\partial A}{\partial u}\right)\cdot B+A\cdot\left(\frac{\partial B}{\partial u}\right)$ *(iv)* $\frac{\partial}{\partial u}(A\times B) = \left(\frac{\partial A}{\partial u}\right)\times B+A\times\left(\frac{\partial B}{\partial u}\right)$.

19. If $\bar{r} = x\bar{i} + y\bar{j} + z\bar{k}$ is a unit vector, then show that

$$x\frac{dx}{dt}+y\frac{dy}{dt}+z\frac{dz}{dt}=0.$$

[Hint: $\bar{r}\,.\,\bar{r} = 1$. Differentiating this w.r.t. *t*, $\frac{d\bar{r}}{dt}\cdot\bar{r} = 0$]

20. The P.V. $\bar{r}$ of a moving particle *P* is $\bar{r} = \cos \omega t\ \bar{i} + \sin \omega t\ \bar{j}$.

Show that the velocity vector of the particle is perpendicular to OP always.

[Hint: Show that $\overline{OP}\,.\,\bar{v} = 0$ i.e., $\bar{r}\cdot\frac{d\bar{r}}{dt} = 0$]

21. If the P.V. of a moving particle is $\bar{r} = a\cos t\,\bar{i} + a\sin t\,\bar{j} + bt\,\bar{k}$, then show that its speed is $\sqrt{a^2+b^2}$.

[**Hint:** $\bar{v} = \dfrac{d\bar{r}}{dt} = -a\sin t\,\bar{i} + a\cos t\,\bar{j} + b\,\bar{k}$. Speed $= |\bar{v}|$]

22. Find the unit tangent vector to the surface $x = t, y = t^2, z = t^3$ at $t = 1$.

[**Hint:** $\bar{r} = x\bar{i} + y\bar{j} + z\bar{k} = t\bar{i} + t^2\bar{j} + t^3\bar{k}$. So the tangent vector is $\dfrac{d\bar{r}}{dt} = \bar{i} + 2t\,\bar{j} + 3t^2\bar{k}$. At $t=1$, it is $\bar{i} + 2\bar{j} + 3\bar{k}$. Unit vector $= (\bar{i} + 2\bar{j} + 3\bar{k})/\sqrt{14}$]

Differentiation Applied to Geometry

1.7. Differential Geometry

Suppose that $\mathbf{f}(u)$ is a vector function defined on the interval $u_1 \le u \le u_2$. Reffering to a fixed point O, a point P in space can be chosen such that $\overline{OP} = \mathbf{f}(u)$. The position of **P** changes as u varies over the interval and consequently **P** traces out a curve (Fig. 1.1). So a vector function of a scalar u can be thought of to specify a curve in space. The direction in

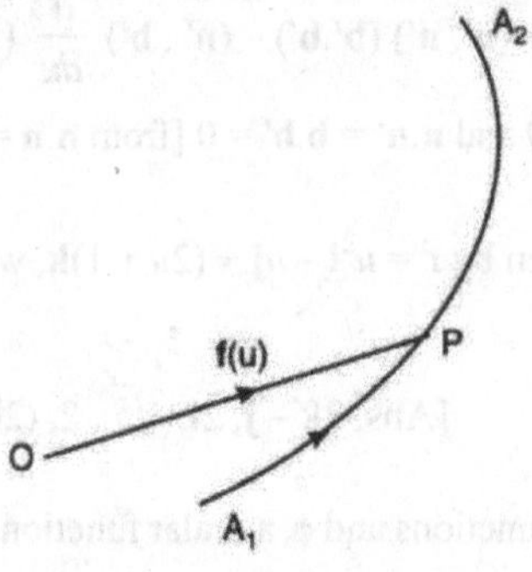

Fig. 1.1

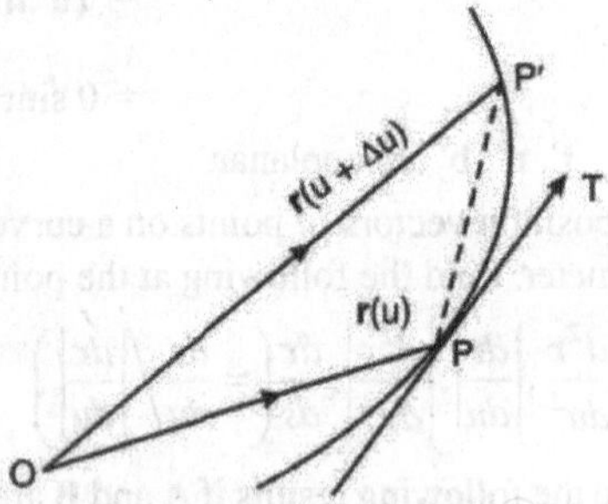

Fig. 1.2

which the curve is traced, as u increases from u_1 to u_2, is defined to be the *sense* of the curve and this sense is usually indicated by an arrow-head on the curve. The end points A_1 and A_2 of the curve correspond to the values $u = u_1$ and $u = u_2$. That is, $\overline{OA_1} = \mathbf{f}(u_1)$ and $\overline{OA_2} = \mathbf{f}(u_2)$. The equation

$$\mathbf{r} = \mathbf{f}(u), \qquad \text{... (1.7.1)}$$

where $\mathbf{r} = \overline{OP}$, is called the parameteric equation of the curve, u being the parameter.

Book Work 1.4.

To obtain Frenet-Serret formulae.

Suppose that C is a space curve whose equation is $\mathbf{r} = \mathbf{f}(u)$. If the position vectors of P and P′ on C are $\mathbf{r}(u)$ and $\mathbf{r}(u + \Delta u)$, then

$$\overline{PP'} = \mathbf{r}(u+\Delta u) - \mathbf{r}(u)$$

and
$$\lim_{\Delta u \to 0} \frac{\overline{PP'}}{\Delta u} = \lim_{\Delta u \to 0} \frac{\mathbf{r}(u+\Delta u) - \mathbf{r}(u)}{\Delta u} = \frac{d\mathbf{r}}{du}$$

In the above limit as Δu tends to zero, the direction of $\overline{PP'}$ becomes the direction of the tangent at P to the curve in the sense in which P moves as u increases. Hence the direction of $\frac{d\mathbf{r}}{du}$ is that of the tangent at P to the curve. If the unit vector in this direction is **T**, then

$$\mathbf{T} = \frac{d\mathbf{r}}{du} \bigg/ \left|\frac{d\mathbf{r}}{du}\right|. \qquad \text{... (1.7.2)}$$

If s is the arcual distance of P measured from a fixed point on C in the sense of the curve, then **r** can be expressed as

$$\mathbf{r} = \mathbf{r}(s),$$

which is called the intrinsic equation of the curve. Here a functional relation between s and u will exist such that, as u increases, s increases, and consequently $\frac{ds}{du}$ is positive.

Now
$$\frac{d\mathbf{r}}{ds} = \lim_{\Delta s \to 0} \frac{\overline{PP'}}{\Delta s} = \lim_{\Delta s \to 0} \left\{ \left(\frac{PP'}{\Delta s} \right) \hat{PP'} \right\} = \mathbf{T} \qquad \text{... (1.7.3)}$$

since
$$\lim_{\Delta s \to 0} \left(\frac{PP'}{\Delta s} \right) = 1 \text{ and } \lim_{\Delta s \to 0} (\hat{P P'}) = \mathbf{T}.$$

Also
$$\frac{d\mathbf{r}}{du} = \frac{d\mathbf{r}}{ds}\frac{ds}{du}; \text{ that is, } \frac{d\mathbf{r}}{du} = \mathbf{T}\frac{ds}{du}.$$

Hence
$$\left|\frac{d\mathbf{r}}{du}\right| = \left|\mathbf{T}\frac{ds}{du}\right| = \frac{ds}{du} \text{ or } \frac{ds}{du} = \left|\frac{d\mathbf{r}}{du}\right|. \qquad \text{... (1.7.4)}$$

Now differentiating $\mathbf{T} \cdot \mathbf{T} = 1$ with respect to s, we get

$$\frac{d\mathbf{T}}{ds} \cdot \mathbf{T} + \mathbf{T} \cdot \frac{d\mathbf{T}}{ds} = 0$$

or

$$\frac{d\mathbf{T}}{ds} \cdot \mathbf{T} = 0$$

That is, $\frac{d\mathbf{T}}{ds}$ is a vector perpendicular to the tangent at P (Fig. 1.2). The direction of this vector is called the direction of the *principal normal* to the curve at P. The unit vector in this direction is denoted by **N**. $\frac{d\mathbf{T}}{ds}$ is written as

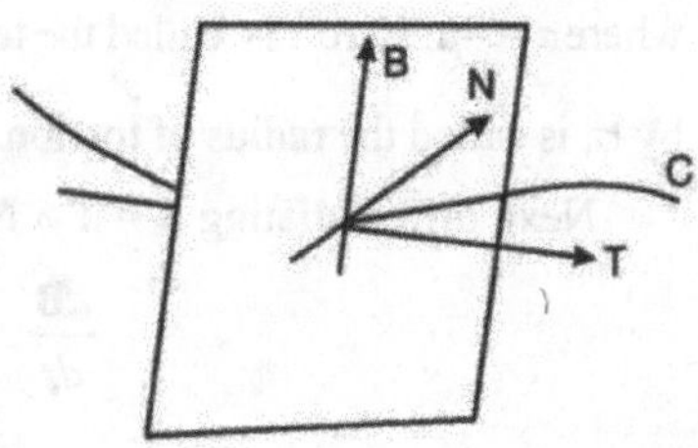

$$\frac{d\mathbf{T}}{ds} = \kappa \mathbf{N}. \qquad \text{... (1.7.5)}$$

Here κ is called the *curvature* of C at P and its reciprocal $\frac{1}{\kappa}$ is called the *radius of curvature* at P. The radius of curvature is usually denoted by ρ.

Next we define another unit vector **B** such that $\mathbf{B} = \mathbf{T} \times \mathbf{N}$. The direction of **B** is called the direction of the *binormal* at P. Now the right-handed rectangular triad **T, N, B** is called the *principal triad at* P. They specify three mutually perpendicular planes through P. Of these the plane perpendicular to the tangent **T** is called the *normal plane*, that perpendicular to the principal normal **N** is called the *rectifying plane* and that perpendicular to the binormal **B** is called the *osculating plane.*

We proceed next to discuss the nature of the derivatives of the principal triad **T**, **N**, **B** with respect to *s*, namely,

$$\frac{d\mathbf{T}}{ds}, \frac{d\mathbf{N}}{ds}, \frac{d\mathbf{B}}{ds}.$$

We already have the result $\frac{d\mathbf{T}}{ds} = \kappa\,\mathbf{N}$. Now differentiating $\mathbf{N} = \mathbf{B} \times \mathbf{T}$ with respect to *s*, we get

$$\frac{d\mathbf{N}}{ds} = \frac{d\mathbf{B}}{ds}\times\mathbf{T}+\mathbf{B}\times\frac{d\mathbf{T}}{ds}. \qquad \text{... (1.7.6)}$$

But differentiating $\mathbf{B}.\mathbf{B} = 1$, we get

$$\frac{d\mathbf{B}}{ds}\cdot\mathbf{B} = 0.$$

That is, $\frac{d\mathbf{B}}{ds}$ is perpendicular to **B**. But **T** and **N** are also perpendicular to **B**. So $\frac{d\mathbf{B}}{ds}$, **T**, **N** are coplanar. So there exist scalars λ and μ such that

$$\frac{d\mathbf{B}}{ds} = \lambda\,\mathbf{T}+\mu\mathbf{N}. \qquad \text{... (1.7.7)}$$

Now, using (1.7.6), (1.7.7) and (1.7.5), we get

$$\begin{aligned}\frac{d\mathbf{N}}{ds} &= (\lambda\,\mathbf{T}+\mu\,\mathbf{N})\times\mathbf{T}+\mathbf{B}\times(\kappa\,\mathbf{N})\\ &= \mu(\mathbf{N}\times\mathbf{T})+\kappa(\mathbf{B}\times\mathbf{N})\\ &= \mu(-\mathbf{B})+\kappa(-\mathbf{T}) = -\mu\mathbf{B}-\kappa\,\mathbf{T}\\ &= \tau\mathbf{B}-\kappa\,\mathbf{T}, \qquad \text{... (1.7.8)}\end{aligned}$$

where $\tau = -\mu$. Here τ is called the torsion of the curve at P. Its reciprocal $\frac{1}{\tau}$ which is usually denoted by σ, is called the radius of torsion.

Next, differentiating $\mathbf{B} = \mathbf{T} \times \mathbf{N}$, we get that

$$\begin{aligned}\frac{d\mathbf{B}}{ds} &= \frac{d\mathbf{T}}{ds}\times\mathbf{N}+\mathbf{T}\times\frac{d\mathbf{N}}{ds}\\ &= (\kappa\mathbf{N})\times\mathbf{N}+\mathbf{T}\times(\tau\mathbf{B}-\kappa\mathbf{T}) = \tau(\mathbf{T}\times\mathbf{B})\\ &= -\tau\mathbf{N}. \qquad \text{... (1.7.9)}\end{aligned}$$

The results (1.7.5), (1.7.8) and (1.7.9) are called Frenet-Serret formulae.

Remark 1. When **r** is given as a function of *u*, we obtain the principal triad **T**, **N**, $\dot{\mathbf{B}}$ and their derivatives $\frac{d\mathbf{T}}{ds}, \frac{d\mathbf{N}}{ds}, \frac{d\mathbf{B}}{ds}$ and κ, τ as follows:

We first obtain **T** from (1.7.2) as

$$\mathbf{T} = \frac{d\mathbf{r}}{du}\bigg/\left|\frac{d\mathbf{r}}{du}\right|$$

Next, $\dfrac{d\mathbf{T}}{ds}$ is found by using

$$\frac{d\mathbf{T}}{du} = \frac{d\mathbf{T}}{du}\Big/\frac{ds}{du}, \text{ where } \frac{ds}{du}=\left|\frac{d\mathbf{r}}{du}\right|.$$

Since $\dfrac{d\mathbf{T}}{ds} = \kappa\mathbf{N}$, we obtain $\kappa = |\kappa\mathbf{N}| = \left|\dfrac{d\mathbf{N}}{ds}\right|$ and hence $\mathbf{N} = \dfrac{1}{\kappa}\dfrac{d\mathbf{T}}{ds}\cdot$ Having found **T** and **N**, **B** is found from $\mathbf{B} = \mathbf{T}\times\mathbf{N}$. Then to find $\dfrac{d\mathbf{N}}{ds}$ and $\dfrac{d\mathbf{B}}{ds}$ we use

$$\frac{d\mathbf{N}}{ds} = \frac{d\mathbf{N}}{du}\Big/\frac{ds}{du} \text{ and } \frac{d\mathbf{B}}{du}=\frac{d\mathbf{B}}{du}\Big/\frac{ds}{du}.$$

Further, using $\dfrac{d\mathbf{B}}{ds} = -\tau\mathbf{N}$, we obtain $\tau = \left|\dfrac{d\mathbf{B}}{ds}\right|$.

As an illustration we now find ρ and σ and the principal triad of the curve, the position vector of whose general points is given by

$$\mathbf{r} = 2\cos u\mathbf{i} + 2\sin u\mathbf{j} + 2u\mathbf{k}.$$

Here $$\frac{d\mathbf{r}}{du} = 2(-\sin u\mathbf{i} + \cos u\mathbf{j} + \mathbf{k}).$$

So $$\frac{ds}{du} = \left|\frac{d\mathbf{r}}{du}\right| = \sqrt{(4\sin^2 u + 4\cos^2 u + 4)} = 2\sqrt{2}\,.$$

Hence $$\mathbf{T} = (-\sin u\mathbf{i} + \cos u\mathbf{j} + \mathbf{k})/\sqrt{2}\cdot$$

Further $$\frac{d\mathbf{T}}{ds} = \frac{d\mathbf{T}}{du}\Big/\frac{ds}{du} = \frac{1}{\sqrt{2}}(-\cos u\mathbf{i} - \sin u\mathbf{j})/2\sqrt{2}$$

$$= \frac{1}{4}(-\cos u\mathbf{i} - \sin u\mathbf{j}).$$

So $$\kappa = \left|\frac{d\mathbf{T}}{ds}\right| = \frac{1}{4}\sqrt{(\cos^2 u + \sin^2 u)} = \frac{1}{4}$$

and $$\rho = \frac{1}{\kappa} = 4.$$

Now $$\mathbf{N} = \frac{1}{\kappa}\frac{d\mathbf{T}}{ds} = 4\cdot\frac{1}{4}(-\cos u\mathbf{i} - \sin u\mathbf{j}) = -\cos u\mathbf{i} - \sin u\mathbf{j}.$$

So $$\mathbf{B} = \mathbf{T}\times\mathbf{N} = \begin{vmatrix} \mathbf{i} & \mathbf{j} & \mathbf{k} \\ -\dfrac{\sin u}{\sqrt{2}} & \dfrac{\cos u}{\sqrt{2}} & \dfrac{1}{\sqrt{2}} \\ -\cos u & -\sin u & 0 \end{vmatrix}$$

$$= \frac{1}{\sqrt{2}}(\sin u\mathbf{i} - \cos u\mathbf{j} + \mathbf{k})$$

Further $$\frac{d\mathbf{N}}{ds} = (\sin u\mathbf{i} - \cos u\mathbf{j})/2\sqrt{2}\,,$$

$$\frac{d\mathbf{B}}{ds} = \frac{1}{\sqrt{2}}(\cos u\mathbf{i} + \sin u\mathbf{j})/2\sqrt{2}$$

$$= \frac{1}{4}(\cos u\mathbf{i} + \sin u\mathbf{j}).$$

Hence $$\tau = \left|\frac{d\mathbf{B}}{ds}\right| = \frac{1}{4} \text{ and } \sigma = \frac{1}{\tau} = 4.$$

Remark 2. Whenever just κ and τ are to be calculated, the formulae found in Ex. 2, § 1.7 and in sum 1, §. 1.8 can be used.

Book Work 1.5. To obtain the equations of the tangent, principal normal, binormal, normal plane, rectifying plane and osculating plane of a given curve.

Let A be a point on the curve. Let its position vector be $\mathbf{r}_1$. To obtaine the equation of the tangent to the curve

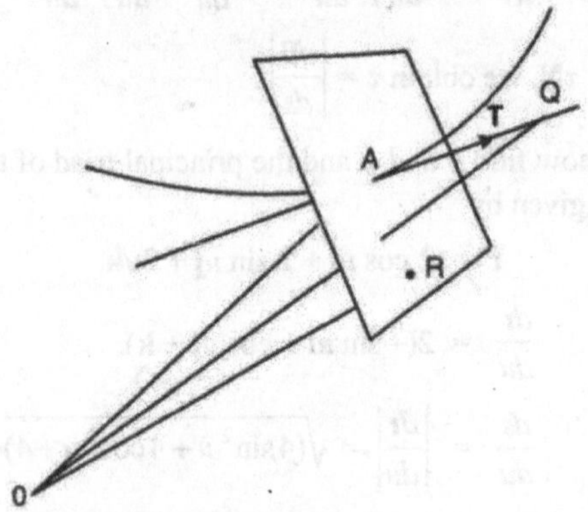

Fig 1.3

at this point take a general point Q on this tangent (Fig. 1.3). Let its position vector be $\mathbf{r}$. Then $\overline{\mathbf{AQ}} \times \mathbf{T} = \mathbf{0}$ gives the equation of the tangent as $(\mathbf{r} - \mathbf{r}_1) \times \mathbf{T} = \mathbf{0}$.

Similarly the equations of the principal normal and the binormal at A are obtained as

$$(\mathbf{r} - \mathbf{r}_1) \times \mathbf{N} = \mathbf{0} \text{ and } (\mathbf{r} - \mathbf{r}_1) \times \mathbf{B} = \mathbf{0}.$$

Also, if **R** is a point in the plane through A and perpendicular to **T**, then $\overline{AR}$ is perpendicular to **T**. Thus $\overline{AR}\,.\,\mathbf{T} = 0$. So the equation of the normal plane is

$$(\mathbf{r} - \mathbf{r}_1).\,\mathbf{T} = 0$$

Similarly the equations of the plane through A and perpendicular to **N** and that through A and perpendicular to **B** are

$$(\mathbf{r} - \mathbf{r}_1).\,\mathbf{N} = 0 \text{ and } (\mathbf{r} - \mathbf{r}_1).\,\mathbf{B} = 0.$$

Solved Examples

Example. 1. A general point on a curve has the position vector $\mathbf{r} = x(s)\mathbf{i} + y(s)\mathbf{j} + z(s)\mathbf{k}$. If the dash denotes differentiation with respect to *s*, then show with the usual meanings that

(*i*) $\rho = \{(x'')^2 + (y'')^2 + (z'')^2\}^{-1/2}$

(*ii*) $[\mathbf{r}', \mathbf{r}'', \mathbf{r}'''] = \dfrac{\tau}{\rho^2}$

(iii) $\tau = \{(x'')^2 + (y'')^2 + (z'')^2\}^{-1} \begin{vmatrix} x' & y' & z' \\ x'' & y'' & z'' \\ x''' & y''' & z''' \end{vmatrix}$

(i) $\dfrac{d\mathbf{r}}{ds} = \mathbf{T} = x'\mathbf{i} + y'\mathbf{j} + z'\mathbf{k}$ and $\dfrac{d\mathbf{T}}{ds} = x''\mathbf{i} + y''\mathbf{j} + z''\mathbf{k}$.

So
$$\kappa = \left|\frac{d\mathbf{T}}{ds}\right| = \{(x'')^2 + (y'')^2 + (z'')^2\}^{1/2}.$$

$$\rho = \frac{1}{\kappa} = \{(x'')^2 + (y'')^2 + (z'')^2\}^{-1/2} \qquad \text{... (1)}$$

(ii)
$$[\mathbf{r}', \mathbf{r}'', \mathbf{r}'''] = \left[\mathbf{T}, \frac{d\mathbf{T}}{ds}, \frac{d^2\mathbf{T}}{ds^2}\right] = \left[\mathbf{T}, \kappa\mathbf{N}, \frac{d}{ds}(\kappa\mathbf{N})\right]$$
$$= \left[\mathbf{T}, \kappa\mathbf{N}, \frac{d\kappa}{ds}\mathbf{N} + \kappa\frac{d\mathbf{N}}{ds}\right]$$
$$= \left[\mathbf{T}, \kappa\mathbf{N}, \frac{dk}{ds}\mathbf{N}\right] + \left[\mathbf{T}, \kappa\mathbf{N}, \kappa\frac{d\mathbf{N}}{ds}\right]$$
$$= 0 + [\mathbf{T}, \kappa\mathbf{N}, \kappa(\tau\mathbf{B} - \kappa\mathbf{T})]$$
$$= [\mathbf{T}, \kappa\mathbf{N}, \kappa\tau\mathbf{B}] + [\mathbf{T}, \kappa\mathbf{N}, -\kappa^2\mathbf{T}]$$
$$= \kappa^2\tau[\mathbf{T}, \mathbf{N}, \mathbf{B}] + 0 = \kappa^2\tau = \frac{\tau}{\rho^2}. \qquad (2)$$

(iii) **The result follows from (1) and (2) since $[\mathbf{r}', \mathbf{r}'', \mathbf{r}''']$ equals the determinant given in the question.**

Example. 2. Show that, in the space curve $\mathbf{r} = \mathbf{r}(t)$,
$$\kappa = |\dot{\mathbf{r}}\times\ddot{\mathbf{r}}|/|\dot{\mathbf{r}}|^3, \rho = |\dot{\mathbf{r}}|^3/|\dot{\mathbf{r}}\times\ddot{\mathbf{r}}|,$$
where t is any scalar, not necessarily the time, and the dot denotes differentiation with respect to t.

$$\dot{\mathbf{r}} = \frac{d\mathbf{r}}{dt} = \frac{d\mathbf{r}}{ds}\frac{ds}{dt} = \mathbf{T}\frac{ds}{dt}.$$

$$\ddot{\mathbf{r}} = \frac{d}{dt}\left(\mathbf{T}\frac{ds}{dt}\right) = \frac{d\mathbf{T}}{dt}\frac{ds}{dt} + \mathbf{T}\frac{d^2s}{ds^2} = \frac{d\mathbf{T}}{ds}\left(\frac{ds}{dt}\right)^2 + \mathbf{T}\frac{d^2s}{dt^2}$$
$$= \kappa\left(\frac{ds}{dt}\right)^2\mathbf{N} + \frac{d^2s}{dt^2}\mathbf{T}.$$

$$\dot{\mathbf{r}}\times\ddot{\mathbf{r}} = \left(\frac{ds}{dt}\mathbf{T}\right)\times\left\{\kappa\left(\frac{ds}{dt}\right)^2\mathbf{N} + \frac{d^2s}{dt^2}\mathbf{T}\right\} = \kappa\left(\frac{ds}{dt}\right)^3\mathbf{B}.$$

So
$$|\dot{\mathbf{r}}\times\ddot{\mathbf{r}}| = \kappa\left|\frac{ds}{dt}\right|^3 \text{ and } |\dot{\mathbf{r}}| = \left|\mathbf{T}\frac{ds}{dt}\right| = \left|\frac{ds}{dt}\right|.$$

Hence the result as stated.

Example. 3. Show that, if the torsion of a curve is zero, then the curve is a plane curve.

$d\mathbf{B}/ds = -\tau\mathbf{N} = 0$ since torsion is zero. So $\mathbf{B}$ is a unit vector in a fixed direction. Let $\overline{OP} = \mathbf{r}$ be the position vector of any point P on the curve. Now $\mathbf{B} = \mathbf{T}\times\mathbf{N}$ and so $\mathbf{r}\,.\,\mathbf{B} = [\mathbf{r}, \mathbf{T}, \mathbf{N}]$. Then

$$\frac{d}{ds}(r.\mathbf{B}) = \frac{d}{ds}[\mathbf{r}, \mathbf{T}, \mathbf{N}] = [\mathbf{T}, \mathbf{T}, \mathbf{N}] + [\mathbf{r}, \kappa\mathbf{N}, \mathbf{T}] + [\mathbf{r}, \mathbf{T}, (\tau\mathbf{B} - \kappa\mathbf{N})]$$

$$= 0 + 0 + \tau[\mathbf{r}. \mathbf{T}, \mathbf{B}] = 0.$$

Hence **r. B** is a constant.

So it is evident from the equation $\mathbf{r.n} = p$ that the given curve lies in a plane which is perpendicular to the vector **B**.

1.8 Examples.

1. Show that in a space curve $\mathbf{r} = \mathbf{r}(t)$, $\tau = \left[\dot{\mathbf{r}}, \ddot{\mathbf{r}}, \dddot{\mathbf{r}}\right] \Big/ \left|\dot{\mathbf{r}} \times \ddot{\mathbf{r}}\right|^2$, where the dot denotes differentiation with respect to t.

   [Hint : $\dot{\mathbf{r}} = \dot{s}\mathbf{T}$ and $\ddot{\mathbf{r}} = \kappa\dot{s}^2\mathbf{N} + \ddot{s}\mathbf{T}$ (Ex. 2, § 1.7.).

   $$\dddot{\mathbf{r}} = \frac{d}{dt}(\kappa\dot{s}^2\mathbf{N} + \ddot{s}\mathbf{T}) = \ldots = \alpha\mathbf{T} + \beta\mathbf{N} + \kappa\tau\dot{s}^3\mathbf{B}, \text{ where } \alpha \text{ and } \beta \text{ are scalars.}$$

   So $\qquad [\dot{r}, \ddot{r}, \dddot{r}] = [\dot{s}\mathbf{T}, \kappa\dot{s}^2\mathbf{N}, \kappa\tau^2\dot{s}^3\mathbf{B}] = \kappa^2\dot{s}^6\tau\,[\mathbf{T}, \mathbf{N}, \mathbf{B}] = \kappa^2\dot{s}^6\tau.$

   But, from Ex. 2, § 1.7.

   $$\kappa = ||\dot{\mathbf{r}} \times \ddot{\mathbf{r}}| \big/ |\dot{\mathbf{r}}|^3 = ||\dot{\mathbf{r}} \times \ddot{\mathbf{r}}| \big/ \dot{s}^3]$$

2. Show that in the following curves the radius of curvature and radius of torsion are as shown against each :

   (*i*) $\mathbf{r} = u\mathbf{i} + u^2\mathbf{j} + \frac{2}{3}u^3\mathbf{k}$, $\rho = \sigma = \frac{1}{2}(1 + 2u^2)^2$.

   (*ii*) $\mathbf{r} = (3u - u^3)\mathbf{i} + 3u^2\mathbf{j} + (3u + u^3)\mathbf{k}$, $\rho = \sigma = 3\,(1 + u^2)^2$.

   (*iii*) $\mathbf{r} = a \cos u\mathbf{i} + a \sin u\mathbf{j} + (a \cot \alpha)\, u\mathbf{k}$, $\rho = a \operatorname{cosec}^2 \alpha$, $\sigma = 2a \operatorname{cosec} 2\alpha$.

   (*iv*) $\mathbf{r} = 3 \cos u\mathbf{i} + 3 \sin u\mathbf{j} + 4\, u\mathbf{k}$, $\rho = \frac{25}{3}$, $\sigma = \frac{25}{4}$.

3. Show that the radius of curvature of the plane curve $y = f(x)$, $z = 0$ is $\frac{(1 + y_1^2)^{3/2}}{|y_2|}$, where y_1 and y_2 are the first and second derivatives of y with respect to x.

   [Hint : $\mathbf{r} = x\mathbf{i} + y\mathbf{j} = x\mathbf{i} + f(x)\mathbf{j}$. So $\frac{d\mathbf{r}}{dx} = \mathbf{i} + y_1\mathbf{j}$, $\frac{d^2\mathbf{r}}{dx^2} = y_2\mathbf{j}$. use Ex. 2, § 1.7]

4. If a plane curve has $f(x, y) = 0$, and $z = 0$ as its equations, shows that $\rho = \frac{\{(x')^2 + (y')^2\}^{3/2}}{|x'y'' - x''y'|}$, where x and y are functions of θ and the dash denotes differentiation with respect to θ.

5. Show that the radius of curvature of the ellipse $\mathbf{r} = a \cos u\mathbf{i} + b \sin u\mathbf{j}$, where a and b are constants, is $\frac{(a^2 \sin^2 u + b^2 \cos^2 u)^{3/2}}{|ab|}$.

6. Show that the curve

$$\mathbf{r} = \frac{2u+1}{u-1}\mathbf{i} + \frac{u^2}{u-1}\mathbf{j} + (u+2)\mathbf{k}$$

is a plane curve by showing that its torsion at every point is zero.

[**Hint :** Show that $[\dot{\mathbf{r}}, \ddot{\mathbf{r}}, \dddot{\mathbf{r}}] = 0$. (Vide sum 1, § 1.8)].

1.9 Partial differentiation applied to Geometry

We know from § 1.9 that a vector function of a single scalar variable specifies a space curve. Similarly a vector function

$$\mathbf{f}(u, v), (u_1 \le u \le u_2;\ v_1 \le v \le v_2)$$

specifies a space surface whose equation is

$$\mathbf{r} = \mathbf{f}(u, v); \qquad \text{... (1.9.1)}$$

for $\mathbf{r} = \mathbf{f}(u, \alpha)$, $u_1 \le u < u_2$, is a space curve which moves generating a surface as α changes from v_1 to v_2. Equation (1.9.1) is called the parametric equation of this surface, u and v being the parameters. Similarly a vector function $\mathbf{f}(u, v, w)$ of three variables specify a volume the position vectors of whose points are

$$\mathbf{r} = \mathbf{f}(u, v, w)$$

Suppose that the parametric values $u = u_0$ and $v = v_0$ specify a point P_0 on the surface

$$\mathbf{r} = \mathbf{f}(u, v), (u_1 \le u \le u_2;\ v_1 \le v \le v_2).$$

Then the equations $\mathbf{r} = \mathbf{f}(u, v_0)$, $(u_1 \le u \le u_2)$ and $\mathbf{r} = \mathbf{f}(u_0, v)$, $(v_1 \le v \le v_2)$ denote two curves on this surface passing through P_0 and as in §1.6.

$$\left\{\frac{d}{du}\mathbf{f}(u, v_0)\right\}_{u=u_0} \quad \text{and} \quad \left\{\frac{d}{dv}\mathbf{f}(u_0, v)\right\}_{v=v_0}$$

are vectors at P_0 tangent to them (Fig. 1.4). So their vector product, namely.

$$\left\{\frac{\partial}{\partial u}\mathbf{f}(u, v) \times \frac{\partial}{\partial v}\mathbf{f}(u, v)\right\}_{u=u_0, v=v_0} \qquad \text{... (1.9.2)}$$

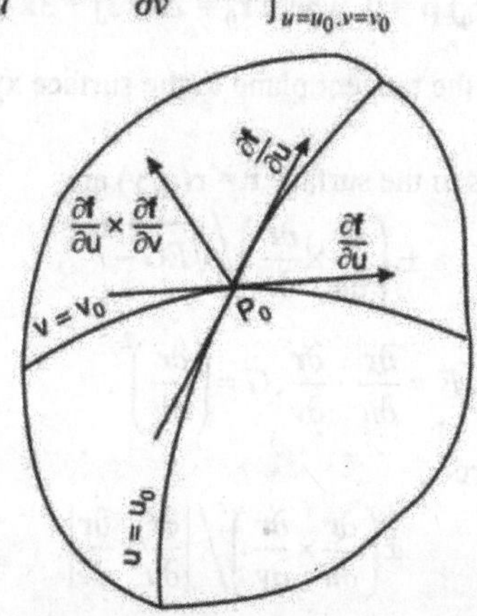

Fig. 1.4

is normal to the surface at P_0. Thus the equation of the plane tangential to the surface at P_0 whose position vector is $\mathbf{f}(u_0, v_0)$, is

$$(\mathbf{r}-\mathbf{r}_0)\cdot\mathbf{p} = 0,$$

where $\mathbf{r}_0 = \mathbf{f}(u_0, v_0)$ and $\mathbf{p} = \left\{\dfrac{\partial \mathbf{f}}{\partial u}\times\dfrac{\partial \mathbf{f}}{\partial v}\right\}_{u=u_0, v=v_0}$

1.10 Solved Examples

Example. 1. Find the equation of the tangent plane to the surface $x^2+y^2-2z=0$ at P (1, – 1, 1).

Supposing that $x = u$ and $y = v$, we get for points on the surface $2z = u^2+v^2$ or $z = \frac{1}{2}(u^2+v^2)$. That is, the parametric equation of the surface is

$$\mathbf{r} = u\mathbf{i}+v\mathbf{j}+\frac{1}{2}(u^2+v^2)\mathbf{k}.$$

Now the position vector of P, say $\mathbf{r}_0$, is given by $u = 1, v = -1$ as $\mathbf{i}-\mathbf{j}+\mathbf{k}$. But

$$\frac{\partial \mathbf{r}}{\partial u} = \mathbf{i}+u\mathbf{k}, \frac{\partial \mathbf{r}}{\partial v} = \mathbf{j}+v\mathbf{k}.$$

These vectors at $P(u = 1, v = -1)$ are $\mathbf{i}+\mathbf{k}$ and $\mathbf{j}-\mathbf{k}$.

So $\mathbf{p} = (\mathbf{i}+\mathbf{k})\times(\mathbf{j}-\mathbf{k}) = -\mathbf{i}+\mathbf{j}+\mathbf{k}$

is normal to the surface at P. So the equation of the tangent plane to the surface at P is

$$(\mathbf{r}-\mathbf{r}_0)\cdot\mathbf{p} = 0$$

i.e., $\{(x\mathbf{i}+y\mathbf{j}+z\mathbf{k})-(\mathbf{i}-\mathbf{j}+\mathbf{k})\}\cdot(-\mathbf{i}+\mathbf{j}+\mathbf{k}) = 0$

which reduces to

$$(x-1)(-1)+(y+1)(1)+(z-1)(1) = 0 \quad \text{or} \quad x-y-z-1 = 0.$$

1.11. Examples

1. Show that the equation of the tangent plane to the surface $\mathbf{r} = u\mathbf{i}+v\mathbf{j}+\frac{1}{2}uv\mathbf{k}$ at the point given by $u = 2$ and $v = 3$ is $(\mathbf{r}-\mathbf{r}_0).\mathbf{p} = 0$, where $\mathbf{r}_0 = 2\mathbf{i}+3\mathbf{j}+3\mathbf{k}$ and $\mathbf{p} = -\frac{3}{2}\mathbf{i}-\mathbf{j}+\mathbf{k}$.
2. Show that the equation of the tangent plane to the surface $xy-2z = 0$ at the point (2, 3, 3) is $3x+2y-2z-6 = 0$.
3. Show that the unit normals to the surface $\mathbf{r} = \mathbf{r}(u, v)$ are

$$\pm\left(\frac{\partial \mathbf{r}}{\partial u}\times\frac{\partial \mathbf{r}}{\partial v}\right)\Big/\sqrt{EG-F^2},$$

where $E = \left(\dfrac{\partial r}{\partial u}\right)^2, F = \dfrac{\partial \mathbf{r}}{\partial u}\cdot\dfrac{\partial \mathbf{r}}{\partial v}, G = \left(\dfrac{\partial r}{\partial v}\right)^2.$

[**Hint:** The unit normals are

$$\pm\left(\frac{\partial \mathbf{r}}{\partial u}\times\frac{\partial \mathbf{r}}{\partial v}\right)\Big/\left|\frac{\partial \mathbf{r}}{\partial u}\times\frac{\partial \mathbf{r}}{\partial v}\right|,$$

where $\left|\dfrac{\partial \mathbf{r}}{\partial u}\times\dfrac{\partial \mathbf{r}}{\partial v}\right|$ can be obtained by using $|\mathbf{a}\times\mathbf{b}|^2 = \mathbf{a}^2\mathbf{b}^2-(\mathbf{a}\cdot\mathbf{b})^2$.]

CHAPTER

2

Gradient of a Scalar Point Function and Divergence and Curl of a Vector Point Function

2.1. Scalar and vector point functions

If, for every point P in a domain D of space, there corresponds a scalar ϕ, then ϕ is said to be a single-valued *scalar point function* defined in the domain D. The value of ϕ at P is denoted by $\phi(P)$ or $\phi(x, y, z)$ if P is (x, y, z). The function ϕ is said to define a *scalar field* in D.

Example. The temperatures of a gas at different points in the region occupied by the gas define a scalar field in that region.

We know that temperature is a scalar. If the temperature of a gas at the point (x, y, z) is the function $T(x, y, z)$, then $T(x, y, z)$ is a scalar point function.

In the same way we define single-valued *vector point functions*. A vector point function defined on a domain D is said to define a *vector field* in D.

Suppose, in a physical situation, for every point (x, y, z), there corresponds a vector.

$$f(x, y, z)\bar{i} + g(x, y, z)\bar{j} + h(x, y, z)\bar{k},$$

then this vector function is said to be a vector point function.

Example. The velocity in a fluid motion is a vector point function.

We know that velocity is a vector. If the velocity of a particle of a fluid in motion, at (x, y, z), is $\bar{v}(x,y,z) = v_1(x,y,z)\bar{i} + v_2(x,y,z)\bar{j} + v_3(x,y,z)\bar{k}$, then $\bar{v}(x, y, z)$ is a vector point function.

2.2. Level surfaces

The surfaces represented by equations $\phi(x, y, z) = c$ for different values of c, are called *level surfaces*. The values of ϕ at all points on a level surface are equal. Also no two level surfaces will intersect each other ; if, however, it is assumed that the two surfaces $\phi = c_1$ and $\phi = c_2$ intersect, P being a common point, then it follows that $\phi(P) = c_1$ and $\phi(P) = c_2$ and that ϕ is not a single-valued function at P which is a contradiction.

Example. Suppose there is a point heating-source at the origin. Then the temperatures at all points equidistant from the origin are the same. Here the temperature is a point function, say $T(x, y, z)$, and the level surfaces $T(x, y, z) = c$ are concentric spheres having centres at the origin.

2.3. Directional derivative of a scalar point function

Suppose $\phi(x, y, z)$ is a scalar point function and $\phi(P)$ is the value of ϕ at P. If P' is any point, close to P, then the limit

$$\lim_{P' \to P} \frac{\phi(P') - \phi(P)}{PP'}$$

is called the directional derivative of ϕ

(i) at the point P

(ii) in the direction from P to P'.

The directional derivative is a scalar. Actually it is the rate of change of ϕ in the given direction. It varies with direction.

Book work 2.1. **The directional derivative of $\phi(x, y, z)$ at the point $P(x, y, z)$ in the direction specified by the direction cosines l, m, n is $l\frac{\partial \phi}{\partial x} + m\frac{\partial \phi}{\partial y} + n\frac{\partial \phi}{\partial z}$.**

Proof: **Let P' be the point reached when one travels from P through a distance s in the given direction. Then P' has the coordinates $(x + ls, y + ms, z + ns)$.**

The directional derivative of ϕ at P in the given direction is

$$\lim_{PP' \to 0} \frac{\phi(P') - \phi(P)}{PP'} = \lim_{s \to 0} \frac{\phi(x + ls, y + ms, z + ns) - \phi(x, y, z)}{s}.$$

But, expanding $\phi(x + ls, y + ms, z + ns)$ in powers of s,

$$\phi(x + ls, y + ms, z + ns) = \phi(x, y, z) + \left(ls\frac{\partial}{\partial x} + ms\frac{\partial}{\partial y} + ns\frac{\partial}{\partial z}\right)\phi(x, y, z)$$

$$+ \frac{1}{2!}\left(ls\frac{\partial}{\partial x} + ms\frac{\partial}{\partial y} + ns\frac{\partial}{\partial z}\right)^2 \phi(x, y, z) + \ldots$$

$$\lim_{PP' \to 0} \frac{\phi(P') - \phi(P)}{PP'} = l\frac{\partial \phi}{\partial x} + m\frac{\partial \phi}{\partial y} + n\frac{\partial \phi}{\partial z} \quad \ldots (2.3.1)$$

since the terms containing $s, s^2, s^3, \ldots$ vanish.

2.4. Gradient of a scalar point function

If ϕ is a scalar point function, then the vector $\frac{\partial \phi}{\partial x}\mathbf{i} + \frac{\partial \phi}{\partial y}\mathbf{j} + \frac{\partial \phi}{\partial z}\mathbf{k}$ is called the *gradient* of ϕ. This vector is written as grad ϕ or $\nabla\phi$, where ∇ (read as del or nabla) is the operator which stands for $\mathbf{i}\frac{\partial}{\partial x} + \mathbf{j}\frac{\partial}{\partial y} + \mathbf{k}\frac{\partial}{\partial z}$. Gradient measures the rate and direction of change in a scalar field. Maps scalar fields to vector fields.

Thus ∇ is an operator whose function is to transform a scalar point function ϕ into a vector point function $\nabla\phi$. Note that ϕ is a scalar and $\nabla\phi$ is a vector.

Book work 2.2. **The directional derivative of ϕ in the direction specified by the unit vector e is $(\nabla\phi).\mathbf{e}$. (2.4.1)**

Proof: **Let the direction cosines of e be l, m, n. Then**

$$\mathbf{e} = l\mathbf{i} + m\mathbf{j} + n\mathbf{k}. \text{ Therefore,}$$

$$(\nabla\phi).\mathbf{e} = (\nabla\phi).(l\mathbf{i} + m\mathbf{j} + n\mathbf{k})$$

$$= \left(\frac{\partial \phi}{\partial x}\mathbf{i} + \frac{\partial \phi}{\partial y}\mathbf{j} + \frac{\partial \phi}{\partial z}\mathbf{k}\right).(l\mathbf{i} + m\mathbf{j} + n\mathbf{k})$$

$$= l\frac{\partial \phi}{\partial x} + m\frac{\partial \phi}{\partial y} + n\frac{\partial \phi}{\partial z}$$

which is the directional derivative of ϕ in the direction whose direction consines are l, m, n, that is, in the direction specified by the unit vector **e**.

Book work 2.3. (*i*) The direction of $\nabla\phi$ at P is normal to the level surface $\phi = c$ through P (*i.e.*, $\nabla\phi$ is a vector along the normal at P to the level surface through P).

(*ii*) Magnitude of $\nabla\phi$ at P is the maximum of the directional derivative of ϕ at P.

Proof: (*i*) Suppose $\phi = c$ is the level surface through P and *C* is an arbitrarily chosen curve through P and on the level surface (Fig. 2.2.1). Now the tangent to *C* at P lies in the tangent plane at P to the level surface. Since ϕ is a constant on C, the directional derivative of ϕ along *C* is zero. So, from (2.4.1), if **T** is the unit vector along the tangent at P to C, then $(\nabla\phi).\mathbf{T} = 0$.

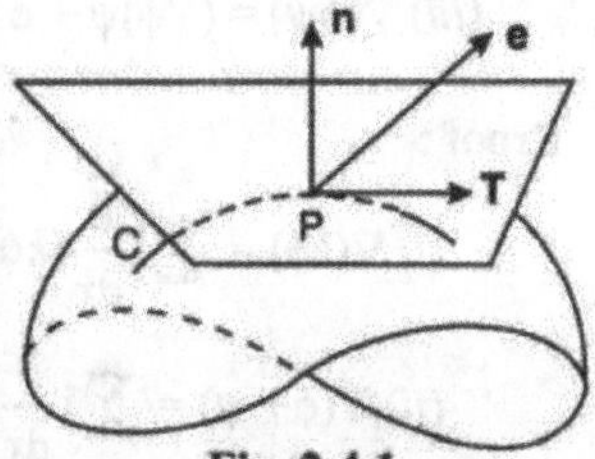

Fig. 2.4.1

Hence $\nabla\phi$ at *P* is perpendicular to **T**. But *C* is arbitrary. Therefore $\nabla\phi$ at P is perpendicular to the tangent plane to the level surface at P. Hence $\nabla\phi$ at P is along the normal to the surface at P.

(*ii*) Let ***n*** be the unit vector along the normal to the level surface at P (in the sense in which ϕ increases). Then $\nabla\phi = |\nabla\phi|\mathbf{n}$. ... (2, 4, 2)

Directional derivative at P in an arbitrary direction $\boldsymbol{e} = (\nabla\phi) . \mathbf{e} = |\nabla\phi|\, \mathbf{n} . \mathbf{e} = |\nabla\phi| \cos\theta$.

where θ is the angle between **n** and **e**.

$\therefore$ Maximum value of the directional derivative at P $= |\nabla\phi|$ ... (2.4.3)

since the maximum value of $\cos\theta = 1$. So the *magnitude of* $\nabla\phi$ is the maximum of the directional derivative.

Remark 1. The directional derivative is a maximum when $\cos\theta = 1$ or $\theta = 0$, *i.e.*, when **e** coincides with **n**. So the directional derivative is a maximum when it is along the normal to the level surface.

Remark 2. The vector $\nabla\phi$ depends only on ϕ and is independent of the choice of x, y, z axes.

Remark 3. Since the unit vector corresponding to $\bar{a}$ is $\dfrac{\bar{a}}{|\bar{a}|}$, the unit vector normal to the surface $\phi(x, y, z) = c$ at any point (x, y, z) is

$$\hat{n} = \frac{\nabla\phi}{|\nabla\phi|}.$$

Angle between two surfaces. The angle between two surfaces at a common point is the angle between the normals to the surfaces at that point. If the equation of the surfaces are

$$\phi_1 = c_1, \phi_2 = c_2$$

and if P is a common point, then the vectors

$$\nabla\phi_1, \nabla\phi_2,$$

at P, are normal to the surfaces and the angle between the surface is

$$\cos^{-1}\frac{(\nabla\phi_1)\cdot(\nabla\phi_2)}{|\nabla\phi_1||\nabla\phi_2|}.$$

2.5. Summation notation for gradient

The following summation notation shortens proofs of results involving gradients:-

$$\nabla\phi = \left(\mathbf{i}\frac{\partial}{\partial x} + \mathbf{j}\frac{\partial}{\partial y} + \mathbf{k}\frac{\partial}{\partial z}\right)\phi = \mathbf{i}\frac{\partial\phi}{\partial x} + \mathbf{j}\frac{\partial\phi}{\partial y} + \mathbf{k}\frac{\partial\phi}{\partial z} = \sum\left(i\frac{\partial\phi}{\partial x}\right).$$

GRADIENT OF SUM AND PRODUCT OF FUNCTIONS

Book work 2.4. If ϕ and ψ are scalar point functions,

(i) $\nabla(k\phi) = k(\nabla\phi)$, where k is a constant (ii) $\nabla(\phi+\psi) = (\nabla\phi + \nabla\psi)$

(iii) $\nabla(\phi\psi) = (\nabla\phi)\psi + \phi(\nabla\psi)$ (iv) $\nabla\left(\frac{\phi}{\psi}\right) = \frac{\psi(\nabla\phi) - \phi(\nabla\psi)}{\psi^2}$

Proof :

(i) $\nabla(k\phi) = \sum \mathbf{i}\frac{\partial}{\partial x}(k\phi) = \sum \mathbf{i}k\frac{\partial\phi}{\partial x} = k\sum \mathbf{i}\frac{\partial\phi}{\partial x} = k(\nabla\phi).$

(ii) $\nabla(\phi+\psi) = \sum \mathbf{i}\frac{\partial}{\partial x}(\phi+\psi) = \sum \mathbf{i}\left[\frac{\partial\phi}{\partial x} + \frac{\partial\psi}{\partial x}\right]$

$$= \sum\left(\mathbf{i}\frac{\partial\phi}{\partial x} + \mathbf{i}\frac{\partial\psi}{\partial x}\right) = \sum \mathbf{i}\frac{\partial\phi}{\partial x} + \sum \mathbf{i}\frac{\partial\psi}{\partial x} = \nabla\phi + \nabla\psi.$$

(iii) $\nabla(\phi\psi) = \sum \mathbf{i}\frac{\partial}{\partial x}(\phi\psi) = \sum \mathbf{i}\left(\frac{\partial\phi}{\partial x}\psi + \phi\frac{\partial\psi}{\partial x}\right) = \left(\sum \mathbf{i}\frac{\partial\phi}{\partial x}\right)\psi + \phi\left(\sum \mathbf{i}\frac{\partial\psi}{\partial x}\right) = (\nabla\phi)\psi + \phi(\nabla\psi).$

(iv) $\nabla\left(\frac{\phi}{\psi}\right) = \sum \mathbf{i}\frac{\partial}{\partial x}\left(\frac{\phi}{\psi}\right) = \sum \mathbf{i}\frac{\psi\frac{\partial\phi}{\partial x} - \phi\frac{\partial\psi}{\partial x}}{\psi^2} = \frac{\psi\sum i\frac{\partial\phi}{\partial x} - \phi\sum i\frac{\partial\psi}{\partial x}}{\psi^2} = \frac{\psi(\nabla\phi) - \phi(\nabla\psi)}{\psi^2}.$

In this the summation ranges over i, j, k and *x*, *y*, *z*.

2.6. Gradient of $f(r)$.

It is conventional to denote the P.V. of any point (x, y, z) in space by $\bar{r}$ so that

$$\bar{r} = x\bar{i} + y\bar{j} + z\bar{k}.$$

Its magnitude and the unit vector in its direction are denoted by r and $\hat{r}$. Since r is the distance of the point (x, y, z) from the origin,

$$r^2 = x^2 + y^2 + z^2,\ \bar{r} = r\hat{r},\ \hat{r} = \frac{\bar{r}}{r}.$$

Theorem. **If $\bar{r} = x\bar{i} + y\bar{j} + z\bar{k}$ and $|\bar{r}| = r$, then**

$$\nabla f(r) = \left[\frac{df(r)}{dr}\right]\hat{r} = f'(r)\hat{r}. \tag{2.6.1}$$

Proof: **Now $r^2 = x^2 + y^2 + z^2$. Differentiating this partially w.r.t. x**

$$2r\frac{\partial r}{\partial x} = 2x \text{ or } \frac{\partial r}{\partial x} = \frac{x}{r}.$$

$$\therefore \quad \frac{\partial r}{\partial x} = \frac{x}{r},\ \frac{\partial r}{\partial y} = \frac{y}{r},\ \frac{\partial r}{\partial z} = \frac{z}{r}.$$

$$\nabla f(r) = \left(\bar{i}\frac{\partial}{\partial x} + \bar{j}\frac{\partial}{\partial y} + \bar{k}\frac{\partial}{\partial z}\right)f(r) = \bar{i}\frac{\partial f(r)}{\partial x} + \bar{j}\frac{\partial f(r)}{\partial y} + \bar{k}\frac{\partial f(r)}{\partial z}$$

$$= \bar{i}f'(r)\frac{\partial r}{\partial x} + \bar{j}f'(r)\frac{\partial r}{\partial y} + \bar{k}f'(r)\frac{\partial r}{\partial z}$$

$$= f'(r)\left[\bar{i}\frac{\partial r}{\partial x} + \bar{j}\frac{\partial r}{\partial y} + \bar{k}\frac{\partial r}{\partial z}\right]$$

$$= f'(r)\left[\bar{i}\frac{x}{r}+\bar{j}\frac{y}{r}+\bar{k}\frac{z}{r}\right] = f'(r) = \frac{\bar{r}}{r}$$

$$= f'(r)\,\hat{r}. \qquad (\because \bar{r} = r\hat{r})$$

Using the formula $\nabla f(r) = f'(r)\,\hat{r}$, we have the following results:

(*i*) If $f(r) = \dfrac{1}{r}$, then

$$\nabla\left(\frac{1}{r}\right) = \left(\frac{d}{dr}\frac{1}{r}\right)\hat{r} = -\frac{1}{r^2}\hat{r} = -\frac{1}{r^2}\frac{\bar{r}}{r} = -\frac{\bar{r}}{r^3}.$$

(*ii*) If $f(r) = r^n$, then

$$\nabla(r^n) = \frac{dr^n}{dr}\hat{r} = nr^{n-1}\hat{r} = nr^{n-1}\frac{\bar{r}}{r} = nr^{n-2}\bar{r}.$$

(*iii*) If $f(r) = \log r$, then

$$\nabla \log r = \frac{d}{dr}\log r\,\hat{r} = \frac{1}{r}\hat{r} = \frac{1}{r}\frac{\bar{r}}{r} = \frac{\bar{r}}{r^2}.$$

(*iv*) If $\nabla f(r) = 5r^3\bar{r}$, then

$$f'(r) = 5r^4.$$

Hence $f(r) = r^5 + c$.

Remark. The above results can also be directly derived. For example,

$$\nabla\frac{1}{r} = \left(\bar{i}\frac{\partial}{\partial x}\frac{1}{r}+\bar{j}\frac{\partial}{\partial y}\frac{1}{r}+\bar{k}\frac{\partial}{\partial z}\frac{1}{r}\right) = \bar{i}\frac{\partial}{\partial x}\left(\frac{1}{r}\right)+\bar{j}\frac{\partial}{\partial y}\left(\frac{1}{r}\right)+\bar{k}\frac{\partial}{\partial z}\left(\frac{1}{r}\right)$$

$$= \bar{i}\left(-\frac{1}{r^2}\frac{\partial r}{\partial x}\right)+\bar{j}\left(-\frac{1}{r^2}\frac{\partial r}{\partial y}\right)+\bar{k}\left(-\frac{1}{r^2}\frac{\partial r}{\partial z}\right)$$

$$= -\frac{1}{r^2}\left[\bar{i}\frac{\partial r}{\partial x}+\bar{j}\frac{\partial r}{\partial y}+\bar{k}\frac{\partial r}{\partial z}\right] = -\frac{1}{r^2}\left[\bar{i}\frac{x}{r}+\bar{j}\frac{y}{r}+\bar{k}\frac{z}{r}\right]$$

$$= -\frac{1}{r^3}(x\bar{i}+y\bar{j}+z\bar{k}) = -\frac{\bar{r}}{r^3}$$

To prove (*iii*) independently, we have

$$\nabla \log r = \left(\bar{i}\frac{\partial}{\partial x}+\bar{j}\frac{\partial}{\partial y}+\bar{k}\frac{\partial}{\partial z}\right)\log r$$

$$= \bar{i}\frac{\partial}{\partial x}\log r+\bar{j}\frac{\partial}{\partial y}\log r+\bar{k}\frac{\partial}{\partial z}\log r$$

$$= \bar{i}\frac{1}{r}\frac{\partial r}{\partial x}+\bar{j}\frac{1}{r}\frac{\partial r}{\partial y}+\bar{k}\frac{1}{r}\frac{\partial r}{\partial z}$$

$$= \frac{1}{r}\left(\bar{i}\frac{x}{r}+\bar{j}\frac{y}{r}+\bar{k}\frac{z}{r}\right) = \frac{1}{r^2}(x\bar{i}+y\bar{j}+z\bar{k})$$

$$= \frac{\bar{r}}{r^2}.$$

grad $(\sin(x^2y)) = \{2xy\cos(x^2y),\, x^2\cos(x^2y)\}$ can be proved.

2.7 Divergence and curl of a vector point function

Divergence. If $\mathbf{V} = V_1\mathbf{i} + V_2\mathbf{j} + V_3\mathbf{k}$ is a vector point function, then the scalar function (or a scalar constant) $\frac{\partial V_1}{\partial x} + \frac{\partial V_2}{\partial y} + \frac{\partial V_3}{\partial z}$ is called the *divergence* of V. Divergence measures the scalar of a source at a given point in a vector field. Maps vector fields to scalar fields.

Notation. Divergence of $\bar{V}$: div $\bar{V}$ or $\nabla . \bar{V}$. Now

$$\nabla . \bar{V} = \left(\bar{i}\frac{\partial}{\partial x} + \bar{j}\frac{\partial}{\partial y} + \bar{k}\frac{\partial}{\partial z}\right)\cdot(V_1\bar{i} + V_2\bar{j} + V_3\bar{k})$$

$$= \frac{\partial V_1}{\partial x} + \frac{\partial V_2}{\partial y} + \frac{\partial V_3}{\partial z}.$$

Note. If $\bar{a}$ is a constant, $\nabla . \bar{a} = 0$.

Curl. If $V = V_1\mathbf{i} + V_2\mathbf{j} + V_3\mathbf{k}$, is a vector point function, then the vector function (or a constant vector)

$$\mathbf{i}\left(\frac{\partial V_3}{\partial y} - \frac{\partial V_2}{\partial z}\right) + \mathbf{j}\left(\frac{\partial V_1}{\partial z} - \frac{\partial V_3}{\partial x}\right) + \mathbf{k}\left(\frac{\partial V_2}{\partial x} - \frac{\partial V_1}{\partial y}\right)$$

is called the *curl* or *rotation of V*. Curl measures the tendency to rotate about a point in a vector field. Maps vector fields to (pseudo) vector fields.

Notation. Curl of $\bar{\mathbf{V}}$ (*i.e.*, rotor operator) of $\bar{\mathbf{V}}$: curl $\bar{V}$ or rot **V** or $\nabla \times \bar{V}$. Now

$$\nabla \times \bar{V} = \begin{vmatrix} \bar{i} & \bar{j} & \bar{k} \\ \frac{\partial}{\partial x} & \frac{\partial}{\partial y} & \frac{\partial}{\partial z} \\ V_1 & V_2 & V_3 \end{vmatrix}.$$

Note. (*i*) If $\bar{a}$ is a constant, $\nabla \times \bar{a} = \bar{0}$.

(*ii*) Curl of a vector field is a *bivector field.*

Solenoidal vector. If $\nabla . V = 0$, then the vector V is said to be *solenoidal* or zero divergence or divergenceless.

Irrotational vector. If $\nabla \times V = \mathbf{0}$, then the vector V is said to be *irrotational*. Such vector fields are curl-less vector fields (cur-free vector field). Since $\nabla \times \nabla\phi = \mathbf{0}$, every conservative vector field is an irrotational vector field.

Scalar Potential

Given a vector point function $\bar{F}$, if there exists a scalar point function ϕ such that

$$\bar{F} = \nabla\phi,$$

then ϕ is called the scalar potential of $\bar{F}$. In such a case,

$$\nabla \times \bar{F} = \nabla\times(\nabla\phi) = \bar{0}$$

that is, $\bar{F}$ is irrotational.

Conversely, if $\bar{F}$ is irrotational, it can be proved that there exists a scalar potential ϕ such that

$$\bar{F} = \nabla\phi.$$

2.8 Summation notation for divergence and curl

The summation notation shortens the proofs of formulae connected with divergence and curl. The notations are

(i) $\nabla.\mathbf{A}=\Sigma\left(\mathbf{i}\cdot\frac{\partial \mathbf{A}}{\partial x}\right)$ (ii) $\nabla\times\mathbf{A}=\Sigma\left(\mathbf{i}\times\frac{\partial \mathbf{A}}{\partial x}\right)$

(iii) $\mathbf{A}.\nabla=\mathbf{A}\cdot\Sigma\left(i\frac{\partial}{\partial x}\right)$ (iv) $(\mathbf{A}.\nabla)\,\mathbf{B}=\left(\mathbf{A}\cdot\Sigma\mathbf{i}\frac{\partial}{\partial x}\right)\mathbf{B}$

Book work 2.5. If **A** and **B** are vector point functions, ϕ a scalar point function and k a constant, then

(i) $\nabla.(\mathbf{A}+\mathbf{B})=\nabla.\mathbf{A}+\nabla.\mathbf{B}$ (ii) $\nabla.(k\mathbf{A})=k(\nabla.\mathbf{A})$

(iii) $\nabla.(\phi\mathbf{A})=(\nabla\phi).\mathbf{A}+\phi(\nabla.\mathbf{A})$ ... (2.8.1)

(iv) $\nabla\times(\mathbf{A}+\mathbf{B})=\nabla\times\mathbf{A}+\nabla\times\mathbf{B}$ (v) $\nabla\times(k\mathbf{A})=k(\nabla\times\mathbf{A})$

(vi) $\nabla\times(\phi\mathbf{A})=(\nabla\phi)\times\mathbf{A}+\phi\,(\nabla\times\mathbf{A})$. ... (2.8.2)

Proof:

$$(i)\ \nabla.(\mathbf{A}+\mathbf{B})=\Sigma\mathbf{i}\cdot\frac{\partial}{\partial x}(\mathbf{A}+\mathbf{B})=\Sigma\mathbf{i}\cdot\left\{\frac{\partial\mathbf{A}}{\partial x}+\frac{\partial\mathbf{B}}{\partial x}\right\}=\Sigma\mathbf{i}\cdot\frac{\partial\mathbf{A}}{\partial x}+\Sigma\mathbf{i}\cdot\frac{\partial\mathbf{B}}{\partial x}=\nabla\cdot\mathbf{A}+\nabla\cdot\mathbf{B}.$$

$$(ii)\ \nabla.(k\mathbf{A})=\Sigma\mathbf{i}\cdot\frac{\partial(k\mathbf{A})}{\partial x}=\Sigma\mathbf{i}\cdot k\frac{\partial\mathbf{A}}{\partial x}=k\Sigma\mathbf{i}\cdot\frac{\partial\mathbf{A}}{\partial x}=k(\nabla\cdot\mathbf{A}).$$

$$(iii)\ \nabla.(\phi\mathbf{A})=\Sigma\mathbf{i}\cdot\frac{\partial(\phi\mathbf{A})}{\partial x}=\Sigma\mathbf{i}\cdot\left\{\frac{\partial\phi}{\partial x}\mathbf{A}+\phi\frac{\partial\mathbf{A}}{\partial x}\right\}=\Sigma\mathbf{i}\cdot\left\{\frac{\partial\phi}{\partial x}\mathbf{A}\right\}+\Sigma\mathbf{i}\cdot\left\{\phi\frac{\partial\mathbf{A}}{\partial x}\right\}$$

$$=\left\{\Sigma\left(\mathbf{i}\frac{\partial\phi}{\partial x}\right)\right\}\cdot\mathbf{A}+\phi\Sigma\mathbf{i}\cdot\frac{\partial\mathbf{A}}{\partial x}=(\nabla\phi).\mathbf{A}+\phi(\nabla.\mathbf{A}).$$

(iv), (v), (vi) : The proofs are same as in cases (i), (ii), (iii), the only differences being instead of dots we have crosses.

Divergence and Curl of $f(r)\bar{r}$.

We have already denoted $x\bar{i}+y\bar{j}+z\bar{k}$ by $\bar{r}$. Thus

$$\nabla.\bar{r}=\frac{\partial(x)}{\partial x}+\frac{\partial(y)}{\partial y}+\frac{\partial(z)}{\partial z}=1+1+1=3,$$

$$\nabla\times\bar{r}=\begin{vmatrix}\bar{i} & \bar{j} & \bar{k}\\ \frac{\partial}{\partial x} & \frac{\partial}{\partial y} & \frac{\partial}{\partial z}\\ x & y & z\end{vmatrix}=\bar{i}(0)-\bar{j}(0)+\bar{k}(0)=\bar{0}.$$

Theorem

If $\bar{r}=x\bar{i}+y\bar{j}+z\bar{k}$ and $\bar{r}=r\hat{r}$, show that

(i) $\nabla.[f(r)\bar{r}]=rf'(r)+3f(r)$ ... (2.8.3) (ii) $\nabla\times[f(r)\bar{r}]=\bar{0}$. ... (2.8.4)

Proof: (i) We know that

$$\nabla.(\phi\bar{A})=(\nabla\phi).\bar{A}+\phi(\nabla.\bar{A}),\ \nabla f(r)=f'(r)\hat{r}.$$

$$\therefore\quad \nabla.[f(r)\bar{r}]=[f'(r)].\bar{r}+f(r)][\nabla.\bar{r}]=[f'(r)\hat{r}].\bar{r}+f(r)\,[3]$$

$$=f'(r)[\hat{r}\cdot(r\hat{r})]+3f(r)=rf'(r)[\hat{r}\cdot\hat{r}]+3f(r)$$

$$=rf'(r)+3f(r).$$

(*ii*) We know that

$$\nabla\times(\phi\bar{A}) = (\nabla\phi)\times\bar{A}+\phi(\nabla\times\bar{A}).$$

$$\therefore\quad \nabla\times[f(r)\bar{r}] = [\nabla f(r)]\times\bar{r}+f(r)[\nabla\times\bar{r}]$$

$$= [f'(r)\hat{r}]\times\bar{r}+f(r)[\bar{0}]$$

$$= f'(r)\,[\hat{r}\times\bar{r}]=\bar{0} \qquad (\because\ \hat{r}\times\bar{r}=\bar{0})$$

Corollaries of (*i*)

$$\nabla\cdot(r^n\bar{r}) = r\frac{d(r^n)}{dr}+3r^n=(n+3)r^n, \text{ by setting } f(r)=r^n \text{ in } (i).$$

$$\nabla\cdot\hat{r} = \nabla\cdot\left(\frac{1}{r}\bar{r}\right)=r\frac{d}{dr}(1/r)+\frac{3}{r}=-\frac{1}{r}+\frac{3}{r}=\frac{2}{r}.$$

Corollaries of (ii)

$$\nabla\times(r^n\bar{r})=\bar{0}. \qquad \text{... (1)}$$

$$\nabla\times\hat{r} = \nabla\times\left(\frac{1}{r}\bar{r}\right)=\bar{0}.$$

Note. To prove (1) independently, using

$$\nabla\times(\phi\bar{A}) = (\nabla\phi)\times\bar{A}+\phi(\nabla\times\bar{A}),$$

we have

$$\nabla\times(r^n\bar{r}) = (\nabla r^n)\times\bar{r}+r^n(\nabla\times\bar{r})=(nr^{n-2}\bar{r})\times\bar{r}+r^n(\bar{0})$$

$$= nr^{n-2}(\bar{r}\times\bar{r})=\bar{0}+\bar{0}=\bar{0}.$$

Book Work 2.6. **If A and B are vector point functions,**

(*i*) $\nabla(\mathbf{A}.\mathbf{B})=\mathbf{A}\times(\nabla\times\mathbf{B})+(\mathbf{A}.\nabla)\mathbf{B}+\mathbf{B}\times(\nabla\times\mathbf{A})+(\mathbf{B}.\nabla)\mathbf{A}.$ **(2.8.5)**

(*ii*) $\nabla.(\mathbf{A}\times\mathbf{B})=(\nabla\times\mathbf{A}).\mathbf{B}-(\nabla\times\mathbf{B}).\mathbf{A}.$ **(2.8.6)**

(*iii*) $\nabla\times(\mathbf{A}\times\mathbf{B})=\{(\mathbf{B}.\nabla)\mathbf{A}-(\nabla.\mathbf{A})\mathbf{B}\}-\{(\mathbf{A}.\nabla)\mathbf{B}-(\nabla.\mathbf{B})\mathbf{A}\}.$ **(2.8.7)**

Proof: (*i*)

$$\nabla(\mathbf{A}.\mathbf{B}) = \Sigma\mathbf{i}\frac{\partial}{\partial x}(\mathbf{A}\cdot\mathbf{B})$$

$$= \Sigma\mathbf{i}\left(\frac{\partial\mathbf{A}}{\partial x}\cdot\mathbf{B}+\mathbf{A}\cdot\frac{\partial\mathbf{B}}{\partial x}\right)$$

$$= \Sigma\mathbf{i}\left(\mathbf{B}\cdot\frac{\partial\mathbf{A}}{\partial x}\right)+\Sigma\mathbf{i}\left(\mathbf{A}\cdot\frac{\partial\mathbf{B}}{\partial x}\right). \qquad \text{... (1)}$$

Note: $\nabla(\mathbf{r}.\mathbf{A})=\mathbf{A}$ **when A is a constant vector.**

In this we simplify the second sum on the right by using $\mathbf{a}\times(\mathbf{b}\times\mathbf{c})=(\mathbf{a}.\mathbf{c})\mathbf{b}-(\mathbf{a}.\mathbf{b})\mathbf{c}$

i.e.,

$$\mathbf{A}\times\left(\mathbf{i}\times\frac{\partial\mathbf{B}}{\partial x}\right) = \left(\mathbf{A}\cdot\frac{\partial\mathbf{B}}{\partial x}\right)\mathbf{i}-(\mathbf{A}\cdot\mathbf{i})\frac{\partial\mathbf{B}}{\partial x}.$$

$$\therefore\quad \Sigma\left(\mathbf{A}\cdot\frac{\partial\mathbf{B}}{\partial x}\right)\mathbf{i}=\Sigma\left\{\mathbf{A}\times\left(\mathbf{i}\times\frac{\partial\mathbf{B}}{\partial x}\right)+(\mathbf{A}\cdot\mathbf{i})\frac{\partial\mathbf{B}}{\partial x}\right\}$$

$$= \Sigma\mathbf{A}\times\left(\mathbf{i}\times\frac{\partial\mathbf{B}}{\partial x}\right)+\Sigma(\mathbf{A}\cdot\mathbf{i})\frac{\partial\mathbf{B}}{\partial x}$$

$$= \mathbf{A}\times\left(\Sigma\mathbf{i}\times\frac{\partial \mathbf{B}}{\partial x}\right)+\left(\Sigma\mathbf{A}\cdot\mathbf{i}\frac{\partial}{\partial x}\right)\mathbf{B}$$

$$= \mathbf{A}\times(\nabla\times\mathbf{B})+\left(\mathbf{A}\cdot\Sigma\mathbf{i}\frac{\partial}{\partial x}\right)\mathbf{B}$$

$$= \mathbf{A}\times(\nabla\times\mathbf{B})+(\mathbf{A}\,.\,\nabla)\mathbf{B}. \qquad ...(2)$$

Interchanging A and B, we get the other sum to be

$$\Sigma\left(\mathbf{B}\,.\,\frac{\partial \mathbf{A}}{\partial x}\right)\mathbf{i} = \mathbf{B}\times(\nabla\times\mathbf{A})+(\mathbf{B}\,.\,\nabla)\mathbf{A}. \qquad ...(3)$$

From (1), (2), (3) we get the result.

(ii)

$$\nabla\,.\,(\mathbf{A}\times\mathbf{B}) = \Sigma\mathbf{i}\cdot\frac{\partial}{\partial x}(\mathbf{A}\times\mathbf{B})$$

$$= \Sigma\mathbf{i}\cdot\left(\frac{\partial \mathbf{A}}{\partial x}\times\mathbf{B}+\mathbf{A}\times\frac{\partial \mathbf{B}}{\partial x}\right)$$

$$= \Sigma\mathbf{i}\cdot\left(\frac{\partial \mathbf{A}}{\partial x}\times\mathbf{B}\right)+\Sigma\mathbf{i}\,.\left(\mathbf{A}\times\frac{\partial \mathbf{B}}{\partial x}\right)$$

$$= \Sigma\mathbf{i}\cdot\left(\frac{\partial \mathbf{A}}{\partial x}\times\mathbf{B}\right)-\Sigma\mathbf{i}\cdot\left(\frac{\partial \mathbf{B}}{\partial x}\times\mathbf{A}\right)$$

$$= \Sigma\left(\mathbf{i}\times\frac{\partial \mathbf{A}}{\partial x}\right)\cdot\mathbf{B}-\Sigma\left(\mathbf{i}\times\frac{\partial \mathbf{B}}{\partial x}\right)\cdot\mathbf{A} \quad \text{interchanging dot and cross}$$

$$= \left(\Sigma\mathbf{i}\times\frac{\partial \mathbf{A}}{\partial x}\right)\cdot\mathbf{B}-\left(\Sigma\mathbf{i}\times\frac{\partial \mathbf{B}}{\partial x}\right)\cdot\mathbf{A}$$

$$= (\nabla\times\mathbf{A})\,.\,\mathbf{B}-(\nabla\times\mathbf{B}).\mathbf{A}.$$

(iii)

$$\nabla\times(\mathbf{A}\times\mathbf{B}) = \Sigma\mathbf{i}\times\frac{\partial}{\partial x}(\mathbf{A}\times\mathbf{B})$$

$$= \Sigma\mathbf{i}\times\left(\frac{\partial \mathbf{A}}{\partial x}\times\mathbf{B}\right)+\Sigma\mathbf{i}\times\left(\mathbf{A}\times\frac{\partial \mathbf{B}}{\partial x}\right)$$

$$= \Sigma\mathbf{i}\times\left(\frac{\partial \mathbf{A}}{\partial x}\times\mathbf{B}\right)-\Sigma\mathbf{i}\times\left(\frac{\partial \mathbf{B}}{\partial x}\times\mathbf{A}\right)$$

We simplify the first sum on the right by using

$$\mathbf{a}\times(\mathbf{b}\times\mathbf{c}) = (\mathbf{a}\,.\,\mathbf{c})\mathbf{b}-(\mathbf{a}\,.\,\mathbf{b})\mathbf{c}.$$

$$\Sigma\mathbf{i}\times\left(\frac{\partial \mathbf{A}}{\partial x}\times\mathbf{B}\right) = \Sigma\left[(\mathbf{i}\cdot\mathbf{B})\frac{\partial \mathbf{A}}{\partial x}-\left(\mathbf{i}\cdot\frac{\partial \mathbf{A}}{\partial x}\right)\mathbf{B}\right]$$

$$= \Sigma(\mathbf{i}\cdot\mathbf{B})\frac{\partial \mathbf{A}}{\partial x}-\Sigma\left(\mathbf{i}\cdot\frac{\partial \mathbf{A}}{\partial x}\right)\mathbf{B}$$

$$= \Sigma(\mathbf{B}\,.\,\mathbf{i})\frac{\partial \mathbf{A}}{\partial x}-\Sigma\left(\mathbf{i}\cdot\frac{\partial \mathbf{A}}{\partial x}\right)\mathbf{B}$$

$$= \left(\mathbf{B}\cdot\Sigma\mathbf{i}\frac{\partial}{\partial x}\right)\mathbf{A}-\left(\Sigma\mathbf{i}\cdot\frac{\partial \mathbf{A}}{\partial x}\right)\mathbf{B}$$

$$= (\mathbf{B}\,.\,\nabla)\,\mathbf{A}-(\nabla\,.\,\mathbf{A})\mathbf{B}.$$

Interchanging **A** and **B**, we get

$$\Sigma \mathbf{i}\times\left(\frac{\partial \mathbf{B}}{\partial x}\times \mathbf{A}\right) = (\mathbf{A}\,.\,\nabla)\,\mathbf{B}-(\nabla\,.\,\mathbf{B})\mathbf{A}.$$

Thus we get the result as stated.

2.9 Laplacian Differential Operator

The operator ∇^2 defined by $\nabla^2=\nabla\,.\,\nabla=\frac{\partial^2}{\partial x^2}+\frac{\partial^2}{\partial y^2}+\frac{\partial^2}{\partial z^2}$ is called Laplacian differential operator.

When it operates on a scalar point function, it results in a scalar. When it operates on a vector point function, it results in a vector.

As illustrations

$$\begin{aligned}\nabla^2 (xy^2z^3) &= \left(\frac{\partial^2}{\partial x^2}+\frac{\partial^2}{\partial y^2}+\frac{\partial^2}{\partial z^2}\right)xy^2z^3\\ &= \frac{\partial^2}{\partial x^2}(xy^2z^3)+\frac{\partial^2}{\partial y^2}(xy^2z^3)+\frac{\partial^2}{\partial z^2}(xy^2z^3)\\ &= 0+2xz^3+6xy^2z\end{aligned}$$

$$\begin{aligned}\nabla^2 (y^2z^2\mathbf{i}+x^2\mathbf{j}+x^2\mathbf{k}) &= \left(\frac{\partial^2}{\partial x^2}+\frac{\partial^2}{\partial y^2}+\frac{\partial^2}{\partial z^2}\right)(y^2z^2\mathbf{i}+x^2\mathbf{j}+x^2\mathbf{k})\\ &= \frac{\partial^2}{\partial x^2}\left(y^2z^2\mathbf{i}+x^2\mathbf{j}+x^2\mathbf{k}\right)\\ &\quad+\frac{\partial^2}{\partial y^2}(y^2z^2\mathbf{i}+x^2\mathbf{j}+x^2\mathbf{k})+\frac{\partial^2}{\partial z^2}(y^2z^2\mathbf{i}+x^2\mathbf{j}+x^2\mathbf{k})\\ &= (2\mathbf{j}+2\mathbf{k})+(2z^2\mathbf{i})+(2y^2\mathbf{i}).\end{aligned}$$

Laplacian $=\nabla^2 f=\nabla.\nabla f$ which is a composition of the gradient and divergence operations. Maps between scalar fields.

Vector Laplacian $\nabla^2\mathrm{F}=\nabla(\nabla.\,\mathrm{F})-\nabla\times(\nabla\times\mathrm{F})$ which is a composition of the gradient and curl operations. Maps between vector fields.

Laplace equation. If ϕ is such that $\nabla^2\phi=0$, then ϕ is said to satisfy the Laplace equation. For example, $\phi=x^2-y^2+z$ satisfies the Laplace equation because $\frac{\partial^2\phi}{\partial x^2}=2,\ \frac{\partial^2\phi}{\partial y^2}=-2,\ \frac{\partial^2\phi}{\partial z^2}=0.$

Harmonic function. A single-valued $f(x, y, z)$ is said to be a harmonic function if its second partial derivatives exist and are continuous and if the function satisfies the Laplace equation

$$\nabla^2 f(x,y,z)=0,\ i.e.,\ \frac{\partial^2 f}{\partial x^2}+\frac{\partial^2 f}{\partial y^2}+\frac{\partial^2 f}{\partial z^2}=0\cdot$$

Note: $\nabla^2\left(\frac{1}{r}\right)=0$

2.10 Other Differential Operators

A . ∇ and **A** × ∇ can also be treated as operators, their operational meaning being given by

$$\mathbf{A}\,.\,\nabla = (A_1\mathbf{i}+A_2\mathbf{j}+A_3\mathbf{k})\,.\,\left(\mathbf{i}\frac{\partial}{\partial x}+\mathbf{j}\frac{\partial}{\partial y}+\mathbf{k}\frac{\partial}{\partial z}\right)$$

$$= A_1\frac{\partial}{\partial x}+A_2\frac{\partial}{\partial y}+A_3\frac{\partial}{\partial z}.$$

Note: $(\mathbf{A}\,.\,\nabla)\,\phi = \mathbf{A}\,.\,(\nabla\,\phi)$

If $\mathbf{A} = x^4\mathbf{i}+y\mathbf{j}-z\mathbf{k}$, then $\mathbf{A}.\nabla = x^4\frac{\partial}{\partial x}+y\frac{\partial}{\partial y}-z\frac{\partial}{\partial z}$.

So
$$(\mathbf{A}\,.\,\nabla)\,xy^2z^3 = \left(x^4\frac{\partial}{\partial x}+y\frac{\partial}{\partial y}-z\frac{\partial}{\partial z}\right)(xy^2z^3)$$
$$= x^4\frac{\partial}{\partial x}(xy^2z^3)+y\frac{\partial}{\partial y}(xy^2z^3)-z\frac{\partial}{\partial z}(xy^2z^3)$$
$$= x^4(y^2z^3)+y(2xyz^3)-z\,(3xy^2z^2).$$
$$(\mathbf{A}.\nabla)\,(y^2z^2\mathbf{i}+x^2\mathbf{j}+x^2\mathbf{k}) = \left(x^4\frac{\partial}{\partial x}+y\frac{\partial}{\partial y}-z\frac{\partial}{\partial z}\right)(y^2z^2\mathbf{i}+x^2\mathbf{j}+x^2\mathbf{k})$$
$$= x^4\frac{\partial}{\partial x}(y^2z^2\mathbf{i}+x^2\mathbf{j}+x^2\mathbf{k})+y\frac{\partial}{\partial y}(y^2z^2\mathbf{i}+x^2\mathbf{j}+x^2\mathbf{k})$$
$$-z\frac{\partial}{\partial z}(y^2z^2\mathbf{i}+x^2\mathbf{j}+x^2\mathbf{k})$$
$$= x^4\,(2x\mathbf{j}+2x\mathbf{k})+y(2yz^2\mathbf{i})-z(2y^2z\mathbf{i}) = 2x^5\,(\mathbf{j}+\mathbf{k}).$$
$$\mathbf{A}\times\nabla = (A_1\,\mathbf{i}+A_2\mathbf{j}+A_3\mathbf{k})\times\left(\mathbf{i}\frac{\partial}{\partial x}+\mathbf{j}\frac{\partial}{\partial y}+\mathbf{k}\frac{\partial}{\partial z}\right)$$

2.11 Divergence and Curl of a Gradient

Book work 2.7. If ϕ is a scalar point function, then

(*i*) divergence of the gradient of ϕ, that is, $\nabla\,.(\nabla\phi) = \nabla^2\phi$ **(2.11.1)**

(*ii*) curl of the gradient of ϕ vanishes, that is, $\nabla\times(\nabla\phi) = \mathbf{0}$ if ϕ has continuous partial derivatives. **(2.11.2).**

Proof:

(*i*)
$$\nabla.(\nabla\phi) = \left(\mathbf{i}\frac{\partial}{\partial x}+\mathbf{j}\frac{\partial}{\partial y}+\mathbf{k}\frac{\partial}{\partial z}\right).\left(\mathbf{i}\frac{\partial\phi}{\partial x}+\mathbf{j}\frac{\partial\phi}{\partial y}+\mathbf{k}\frac{\partial\phi}{\partial z}\right)$$
$$= \frac{\partial^2\phi}{\partial x^2}+\frac{\partial^2\phi}{\partial y^2}+\frac{\partial^2\phi}{\partial z^2} = \nabla^2\phi.$$

(*ii*)
$$\nabla\times(\nabla\phi) = \nabla\times\left(\mathbf{i}\frac{\partial\phi}{\partial x}+\mathbf{j}\frac{\partial\phi}{\partial y}+\mathbf{k}\frac{\partial\phi}{\partial z}\right)$$

curl of a gradient vanishes

$$= \begin{vmatrix}\mathbf{i} & \mathbf{j} & \mathbf{k}\\ \frac{\partial}{\partial x} & \frac{\partial}{\partial y} & \frac{\partial}{\partial z}\\ \frac{\partial\phi}{\partial x} & \frac{\partial\phi}{\partial y} & \frac{\partial\phi}{\partial z}\end{vmatrix} = \Sigma\mathbf{i}\left(\frac{\partial^2\phi}{\partial y\partial z}-\frac{\partial^2\phi}{\partial z\partial y}\right) = 0$$

because $\frac{\partial^2\phi}{\partial y\delta z} = \frac{\partial^2\phi}{\partial z\partial y}$ since the second partial derivatives of ϕ are continuous.

Note: It is noteworthy that, for every scalar point function, having continuous second partials, $\nabla \times (\nabla\phi) = \overline{0}$. In words, this result is *curl of a gradient vanishes.*

Remark. Now it follows that, if $\nabla^2\phi = 0$, then $\nabla\phi$ is both solenoidal and irrotational.

(*i*) Solenoidal, since $\nabla \cdot \nabla\phi = \nabla^2\phi = 0$ (given)

(*ii*) Irrotational, since $\nabla \times (\nabla\phi) = \overline{0}$ always.

2.12 Divergence and Curl of a Curl

Book work 2.8. If a vector point function $\mathbf{A} = A_1\mathbf{i} + A_2\mathbf{j} + A_3\mathbf{k}$, where A_1, A_2, A_3 have continuous second partials, then

(*i*) $\nabla.(\nabla \times \mathbf{A}) = 0$ (*i.e.*, divergence of a curl of a vector vanishes). **(2.12.1)**

(*ii*) $\nabla \times (\nabla \times \mathbf{A}) = \nabla(\nabla \cdot \mathbf{A}) - \nabla^2 \mathbf{A}$. **(2.12.2)**

Proof:

(*i*) $$\nabla \cdot (\nabla \times \mathbf{A}) = \nabla \cdot \begin{vmatrix} \mathbf{i} & \mathbf{j} & \mathbf{k} \\ \frac{\partial}{\partial x} & \frac{\partial}{\partial y} & \frac{\partial}{\partial z} \\ A_1 & A_2 & A_3 \end{vmatrix}$$

$$= \nabla \cdot \Sigma \mathbf{i}\left(\frac{\partial A_3}{\partial y} - \frac{\partial A_2}{\partial z}\right) = \left(\mathbf{i}\frac{\partial}{\partial x} + \mathbf{j}\frac{\partial}{\partial y} + \mathbf{k}\frac{\partial}{\partial z}\right) \cdot \Sigma \mathbf{i}\left(\frac{\partial A_3}{\partial y} - \frac{\partial A_2}{\partial z}\right)$$

$$= \Sigma \frac{\partial}{\partial x}\left(\frac{\partial A_3}{\partial y} - \frac{\partial A_2}{\partial z}\right) = \Sigma\left(\frac{\partial^2 A_3}{\partial x \partial y} - \frac{\partial^2 A_2}{\partial x \partial z}\right) = 0$$

because $\frac{\partial^2 A_3}{\partial x \partial y} = \frac{\partial^2 A_3}{\partial y \partial x}$, etc., since the second partial derivatives of A_1, A_2, A_3 are continous.

Note. It is noteworthy that, for every vector point function $\bar{A}$, having continuous second partials, $\nabla.(\nabla \times \bar{A}) = 0$. In words, the result is

Divergence of a curl vanishes.

(*ii*) $$\nabla \times (\nabla \times \mathbf{A}) = \nabla \times \begin{vmatrix} \mathbf{i} & \mathbf{j} & \mathbf{k} \\ \frac{\partial}{\partial x} & \frac{\partial}{\partial y} & \frac{\partial}{\partial z} \\ A_1 & A_2 & A_3 \end{vmatrix}$$

$$= \begin{vmatrix} \mathbf{i} & \mathbf{j} & \mathbf{k} \\ \frac{\partial}{\partial x} & \frac{\partial}{\partial y} & \frac{\partial}{\partial z} \\ \frac{\partial A_3}{\partial y} - \frac{\partial A_2}{\partial z} & \frac{\partial A_1}{\partial z} - \frac{\partial A_3}{\partial x} & \frac{\partial A_2}{\partial x} - \frac{\partial A_1}{\partial y} \end{vmatrix}$$

$$= \Sigma \mathbf{i}\left[\frac{\partial^2 A_2}{\partial y \partial x} - \frac{\partial^2 A_1}{\partial y^2} - \frac{\partial^2 A_1}{\partial z^2} + \frac{\partial^2 A_3}{\partial x \partial z}\right]$$

$$= \Sigma \mathbf{i}\left[\frac{\partial}{\partial x}\left(\frac{\partial A_2}{\partial y}+\frac{\partial A_3}{\partial z}\right)-\left(\frac{\partial^2 A_1}{\partial y^2}+\frac{\partial^2 A_1}{\partial z^2}\right)\right]$$

$$= \Sigma \mathbf{i}\left[\frac{\partial}{\partial x}\left(\frac{\partial A_1}{\partial x}+\frac{\partial A_2}{\partial y}+\frac{\partial A_3}{\partial z}\right)-\left(\frac{\partial^2 A_1}{\partial x^2}+\frac{\partial^2 A_1}{\partial y^2}+\frac{\partial^2 A_1}{\partial z^2}\right)\right], \text{ adding and subtracting } \mathbf{i}\frac{\partial^2 A_1}{\partial x^2}.$$

$$= \Sigma \mathbf{i}\left[\frac{\partial}{\partial x}(\nabla\cdot \mathbf{A})-\nabla^2 A_1\right]$$

$$= \Sigma \mathbf{i}\frac{\partial}{\partial x}(\nabla\cdot \mathbf{A})-\Sigma\nabla^2(A_1\mathbf{i})$$

$$= \nabla(\nabla\,.\,\mathbf{A})-\nabla^2\mathbf{A}.$$

Non-trivial calculation rules

1. div grad $f = \nabla\,.\,\nabla f$
2. curl grad $f = \nabla\times\nabla f = 0$
3. div curl $\mathbf{A} = \nabla\,.\,(\nabla\times\mathbf{A}) = 0$
4. curl curl $\mathbf{A} = \nabla\times(\nabla\times\mathbf{A}) = \nabla(\nabla\,.\,\mathbf{A})-\nabla^2\mathbf{A}$

2.13. Examples

2.3. Directional derivative

1. Solved Sum

Find the directional derivative of the function $x^2+y^2+z^2$ at (3, 6, 9) in the direction whose d.c.'s are 1/3, 2/3, 2/3.

Method 1. Directional derivative of $x^2+y^2+z^2$ at (x, y, z) in the direction whose d.c.'s are l, m, n is

$$l\frac{\partial\phi}{\partial x}+m\frac{\partial\phi}{\partial y}+n\frac{\partial\phi}{\partial z} = \frac{1}{3}\frac{\partial}{\partial x}(x^2+y^2+z^2)+\frac{2}{3}\frac{\partial}{\partial y}(x^2+y^2+z^2)+\frac{2}{3}\frac{\partial}{\partial z}(x^2+y^2+z^2)$$

$$= \frac{2x+4y+4z}{3}.$$

$$\therefore \text{ D.D. at } (3, 6, 9) = \frac{2\times3+4\times6+4\times9}{3} = 22.$$

Method 2. The unit vector in the given direction is

$$\hat{e} = (1/3)\,\bar{i}+(2/3)\,\bar{j}+(2/3)\,\bar{k}.$$

The D.D. of $x^2+y^2+z^2$ at (x, y, z) in the direction of $\hat{e}$ is

$$\nabla\phi.\hat{e} = \left[\left(\bar{i}\frac{\partial}{\partial x}+\bar{j}\frac{\partial}{\partial y}+\bar{k}\frac{\partial}{\partial z}\right)(x^2+y^2+z^2)\right]\cdot\frac{\bar{i}+2\bar{j}+2\bar{k}}{3}$$

$$= (2x\bar{i}+2y\bar{j}+2z\bar{k})\cdot\frac{\bar{i}+2\bar{j}+2\bar{k}}{3} = \frac{2x+4y+4z}{3}.$$

$$\therefore \quad [(\nabla\phi)\cdot\hat{e}]_{(3,6,9)} = \frac{2\times3+4\times6+4\times9}{3} = 22.$$

2. Find the directional derivative of $x + xy^2 + yz^3$ at the point (0, 1, 1) in the direction whose direction consines are 2/3, 2/3, –1/3.

[*Hint* : **Method 1.** Use 2.3.1; d.d = $\frac{2}{3}(1+y^2)+\frac{2}{3}(2xy+z^3) - \frac{1}{3}(3yz^2)$, where $x = 0, y = 1, z = 1$, d.d. = 1.

Method 2. The unit vector in the given direction is $\hat{e} = 2/3\bar{i} + 2/3\bar{j} - 1/3\bar{k}$.

$\therefore$ $(D.D.)_{(0,1,1)} = [(1+y^2)\bar{i} + (2xy+z^3)\bar{j} + (3yz^2)\bar{k} \cdot \hat{e} = 1]$

2.4. Gradient of a scalar point function

3. Find $\nabla\phi$ if

(*i*) $\phi = x^2y^3z^2$ (*ii*) $\phi = xyz - x^2$

[**Hint:** (*i*) $\nabla\phi = \bar{i}\frac{\partial}{\partial x}(x^2y^3z^2) + \bar{j}\frac{\partial}{\partial y}(x^2y^3z^2) + \bar{k}\frac{\partial}{\partial z}(x^2y^3z^2)$

$$= 2xy^3z^2\bar{i} + 3x^2y^2z^2\bar{j} + 2x^2y^3z\bar{k}$$

(*ii*) $\nabla\phi = (yz - 2x)\bar{i} + (xz)\bar{j} + (xy)\bar{k}$]

4. Find $\nabla\phi$ in the following cases at the points specified:

(*i*) $\phi(x, y, z) = 2xz - y^2$ at (1, 3, 2).

(*ii*) $\phi(x, y, z) = x + xy^2 + yz^2$ at (1, 0, 0)

(*iii*) $\phi(x, y, z) = y^2(x - z)$ at (1, 1, 2)

[**Hint:** (*i*) $\nabla\phi = \left[\bar{i}\frac{\partial}{\partial x} + \bar{j}\frac{\partial}{\partial y} + \bar{k}\frac{\partial}{\partial z}\right](2xz - y^2) = 2z\bar{i} - 2y\bar{j} + 2x\bar{k}.$

$$(\nabla\phi)_{(1,3,2)} = 4\bar{i} - 6\bar{j} + 2\bar{k}.$$

(*ii*) $(\nabla\phi)_{(1,0,0)} = [(1 + y^2)\bar{i} + (2xy + z^2)\bar{j} + 2yz\,\bar{k}]_{(1,0,0)} = \bar{i}$

(*iii*) $(\nabla\phi)_{(1,1,2)} = [y^2\bar{i} + 2y(x - z)\bar{j} - y^2\bar{k}]_{(1,1,2)} = \bar{i} - 2\bar{j} - \bar{k}$]

5. Find $\nabla\phi$ at (x, y, z) if $\phi = x + xy^2 + yz^3$. [Ans. $(1 + y^2)\mathbf{i} + (2xy + z^3)\mathbf{j} + 3yz^2\mathbf{k}$.]

6. *Solved sum.* Find the directional derivative of $\phi = x + xy^2 + yz^3$ at (0, 1, 1) in the direction of the vector $2\mathbf{i} + 2\mathbf{j} - \mathbf{k}$.

The unit vector corresponding to $2\mathbf{i} + 2\mathbf{j} - \mathbf{k}$ is

$$\frac{2\mathbf{i}+2\mathbf{j}-\mathbf{k}}{\sqrt{2^2+2^2+(-1)^2}} = \frac{2\mathbf{i}+2\mathbf{j}-\mathbf{k}}{3} = \mathbf{e} \text{ say.}$$

$$\nabla\phi = \left(\mathbf{i}\frac{\partial}{\partial x} + \mathbf{j}\frac{\partial}{\partial y} + \mathbf{k}\frac{\partial}{\partial z}\right)(x + xy^2 + yz^3)$$

$$= \mathbf{i}(1 + y^2) + \mathbf{j}(2xy + z^3) + \mathbf{k}(3yz^2)$$

$\therefore$ $\nabla\phi$ at (0, 1, 1) = $2\mathbf{i} + \mathbf{j} + 3\mathbf{k}$.

The directional derivatives of ϕ in the direction of $\mathbf{e}$ is

$$(\nabla\phi)\cdot\mathbf{e} = (2\mathbf{i} + \mathbf{j} + 3\mathbf{k})\cdot\frac{2\mathbf{i}+2\mathbf{j}-\mathbf{k}}{3} = \frac{4+2-3}{3} = 1.$$

7. *Solved sum.* Find the directional derivatives of the following scalar point functions at the given points in the given directions:

Function ϕ	Point	Direction
(*i*) $xy+yz+zx$	(1, 1, 3)	$\bar{i}+2\bar{j}+2\bar{k}$
(*ii*) xyz	(2, 1, 1)	$\bar{j}+\bar{k}$
(*iii*) $x^2yz+4xz^2$	(1, −2, − 1)	$2\bar{i}-\bar{j}-2\bar{k}$
(*iv*) $2xy+5yz+zx$	(1, 2, 3)	$3\bar{i}-5\bar{j}+4\bar{k}$

Formula. D.D. in the direction of $\hat{e}=(\nabla\phi).\hat{e}$.

(*i*) The unit vector in the direction of the given vector is

$$\hat{e} = \frac{\bar{i}+2\bar{j}+2\bar{k}}{\sqrt{1^2+2^2+2^2}}=\frac{\bar{i}+2\bar{j}+2\bar{k}}{3}.$$

$$\nabla\phi = \left(\bar{i}\frac{\partial}{\partial x}+\bar{j}\frac{\partial}{\partial y}+\bar{k}\frac{\partial}{\partial z}\right)[xy+yz+zx]$$

$$= \bar{i}(y+0+z)+\bar{j}(x+z+0)+\bar{k}(0+y+x)$$

$$= \bar{i}(y+z)+\bar{j}(x+z)+\bar{k}(y+x).$$

$$\therefore \quad (\nabla\phi)_{(1,1,3)} = \bar{i}(1+3)+\bar{j}(1+3)+\bar{k}(1+1) = 4\bar{i}+4\bar{j}+2\bar{k}.$$

So the directional derivative of ϕ at (1, 1, 3) in the given direction is

$$(\nabla\phi).\hat{e} = (4\bar{i}+4\bar{j}+2\bar{k}).\frac{\bar{i}+2\bar{j}+2\bar{k}}{3}$$

$$= \frac{4(1)+4(2)+2(2)}{3}=\frac{16}{3}.$$

(*ii*)

$$\hat{e} = \frac{\bar{j}+\bar{k}}{\sqrt{1^2+1^2}}=\frac{\bar{j}}{\sqrt{2}}+\frac{\bar{k}}{\sqrt{2}}.$$

$$\nabla\phi = \bar{i}yz+\bar{j}zx+\bar{k}xy.$$

$$\therefore \quad (\nabla\phi)_{(2,1,1)} = \bar{i}+2\bar{j}+2\bar{k}.$$

$$\therefore \quad (\nabla\phi).\hat{e} = 0+\frac{2}{\sqrt{2}}+\frac{2}{\sqrt{2}}=2\sqrt{2}.$$

(*iii*)

$$\hat{e} = \frac{2\bar{i}-\bar{j}-2\bar{k}}{\sqrt{2^2+(-1)^2+(-2)^2}}=\frac{2\bar{i}-\bar{j}-2\bar{k}}{3}.$$

$$\nabla\phi = \bar{i}(2xyz+4z^2)+\bar{j}(x^2z)+\bar{k}(x^2y+8xz).$$

$$\therefore \quad (\nabla\phi)_{(1,-2,-1)} = 8\bar{i}-\bar{j}-10\bar{k}.$$

$$\therefore \quad (\nabla\phi).\hat{e} = \frac{16+1+20}{3}=\frac{37}{3}.$$

(*iv*)

$$\hat{e} = \frac{3\bar{i}-5\bar{j}+4\bar{k}}{\sqrt{3^2+(-5)^2+4^2}}=\frac{3\bar{i}-5\bar{j}+4\bar{k}}{5\sqrt{2}}.$$

$$\nabla\phi = \bar{i}(2y+z)+\bar{j}(2x+5z)+\bar{k}(5y+x).$$

$$\therefore \quad (\nabla\phi)_{(1,2,3)} = 7\bar{i}+17\bar{j}+11\bar{k}.$$

$$\therefore \quad (\nabla\phi).\hat{e} = \frac{21-85+44}{5\sqrt{2}}=-\frac{20}{5\sqrt{2}}=-2\sqrt{2}.$$

8. Find the directional derivative of ϕ at the given point in the direction of the given vector.

	Function ϕ	Point	Vector
(i)	$2x^2+3y^2+z^2$	(2, 1, 3)	$\bar{i}-2\bar{k}$
(ii)	xy^2	(1, 1, 0)	$\bar{i}+\bar{j}+\bar{k}$
(iii)	yz^2	(0, 1, 1)	$\bar{i}+\bar{j}+\bar{k}$
(iv)	$3xy^2-x^2yz$	(1, 2, 3)	$\bar{i}-2\bar{j}+2\bar{k}$
(v)	$xyz-xy^2z^3$	(1, 2, –1)	$\bar{i}-\bar{j}-3\bar{k}$
(vi)	$x^3+y^3+z^3$	(1, –1, 2)	$\bar{i}+2\bar{j}+\bar{k}$
(vii)	$4xz^2+2xyz$	(1, 2, 3)	$2\bar{i}+\bar{j}-\bar{k}$
(viii)	x^2y^2z	(2, 1, 4)	$\bar{i}+2\bar{j}+2\bar{k}$
(ix)	$x^2yz+4xz^2+xyz$	(1, 2, 3)	$2\bar{i}+\bar{j}-\bar{k}$

[**Hint:** $\nabla\phi$ at the point, $\hat{e}$ and the required directional derivatives $(\nabla\phi).\hat{e}$ at the point are

(i) $8\bar{i}+6\bar{j}+6\bar{k}$, $\quad \dfrac{\bar{i}-2\bar{k}}{\sqrt{5}}$, $\quad -\dfrac{4}{\sqrt{5}}$

(ii) $\bar{i}+2\bar{j}$, $\quad \dfrac{\bar{i}+\bar{j}+\bar{k}}{\sqrt{3}}$, $\quad \dfrac{3}{\sqrt{3}}$

(iii) $\bar{j}+2\bar{k}$, $\quad \dfrac{\bar{i}+\bar{j}+\bar{k}}{\sqrt{3}}$, $\quad \dfrac{3}{\sqrt{3}}$

(iv) $0\bar{i}+9\bar{j}-2\bar{k}$, $\quad \dfrac{\bar{i}-2\bar{j}+2\bar{k}}{3}$, $\quad \dfrac{-22}{3}$

(v) $2\bar{i}+3\bar{j}-10\bar{k}$, $\quad \dfrac{\bar{i}-\bar{j}-3\bar{k}}{\sqrt{11}}$, $\quad \dfrac{29}{\sqrt{11}}$

(vi) $3\bar{i}+3\bar{j}+12\bar{k}$, $\quad \dfrac{\bar{i}+2\bar{j}+\bar{k}}{\sqrt{6}}$, $\quad \dfrac{21}{\sqrt{6}}$

(vii) $48\bar{i}+6\bar{j}+28\bar{k}$, $\quad \dfrac{2\bar{i}+\bar{j}-\bar{k}}{\sqrt{6}}$, $\quad \dfrac{74}{\sqrt{6}}$

(viii) $16\bar{i}+32\bar{j}+4\bar{k}$, $\quad \dfrac{\bar{i}+2\bar{j}+2\bar{k}}{3}$, $\quad \dfrac{88}{3}$

(ix) $54\bar{i}+6\bar{j}+28\bar{k}$, $\quad \dfrac{2\bar{i}+\bar{j}-\bar{k}}{\sqrt{6}}$, $\quad \dfrac{86}{\sqrt{6}}$]

9. Solved Sum

Prove that the directional derivative of $\phi = x^3y^2z$ at (1, 2, 3) is a maximum along the direction $9\bar{i}+3\bar{j}+\bar{k}$. Find this maximum directional derivative.

We know that the direction in which the directional derivative of ϕ is maximum is the direction of the vector $\nabla\phi$ and this maximum is $|\nabla\phi|$. Now

$$\nabla\phi = \nabla(x^3y^2z) = \left(\bar{i}\frac{\partial}{\partial x}+\bar{j}\frac{\partial}{\partial y}+\bar{k}\frac{\partial}{\partial z}\right)(x^3y^2z)$$

$$= \bar{i}(3x^2y^2z)+\bar{j}(2x^3yz)+\bar{k}(x^3y^2).$$

$$(\nabla\phi)_{(1,2,3)} = \bar{i}(3.1.4.3)+\bar{j}(2.1.2.3)+\bar{k}(1.4)$$
$$= 4(9\bar{i}+3\bar{j}+\bar{k}).$$

So, in the direction of $4(9\bar{i}+3\bar{j}+\bar{k})$ or $9\bar{i}+3\bar{j}+\bar{k}$, the directional derivative is a maximum.

This maximum is $|4(9\bar{i}+3\bar{j}+\bar{k})| = 4\sqrt{9^2+3^2+1} = 4\sqrt{91}.$

10. *Solved sum.* Find the directional derivative of $\phi = 3x^2 + 2y - 3z$ at the point (1, 1, 1) in the direction specified by $2\mathbf{i} + 2\mathbf{j} - \mathbf{k}$. Also find the maximum value of the directional derivative at that point and the unit vector of the direction pertaining to this maximum.

The gradient of ϕ at (x, y, z) is

$$\nabla\phi = \left(\mathbf{i}\frac{\partial}{\partial x}+\mathbf{j}\frac{\partial}{\partial y}+\mathbf{k}\frac{\partial}{\partial z}\right)(3x^2+2y-3z)$$

$$= \mathbf{i}\frac{\partial}{\partial x}(3x^2+2y-3z)+\mathbf{j}\frac{\partial}{\partial y}(3x^2+2y-3z)+\mathbf{k}\frac{\partial}{\partial z}(3x^2+2y-3z)$$

$$= 6x\mathbf{i} + 2\mathbf{j} - 3\mathbf{k}.$$

So at (1, 1, 1), $\nabla\phi = 6\mathbf{i}+2\mathbf{j}-3\mathbf{k}$. Now the unit vector in the direction of the vector $2\mathbf{i}+2\mathbf{j}-\mathbf{k}$ is

$$\frac{2\mathbf{i}+2\mathbf{j}-\mathbf{k}}{\sqrt{\{2^2+2^2+(-1)^2\}}} = \frac{2}{3}\mathbf{i}+\frac{2}{3}\mathbf{j}-\frac{1}{3}\mathbf{k}.$$

So the directional derivative in this direction at (1, 1, 1) is $(\nabla\phi).\left(\frac{2}{3}\mathbf{i}+\frac{2}{3}\mathbf{j}-\frac{1}{3}\mathbf{k}\right)$

$$= (6\mathbf{i}+2\mathbf{j}-3\mathbf{k}).\left(\frac{2}{3}\mathbf{i}+\frac{2}{3}\mathbf{j}-\frac{1}{3}\mathbf{k}\right) = 6\frac{1}{3}.$$

The maximum value and the respective unit vector are respectively $|\nabla\phi|$ and $\frac{\nabla\phi}{|\nabla\phi|}$.

But, at (1, 1, 1),

$$|\nabla\phi| = \sqrt{6^2+2^2+(-3)^2} = 7$$

and

$$\frac{\nabla\phi}{|\nabla\phi|} = \frac{6}{7}\mathbf{i}+\frac{2}{7}\mathbf{j}+\frac{3}{7}\mathbf{k}.$$

11. If $\mathbf{r} = x\mathbf{i} + y\mathbf{j} + z\mathbf{k}$ and $|\mathbf{r}| = r$, that is, if $\mathbf{r}$ is the position vector of a variable point (x, y, z), then show that

(*i*) $\nabla r^n = nr^{n-1}\hat{\mathbf{r}} = nr^{n-2}\mathbf{r}$. **(1977)**

(*ii*) $\nabla(\mathbf{r}.\mathbf{a}) = \mathbf{a}$ and $\nabla[\mathbf{r}, \mathbf{a}, \mathbf{b}] = \mathbf{a} \times \mathbf{b}$, where $\mathbf{a}$ and $\mathbf{b}$ are constant vectors.

(*iii*) $\nabla(\mathbf{a}.\mathbf{r}) = 2\mathbf{a}$ if $\mathbf{a} = \alpha x\mathbf{i} + \beta y\mathbf{j} + \gamma z\mathbf{k}$, where α, β, γ are constants.

(*iv*) $\frac{dF}{dt} = \frac{\partial F}{\partial t}+(\nabla F)\cdot\frac{d\mathbf{r}}{dt}$, where F is a scalar function of t and of x, y, z which are functions of t.

[**Hint:** (*i*) Let $\mathbf{a} = a_1\mathbf{i} + a_2\mathbf{j} + a_3\mathbf{k}$. Then $\mathbf{r}.\mathbf{a} = a_1x + a_2y + a_3z$

$$\therefore\ \nabla(\bar{r}\cdot\bar{a}) = \nabla(a_1x+a_2y+a_3z) = a_1\bar{i}+a_2\bar{j}+a_3\bar{k} = \bar{a} \quad (1)$$

$$\nabla[\bar{r},\bar{a},\bar{b}] = \nabla[\bar{r}\cdot(\bar{a}\times\bar{b})] = \bar{a}\times\bar{b} \text{ by (1)}$$

(*iii*) $\nabla(\bar{a}\cdot\bar{r}) = \nabla(\alpha x^2+\beta y^2+\gamma z^2) = 2\bar{a}$

(*iv*) $\frac{dF}{dt} = \frac{\partial F}{\partial x}\frac{dx}{dt}+\frac{\partial F}{\partial y}\frac{dy}{dt}+\frac{\partial F}{\partial z}\frac{dz}{dt}+\frac{\partial F}{\partial t}$.]

12. If $\nabla\phi = 5r^3\mathbf{r}$, find ϕ.

[Hint: $\nabla\phi = 5r^4\,\hat{\mathbf{r}}$. But $\nabla\phi(r) = \phi'(r)\,\hat{\mathbf{r}}$. Therefore, $\phi'(r) = 5r^4$ and $\phi = r^5 + C$.**]**

13. If $\nabla\phi = (6r - 8r^2)\,\mathbf{r}$ and $\phi(2) = 4$, find ϕ. *[Ans.:* $2(r^3 - r^4 + 10)$*]*

14. *Solved Sum.* If

$\nabla\phi = (y + y^2 + z^2)\mathbf{i} + (x + z + 2xy)\mathbf{j} + (y + 2zx)\mathbf{k}$ and if $\phi(1, 1, 1) = 3$, find ϕ.

Since LHS is $\mathbf{i}\dfrac{\partial\phi}{\partial x} + \mathbf{j}\dfrac{\partial\phi}{\partial y} + \mathbf{k}\dfrac{\partial\phi}{\partial z}$, we get

$$\frac{\partial\phi}{\partial x} = y + y^2 + z^2 \quad \text{... (1)}$$

$$\frac{\partial\phi}{\partial y} = x + z + 2xy \quad \text{... (2)}$$

$$\frac{\partial\phi}{\partial z} = y + 2zx. \quad \text{... (3)}$$

Integrating (1) with respect to x,

$$\phi = xy + xy^2 + xz^2 + f(y, z), \quad \text{... (4)}$$

where the arbitrary constant of integration, $f(y, z)$, is free from x. Differentiating (4) partially with respect to y and using (2).

$$\frac{\partial\phi}{\partial y} = x + 2xy + \frac{\partial f(y,z)}{\partial y} = x + z + 2xy.$$

$$\therefore \quad \frac{\partial f(y,z)}{\partial y} = z. \quad \text{... (5)}$$

Now integrating this with respect to y, $f(y, z) = yz + g(z)$. So (4) becomes

$$\phi = xy + xy^2 + xz^2 + yz + g(z). \quad \text{... (6)}$$

Differentiating this partially with respect to z and using (3)

$$\frac{\partial\phi}{\partial z} = 2zx + y + \frac{dg(z)}{dz} = y + 2zx.$$

$$\therefore \quad \frac{dg(z)}{dz} = 0 \text{ or } g(z) = C.$$

So (6) becomes $\phi = xy + xy^2 + xz^2 + yz + C$. Here $\phi(1, 1, 1) = 4 + C$. But $\phi(1, 1, 1)$ is given to be be 3. So $4 + C = 3$ or $C = -1$. Thus $\phi = xy + xy^2 + xz^2 + yz - 1$.

Remark. Integrating $\dfrac{\partial\phi}{\partial x}, \dfrac{\partial\phi}{\partial y}, \dfrac{\partial\phi}{\partial z}$ with respect to x, y, z respectively,

$$\phi = xy + xy^2 + xz^2 + f_1(y, z),$$

$$\phi = xy + yz^2 + xy^2 + f_2(x, z),$$

$$\phi = yz + xz^2 + f_3(x, y).$$

From these results ϕ can be obtained by adding all the terms and an arbitrary constant, but omitting

$$f_1(y, z), \quad f_2(x, z), \quad f_3(x, y).$$

and choosing only one from the repeated terms. Thus

$$\phi = xy + xy^2 + xz^2 + yz + C.$$

Since $\phi(1, 1, 1) = 3, C = -1.$

$\therefore$ $\phi = xy + xy^2 + xz^2 + yz - 1.$

15. Find ϕ if $\nabla\phi$ is

(i) $(6xy + z^3)\mathbf{i} + (3x^2 - z)\mathbf{j} + (3xz^2 - y)\mathbf{k}.$ **(1977)**

(ii) $(2x + 3y + 4xy^2)\mathbf{i} + (3x + 3y^2 + 4x^2y + z + 1)\mathbf{j} + (y + 2)\mathbf{k}.$

(iii) $x(2yz + 1)\mathbf{i} + x^2z\,\mathbf{j} + x^2y\,\mathbf{k}.$ **(1978)**

(iv) $(y + \sin z)\mathbf{i} + x\,\mathbf{j} + x\cos z\,\mathbf{k}.$ **(1985)**

(v) $(y^2\cos x + z^3)\bar{i} + (2y\sin x - 4)\bar{j} + 3xz^2\bar{k}$

[Ans.: (i) $3x^2y + xz^3 - yz + C.$

(ii) $x^2 + 3xy + 2x^2y^2 + yz + y^3 + y + 2z + C.$

(iii) $x(xyz + \frac{1}{2}x) + C.$

(iv) $x(y + \sin z) + C.$

(v) $y^2\sin x - 4y + xz^3 + C$]

16. State the unit vector normal to the surface $\phi = c$. [Ans. $\frac{\nabla\phi}{|\nabla\phi|}$]

17. *Solved Sum.* Find the unit vector normal to the surface $x^2 + y^2 + 2z^2 = 4$ at the point (1, 1, 1).

Formula. Unit vector normal to $\phi = c$ is $\frac{\nabla\phi}{|\nabla\phi|}$.

$$\nabla\phi = \nabla(x^2 + y^2 + 2z^2 - 4)$$

$$= \left(\bar{i}\frac{\partial}{\partial x} + \bar{j}\frac{\partial}{\partial y} + \bar{k}\frac{\partial}{\partial z}\right)(x^2 + y^2 + 2z^2 - 4)$$

$$= \bar{i}(2x) + \bar{j}(2y) + \bar{k}(4z).$$

$\therefore$ $(\nabla\phi)_{(1,1,1)} = 2\bar{i} + 2\bar{j} + 4\bar{k}.$

This vector is normal to the surface at (1, 1, 1). So the corresponding required unit vector is

$$\frac{2\bar{i} + 2\bar{j} + 4\bar{k}}{\sqrt{2^2 + 2^2 + 4^2}} \text{ or } \frac{\bar{i} + \bar{j} + 2\bar{k}}{\sqrt{6}}.$$

18. Find the unit vectors $\hat{n}$ normal to the following surfaces at the specified points:

(i) $x^2 + 2y^2 + z^2 = 7$ at (1, – 1, 2) (ii) $x^2 + y^2 - z^2 = 1$ at (1, 1, 1)

(iii) $x^2 + 3y^2 + 2z^2 = 6$ at (2, 0, 1) (iv) $x^2 + y^2 - z = 1$ at (1, 1, 1)

(v) $x^3 - xyz + y^3 = 1$ at (1, 1, 1)

[Ans: (i) $\frac{\bar{i} - 2\bar{j} + 2\bar{k}}{3}$ (ii) $\frac{\bar{i} + \bar{j} - \bar{k}}{\sqrt{3}}$ (iii) $\frac{\bar{i} + \bar{k}}{\sqrt{2}}$ (iv) $\frac{2\bar{i} + 2\bar{j} - \bar{k}}{3}$ (v) $\frac{2\bar{i} + 2\bar{j} - \bar{k}}{3}$]

19. Solved Sum. Find the angle between the surfaces

$$x^2 + y^2 + z^2 = 9, z = x^2 + y^2 - 3$$

at the point (2, – 1, 2).

The required angle is the angle between the tangent planes at (2, –1, 2). which is the angle between the normals at (2, –1, 2). The normal to the first surface is parallel to

$$\nabla(x^2 + y^2 + z^2 - 9) = 2x\bar{j} + 2y\bar{j} + 2z\bar{k} \text{ or } x\bar{i} + y\bar{j} + z\bar{k}.$$

At (2, –1, 2) this vector is $2\bar{i} - \bar{j} + 2\bar{k}$.

The normal to the second surface is parallel to

$$\nabla(x^2 + y^2 - z - 3) = 2x\bar{i} + 2y\bar{j} - \bar{k}.$$

At (2, –1, 2) this vector is $4\bar{i} - 2\bar{j} - \bar{k}$.

Thus, if θ is the required angle, then

$$\cos\theta = \frac{(2\bar{i} - \bar{j} + 2\bar{k})\cdot(4\bar{i} - 2\bar{j} - \bar{k})}{\sqrt{2^2 + 1^2 + 2^2}\sqrt{4^2 + 2^2 + 1^2}} = \frac{8 + 2 - 2}{3\sqrt{21}} = \frac{8}{3\sqrt{21}}.$$

$$\therefore \quad \theta = \cos^{-1}\frac{8}{3\sqrt{21}}.$$

20. *Solved sum.* **Find the equation of the tangent plane to the ellipsoid $x^2 + 2y^2 + 3z^2 = 6$ at the point (1, –1, 1).**

The vector normal to the surface at (x, y, z) is $\nabla(x^2 + 2y^2 + 3z^2 - 6) = 2x\mathbf{i} + 4y\mathbf{j} + 6z\mathbf{k}$.

$$\nabla\phi \text{ at } (1, 1, 1) = 2\mathbf{i} + 4\mathbf{j} + 6\mathbf{k}.$$

This vector at (1, –1, 1) is $2\mathbf{i} - 4\mathbf{j} + 6\mathbf{k}$. $\nabla\phi$ is normal to the ellipsoid at (1, 1, 1). So the d.r's of the normal at (1, 1, 1) are 2, 4, 6. So, by $(\mathbf{r} - \mathbf{r}_1).\mathbf{p} = 0$, the vector equation of the tangent plane at (1, – 1, 1) is

$$[(x\mathbf{i} + y\mathbf{j} + z\mathbf{k}) - (\mathbf{i} - \mathbf{j} + \mathbf{k})] . (2\mathbf{i} - 4\mathbf{j} + 6\mathbf{k}) = 0$$

$$\text{or } x - 2y + 3z - 6 = 0.$$

21. **Find the equation of the tangent plane to the surface $x^2 + 2y^2 + 3z^2 = 6$ at (1, –1, 1).**

[Hint: $[\nabla(x^2 + 2y^2 + 3z^2)]_{(1,-1,1)} = 2\bar{i} - 4\bar{j} + 6\bar{k}$. So the d.r's of the normal to the plane are 2, –4, 6. The equation of the plane is $2(x - 1) - 4(y + 1) + 6(z - 1) = 0$ or $x - 2y + 3z = 6$]

22. **Find the equation of the tangent plane to the surface $x^2 - 4y^2 + 3z^2 + 4 = 0$ at (3, 2, 1).**

[Ans: $3x - 8y + 3z + 4 = 0$.]

23. **Find the angle between the normals to the surface $xy - z^2 = 0$ at the points (1, 4, – 2) and (–3, –3, 3). (1983)**

[Hint: If θ is the angle between the normals, then

$$\cos\theta = \frac{(4\mathbf{i} + \mathbf{j} + 4\mathbf{k})\cdot(-3\mathbf{i} - 3\mathbf{j} - 6\mathbf{k})}{\sqrt{16 + 1 + 16}\sqrt{9 + 9 + 36}} = \frac{-13}{3\sqrt{22}}].$$

24. **Show that the surfaces $5x^2 - 2yz - 9x = 0$ and $4x^2y + z^3 - 4 = 0$ are orthogonal at (1, –1, 2).**

[Hint: $(\bar{i} - 4\bar{j} + 2\bar{k})\cdot(-8\bar{i} + 4\bar{j} + 12\bar{k}) = 0$. So the normals are perpendicular. Hence the tangent planes are perpendicular].

25. **Find the angle between the surfaces $x^2 + yz = 2$ and $x + 2y - z = 2$ at (1, 1, 1).**

[Hint: If θ is the angle between the surfaces, then

$$\cos\theta = \frac{(2\bar{i} + \bar{j} + \bar{k})\cdot(\bar{i} + 2\bar{j} - \bar{k})}{\sqrt{2^2 + 1^2 + 1^2}\sqrt{1^2 + 2^2 + (-1)^2}} = \frac{1}{2}. \text{ So } \theta = \frac{\pi}{3}].$$

26. **Find the angle between the surfaces $x^2 - y^2 - z^2 - 11 = 0$, $xy + yz - zx - 18 = 0$ at (6, 4, 3).**

[Hint: If θ is the angle between the surfaces, then

$$\cos\theta = \frac{(12\bar{i} - 8\bar{j} - 6\bar{k})\cdot(\bar{i} + 9\bar{j} - 2\bar{k})}{\sqrt{12^2 + (-8)^2 + (-6)^2}\sqrt{1^2 + 9^2 + (-2)^2}} = -\frac{24}{\sqrt{61}\sqrt{86}}].$$

27. Find the angle between the surfaces $x^2 + y^2 = 4 - 5z$ and $x^2 + y^2 + 3z^2 = 104$ at $(5, 2, -5)$.
[Hint: If θ is the angle between the surfaces, then

$$\cos\theta = \frac{(10\bar{i} + 4\bar{j} + 5\bar{k})\cdot(10\bar{i} + 4\bar{j} - 30\bar{k})}{\sqrt{10^2 + 4^2 + 5^2}\sqrt{10^2 + 4^2 + (-30)^2}} = \frac{17}{\sqrt{141}\sqrt{254}}.]$$

28. Find the angle between the normals to the intersecting surfaces $xy - z^2 - 1 = 0$ and $y^2 - 3z - 1 = 0$ at (1, 1, 0). Also find a unit vector along the tangent to the curve of intersection of the surfaces at (1, 1, 0).

[Hint: Vectors along the normals: $\mathbf{a} = \mathbf{i} + \mathbf{j}$, $\mathbf{b} = 2\mathbf{j} - 3\mathbf{k}$; angle: $\cos^{-1}(2/\sqrt{26})$; unit vector along the tangent : $\mathbf{a} \times \mathbf{b} / |\mathbf{a} \times \mathbf{b}| = (-3\mathbf{i} + 3\mathbf{j} + 2\mathbf{k}) / \sqrt{22}$.]

29. *Solved Sum.* The temperature at any point (x, y, z) in space is given by

$$T = xy + yz + 2zx.$$

Find the direction in which the rate of increase in temperature at (1, 1, 1), w.r.t distance, is the greatest. Find this rate.

$$\nabla T = \left(\bar{i}\frac{\partial}{\partial x} + \bar{j}\frac{\partial}{\partial y} + \bar{k}\frac{\partial}{\partial z}\right)(xy + yz + 2zx)$$

$$= \bar{i}(y+2z) + \bar{j}(x+z) + \bar{k}(y+2x).$$

$$\therefore \quad (\nabla T)_{(1,1,1)} = 3\bar{i} + 2\bar{j} + 3\bar{k}. \qquad \text{... (1)}$$

The direction of this vector (1) is the direction in which the rate of increase of temperature at (1, 1, 1) is the greatest. This rate is

$$\sqrt{3^2 + 2^2 + 3^2} \text{ or } \sqrt{22}.$$

Note. The level surface of temperature through (1, 1, 1) is $xy + yz + 2zx = 4$. The vector (1) is normal to this surface at (1, 1, 1).

30. Find the maximum value of directional derivative of the function $\phi = 2x^2 + 3y^2 + 5z^2$ at the point $(1, 1, -4)$.

[Hint: Max. D.D. at $(1, 1, -4) = |\nabla\phi|$ at $(1, 1, -4) = |4x\bar{i} + 6y\bar{j} + 10z\bar{k}|_{(1,1,-4)} = \sqrt{4^2 + 6^2 + 40^2}$]

31. The temperature T at the point (x, y, z) in space is given by $T = xy^2z^3$. Find the direction in which the rate of increase of temperature at (1, 1, 1) is the greatest. Find this maximum rate.
[Hint: $[\nabla(xy^2z^3)]_{(1,1,1)} = \bar{i} + 2\bar{j} + 3\bar{k}$. The direction of this vector is the required direction. Maximum rate of increase is $|\bar{i} + 2\bar{j} + 3\bar{k}|$ or $\sqrt{14}$]

32. Find the direction in which $\phi = xy^2 + yz^2 + zx^2$ increases most rapidly at the point $(1, 2, -3)$.
[Hint: Direction of $\nabla(xy^2 + yz^2 + zx^2) = -2\mathbf{i} + 13\mathbf{j} - 11\mathbf{k}$.]

33. In a fluid which is in motion the temperature at (x, y, z) is $T(x, y, z, t) = x^2 + y^2 + tz^2$. Find the time-rate of change of temperature at (1, 2, 3) at time $t = 2$ if the velocity of the fluid particle there is $\mathbf{i} + \mathbf{j} + \mathbf{k}$.

[Hint: $\dfrac{dT}{dt} = \dfrac{\partial T}{\partial x}\dfrac{dx}{dt} + \dfrac{\partial T}{\partial y}\dfrac{dy}{dt} + \dfrac{\partial T}{\partial z}\dfrac{dz}{dt} + \dfrac{\partial T}{\partial t}$

$$= (\nabla T).v + \frac{\partial T}{\partial t} = (2x\mathbf{i} + 2y\mathbf{j} + 2zt\mathbf{k})\cdot(\mathbf{i} + \mathbf{j} + \mathbf{k}) + z^2 = 27$$

since $x = 1, y = 2, z = 3, t = 2$.]

34. Show that, if A is a fixed point and P, a variable point, then $\nabla(AP) = \nabla\hat{AP}$. Use this result to show that the focal chords passing through a point P on an ellipse are equally inclined to the tangent to the ellipse at P.

[**Hint:** If $\mathbf{a} = a_1\mathbf{i} + a_2\mathbf{j} + a_3\mathbf{k}$ and $\mathbf{r} = x\mathbf{i} + y\mathbf{j} + z\mathbf{k}$ are the position vectors of A and P,

$$\nabla(AP) = \nabla|\mathbf{r}-\mathbf{a}| = \nabla\{(x-a_1)^2 + (y-a_2)^2 + (z-a_3)^2\}^{1/2}$$

$$= \frac{(x-a_1)\mathbf{i}+(y-a_2)\mathbf{j}+(z-a_3)\mathbf{k}}{\{(x-a_1)^2+(y-a_2)^2+(z-a_3)^2\}^{1/2}}$$

$$= \frac{\overline{AP}}{AP} = \hat{AP}$$

Let S_1 and S_2 be the foci. Then $S_1P + S_2P$ is a constant. That is, the equation of the ellipse is $\phi = S_1P + S_2P = c$. So $\nabla(S_1P + S_2P)$ is normal to the ellipse. Thus

$\nabla(S_1P + S_2P) \cdot \mathbf{T} = 0$ or $(\hat{S_1P} + \hat{S_2P}) \cdot \mathbf{T} = 0$, where **T** is the unit tangent vector to the ellipse at P. So $\hat{S_1P} \cdot \mathbf{T} + \hat{S_2P} \cdot (-\mathbf{T})$].

2.7. Divergence of a vector point function

35. *Solved sum.* Find the divergence of $x^2\bar{i} + y^2\bar{j} + z^2\bar{k}$.

$$\text{Divergence} = \frac{\partial}{\partial x}(x^2) + \frac{\partial}{\partial y}(y^2) + \frac{\partial}{\partial z}(z^2) = 2(x+y+z).$$

36. Find the divergence of the vector point function

$$\bar{A} = x^2yz\bar{i} + xyz^2\bar{j} + \bar{k}.$$

[**Hint:** $\nabla.\ \bar{A} = \frac{\partial}{\partial x}(x^2yz) + \frac{\partial}{\partial y}(xyz^2) + \frac{\partial(1)}{\partial z} = 2xyz + xz^2$]

37. *Solved sum.* Show that the vector $\mathbf{A} = x^2z^2\mathbf{i} + xyz^2\mathbf{j} - xz^3\mathbf{k}$ is solenoidal.

$$\nabla.\mathbf{A} = \left(\mathbf{i}\frac{\delta}{\delta x} + \mathbf{j}\frac{\delta}{\delta y} + \mathbf{k}\frac{\delta}{\delta z}\right).(x^2z^2\mathbf{i} + xyz^2\mathbf{j} - xz^3\mathbf{k})$$

$$= \frac{\delta}{\delta x}(x^2z^2) + \frac{\delta}{\delta y}(xyz^2) - \frac{\delta}{\delta z}(xz^3)$$

$$= 2xz^2 + xz^2 - 3xz^2 = 0.$$

Since $\nabla.\ \mathbf{A} = 0$, **A** is solenoidal.

38. Define solenoidal vector. Show that the vector $3x^2y\mathbf{i} - 4xy^2\mathbf{j} + 2xyz\ \mathbf{k}$ is solenoidal. **(1979)**

39. Show that $x^2\bar{\mathbf{i}} + 2xy\bar{\mathbf{j}} - 4xz\bar{\mathbf{k}}$ is solenoidal.

[**Hint:** Show that divergence is zero].

40. Find the values of m if the following vectors are solenoidal:

(*i*) $(x+2y)\bar{i} + (my+4z)\bar{j} + (5z+6x)\bar{k}$

(*ii*) $(2x+z)\bar{i} + (4x-11y+3z)\bar{j} + (3x+mz)\bar{k}$

(*iii*) $(x+3y)\bar{i} + (y-2z)\bar{j} + (x+mz)\bar{k}$

[**Hint:** "Divergence = 0" gives (*i*) $1 + m + 5 = 0, m = -6$. (*ii*) $2 - 11 + m = 0, m = 9$. (*iii*) $1 + 1 + m = 0, m = -2$]

41. If $ax^2z^2\bar{i} + xyz^2\bar{j} - xz^3\bar{k}$ is solenoidal, find the value of a.

[Hint: "Divergence = 0" gives $a = 1$**]**

42. *Solved Sum.* Determine the constant a so that the vector

$$\overline{F} = (x+2y)\,\overline{i} + (y-2z)\,\overline{j} + (x+az)\,\overline{k}$$

is solenoidal.

$\overline{F}$ is solenoidal. Therefore, $\nabla.\overline{F} = 0$. So

$$\frac{\partial}{\partial x}(x+2y)+\frac{\partial}{\partial y}(y-2z)+\frac{\partial}{\partial z}(x+az)=0$$

i.e., $1+1+a = 0. \quad \therefore\ a=-2$

43. If the vector $3x\mathbf{i} + (x+y)\mathbf{j} - az\mathbf{k}$ is solenoidal, find a. *[Ans: 4.]*

44. *Solved Sum.* If $\overline{r} = x\overline{i} + y\overline{j} + z\overline{k}$, find n so that $r^n\overline{r}$ is solenoidal.

Since $r^n\overline{r}$ is solenoidal, $\nabla.(r^n\overline{r}) = 0$.

But, from (*i*) of Corollaries of (*i*), $\nabla.(r^n\overline{r}) = (n+3)r^n$.

$$(n+3)r^n = 0 \text{ or } n = -3.$$

45. *Solved Sum.* Show that $\nabla.\dfrac{\overline{r}}{r^3} = 0$.

We shall find $\nabla.\dfrac{\overline{r}}{r^3}$ directly using the results

(*i*) $\nabla.(\phi\,\overline{\mathrm{A}}) = (\nabla\phi).\,\overline{\mathrm{A}} + \phi(\nabla.\,\overline{\mathrm{A}})$ (*ii*) $\nabla f(r) = f'(r)\hat{r}$.

$$\nabla\cdot\frac{\overline{r}}{r^3} = \nabla\cdot\left(\frac{1}{r^3}\overline{r}\right) = \left[\nabla\frac{1}{r^3}\right]\cdot\overline{r} + \frac{1}{r^3}[\nabla\cdot\overline{r}]$$

$$= \left[\left(\frac{d}{dr}\frac{1}{r^3}\right)\hat{r}\right]\cdot\overline{r} + \frac{1}{r^3}[3] = \left[-\frac{3}{r^4}\hat{r}\right]\cdot\overline{r} + \frac{3}{r^3}$$

$$= -\frac{3}{r^4}[r] + \frac{3}{r^3} = 0.$$

46. If $\mathbf{r} = x\mathbf{i} + y\mathbf{j} + z\mathbf{k}$, show that $\nabla.\mathbf{r} = 3$. **(1980, 81)**

$$\left[\textbf{Hint: } \nabla\cdot\overline{r} = \frac{\partial x}{\partial x} + \frac{\partial y}{\partial y} + \frac{\partial z}{\partial z} = 1+1+1\right]$$

47. *Solved sum.* If $\mathbf{r} = x\mathbf{i} + y\mathbf{j} + z\mathbf{k}$ and $r = |\,\mathbf{r}\,|$, show that $\nabla.(\mathbf{r}^n\mathbf{r}) = (n+3)r^n$. **(1976, 79, 81)**

$$\begin{aligned}\nabla.(\mathbf{r}^n\mathbf{r}) &= (\nabla\,\mathbf{r}^n).\mathbf{r} + r^n(\nabla.\mathbf{r}) \quad \text{by (2.8.1)}\\ &= (nr^{n-1}\hat{\mathbf{r}}).\mathbf{r} + r^n(3) \text{ since } \nabla f(r) = f'(r)\hat{\mathbf{r}}\\ &= (nr^{n-1}\hat{\mathbf{r}}).(r\hat{\mathbf{r}}) + 3r^n = nr^n(\hat{\mathbf{r}}.\hat{\mathbf{r}}) + 3r^n = (n+3)r^n.\end{aligned}$$

48. Show that $\nabla.\left(\dfrac{1}{r^3}\mathbf{r}\right) = 0,\ \nabla.\hat{\mathbf{r}} = \dfrac{2}{r}$.

49. *Solved sum.* If $\mathbf{r} = x\mathbf{i} + y\mathbf{j} + z\mathbf{k}$ and $|\,\mathbf{r}\,| = r$, show that $\nabla.\{\mathrm{f}(r)\mathbf{r}\} = rf'(r) + 3f(r)$. **(1978, 80, 82)**

Also, if $\nabla.\{\mathrm{f}(r)\mathbf{r}\} = 0$, show that $f(r) = \dfrac{C}{r^3}$, where C is an arbitrary constant.

$$\begin{aligned}\nabla.\{f(r)\mathbf{r}\} &= \{\nabla f(r)\}.\mathbf{r} + f(r)\,(\nabla.\mathbf{r}) \text{ by (2.8.1)}\\ &= \{f'(r)\hat{\mathbf{r}}\}.\mathbf{r} + f(r)\,(3)\end{aligned}$$

$$= \{f'(r)\hat{\mathbf{r}}\}.(r\hat{\mathbf{r}}) + 3f(r) = rf'(r) + 3f(r).$$

If $\nabla.\{f(r)\mathbf{r}\} = 0$, then $rf'(r) + 3f(r) = 0$. That is, $\dfrac{f'(r)}{f(r)} = -\dfrac{3}{r}$.

Integrating, $\log f(r) = -3\log r + \log C = \log \dfrac{C}{r^3}$ or $f(r) = \dfrac{C}{r^3}$.

50. *Solved sum.* **If a is a constant vector and r = xi + yj + zk, then show that**

$$\nabla.\{(\mathbf{a.r})\mathbf{r}\} = 4\mathbf{a.r.}$$ **(1975, 81)**

Let $\mathbf{a} = \mathbf{a_1 i} + \mathbf{a_2 j} + \mathbf{a_3 k}$. Then

$$\nabla.\{(\mathbf{a.r})\mathbf{r}\} = \nabla.\{(a_1x + a_2y + a_3z)\mathbf{r}\}$$
$$= \{\nabla(a_1x + a_2y + a_3z)\}.\mathbf{r} + (a_1x + a_2y + a_3z)(\nabla.\mathbf{r}) \text{ by (2.8.1)}$$
$$= (a_1\mathbf{i} + a_2\mathbf{j} + a_3\mathbf{k}).\mathbf{r} + (a_1x + a_2y + a_3z)(3)$$
$$= \mathbf{a.r} + 3\mathbf{a.r} = 4\mathbf{a.r}.$$

2.7. Curl of a Vector Point Function

51. **If r = xi + yj + zk, show that ∇ × r = 0.** **(1981)**

52. Find $\nabla.\overline{F}$ and $\nabla\times\overline{F}$ at the point (1, –1, 1) if $\overline{F} = xz^3\overline{i} - 2x^2yz\overline{j} + 2yz^4\overline{k}$.

[**Hint:** $[\nabla.\overline{F}]_{(1,-1,1)} = \dfrac{\partial}{\partial x}(xz^3) - \dfrac{\partial}{\partial y}(2x^2yz) + \dfrac{\partial}{\partial z}(2yz^4)$

$= [z^3 - 2x^2z + 8yz^3]_{(1,-1,1)} = -9.$

$$\nabla\times\overline{F} = \begin{vmatrix} \overline{i} & \overline{j} & \overline{k} \\ \dfrac{\partial}{\partial x} & \dfrac{\partial}{\partial y} & \dfrac{\partial}{\partial z} \\ xz^3 & 2x^2yz & 2yz^4 \end{vmatrix} = \overline{i}(2z^4 - 2x^2y) - \overline{j}(-3xz^2) + \overline{k}(4xyz)$$

$\therefore\ (\nabla\times\overline{F})_{(1,-1,1)} = 4\overline{i} + 3\overline{j} - 4\overline{k}$]

53. Find $\nabla\times\overline{F}$ if $\overline{F} = xz^3\overline{i} - 2x^2yz\overline{j} + 2yz^4\overline{k}$.

54. *Solved Sum.* Find curl $\overline{F}$ in the following cases:

(*i*) $\overline{F} = x^2y\overline{i} + y^2z\overline{j} + z^2x\overline{k}$

(*ii*) $\overline{F} = xyz\,\overline{i} + xyz^2\overline{j} + x^2yz\overline{k}$

(*i*)
$$\nabla\times\overline{F} = \begin{vmatrix} \overline{i} & \overline{j} & \overline{k} \\ \dfrac{\partial}{\partial x} & \dfrac{\partial}{\partial y} & \dfrac{\partial}{\partial z} \\ x^2y & y^2z & z^2x \end{vmatrix}$$
$$= \overline{i}(0 - y^2) - \overline{j}(z^2 - 0) + \overline{k}(0 - x^2)$$
$$= -(y^2\overline{i} + z^2\overline{j} + x^2\overline{k}).$$

(*ii*)
$$\nabla\times\overline{F} = \begin{vmatrix} \overline{i} & \overline{j} & \overline{k} \\ \dfrac{\partial}{\partial x} & \dfrac{\partial}{\partial y} & \dfrac{\partial}{\partial z} \\ xyz & xyz^2 & x^2yz \end{vmatrix}$$

$= \bar{i}\,(x^2z - 2xyz) - \bar{j}\,(2xyz - xy) + \bar{k}\,(yz^2 - xz).$

$= \bar{i}\,(x^2z - 2xyz) + \bar{j}(xy - 2xyz) + \bar{k}(yz^2 - xz).$

55. Find the curl of the following functions at the specified points:

(*i*) $x^2z\,\bar{i} - 2y^3z^2\,\bar{j} + xy^2z\,\bar{k}$ at $(1, -1, 1)$

(*ii*) $xyz\,\bar{i} + 3x^2y\,\bar{j} + (xz^2 - y^2z)\,\bar{k}$ at $(1, 2, -1)$

[**Hint:** (*i*) Curl $= \bar{i}\,(2xyz + 4y^3z) - \bar{j}\,(y^2z - x^2) + \bar{k}\,(0).$

$(\text{Curl})_{(1,-1,1)} = \bar{i}\,(-2-4) - \bar{j}\,(1-1) = -6\,\bar{i}$

(*ii*) Curl $= \bar{i}\,(-2yz - 0) - \bar{j}\,(z^2 - xy) + \bar{k}\,(6xy - xz).$

$(\text{Curl})_{(1,2,-1)} = 4\,\bar{i} + \bar{j} + 13\,\bar{k}$

56. Define irrotational vector.

57. ***Solved Sum:*** **Show that $\bar{F} = yz\,\bar{i} + zx\,\bar{j} + xy\,\bar{k}$ is irrotational.**

$$\nabla \times \bar{F} = \begin{vmatrix} \bar{i} & \bar{j} & \bar{k} \\ \frac{\partial}{\partial x} & \frac{\partial}{\partial y} & \frac{\partial}{\partial z} \\ yz & zx & xy \end{vmatrix} = \bar{i}\,(x - x) - \bar{j}\,(y - y) + \bar{k}\,(z - z) = \bar{0}.$$

Since $\nabla \times \bar{F} = \bar{0}$, $\bar{F}$ is irrotational.

58. ***Solved Sum:*** **If $\bar{A} = axy\,\bar{i} + (x^2 + 2yz)\,\bar{j} + y^2\,\bar{k}$ is irrotational, find the value of a.**

Since $\bar{A}$ is irrotational, $\nabla \times \bar{A} = \bar{0}$.

Hence,

$$\begin{vmatrix} \bar{i} & \bar{j} & \bar{k} \\ \frac{\partial}{\partial x} & \frac{\partial}{\partial y} & \frac{\partial}{\partial z} \\ axy & x^2 + 2yz & y^2 \end{vmatrix} = \bar{0}$$

i.e., $\bar{i}\,(2y - 2y) - \bar{j}\,(0 - 0) + \bar{k}\,(2x - ax) = \bar{0}.$

i.e., $(2 - a)x = 0. \;\therefore\; 2 - a = 0$ or $a = 2$

59. Show that $\bar{f} = (\sin y + z)\,\bar{i} + (x\cos y - z)\,\bar{j} + (x - y)\,\bar{k}$ is irrotational.

[**Hint:** $\nabla \times \bar{f} = \bar{i}\,[-1 - (-1)] - \bar{j}\,(1 - 1) + \bar{k}\,(\cos y - \cos y) = \bar{0}$]

60. Prove that the following vector is irrotational.

$(y^2\cos x + z^3)\,\bar{i} + (2y\sin x - 4)\,\bar{j} + 3xz^2\,\bar{k}.$

61. ***Solved Sum:*** **Find the value of the constants a, b, c so that**

$\bar{F} = (x + 2y + az)\,\bar{i} + (bx - 3y - z)\,\bar{j} + (4x + cy + 2z)\,\bar{k}$ **is irrotational.**

Since $\bar{F}$ is irrotational $\nabla \times \bar{F} = \bar{0}$. Therefore,

$$\nabla \times \bar{F} = \begin{vmatrix} \bar{i} & \bar{j} & \bar{k} \\ \frac{\partial}{\partial x} & \frac{\partial}{\partial y} & \frac{\partial}{\partial z} \\ x + 2y + az & bx - 3y - z & 4x + cy + 2z \end{vmatrix} = \bar{0}.$$

$$= \bar{i}\,(c+1) - \bar{j}\,(a-4) + \bar{k}\,(b-2) = \bar{0}$$

$$\therefore \quad c+1=0,\quad a-4=0,\quad b-2=0 \quad \text{or} \quad a=4, b=2, c=-1.$$

62. *Solved Sum:* If $\bar{F} = \nabla(x^3+y^3+z^3-3xyz)$ find $\nabla \times \bar{F}$.

We know that $\nabla \times \nabla\phi = \bar{0}$. Now $\nabla \times \bar{F} = \nabla \times \nabla\phi$, where $\phi = x^3+y^3+z^3-3xyz$ is $\bar{0}$.

63. If $\bar{a}$ is a constant vector and $\bar{r} = x\bar{i} + y\bar{j} + z\bar{k}$, then show that

$$\nabla\times[(\bar{a}\cdot\bar{r})\bar{r}] = \bar{a}\times\bar{r}.$$

[**Hint:** Use $\nabla\times[\phi\bar{r}] = (\nabla\phi)\times\bar{r} + \phi(\nabla\times\bar{r})$]

64. *Solved sum.* Find the value of a if

$\mathbf{A} = (axy - z^2)\mathbf{i} + (x^2+2yz)\mathbf{j} + (y^2 - axz)\mathbf{k}$ is irrotational.

$$\nabla\times\mathbf{A} = \left(\mathbf{i}\frac{\partial}{\partial x} + \mathbf{j}\frac{\partial}{\partial y} + \mathbf{k}\frac{\partial}{\partial z}\right) \times \{(axy-z^2)\mathbf{i} + (x^2+2yz)\mathbf{j} + (y^2-axz)\mathbf{k}\}$$

$$= \begin{vmatrix} \mathbf{i} & \mathbf{j} & \mathbf{k} \\ \dfrac{\partial}{\partial x} & \dfrac{\partial}{\partial y} & \dfrac{\partial}{\partial z} \\ axy-z^2 & x^2+2yz & y^2-axz \end{vmatrix}$$

$$= \mathbf{i}\left\{\frac{\partial}{\partial y}(y^2-axz) - \frac{\partial}{\partial z}(x^2+2yz)\right\} + \mathbf{j}\left\{\frac{\partial}{\partial z}(axy-z^2) - \frac{\partial}{\partial x}(y^2-axz)\right\}$$

$$+ \mathbf{k}\left\{\frac{\partial}{\partial x}(x^2+2yz) - \frac{\partial}{\partial y}(axy-z^2)\right\}$$

$$= \mathbf{i}(2y-2y) + \mathbf{j}(-2z+az) + \mathbf{k}(2x-ax).$$

If $\mathbf{A}$ is irrotational, then $\nabla\times\mathbf{A} = \mathbf{0}$. It is so when $a = 2$.

65. Find a if $\bar{A} = (4xy - z^3)\bar{i} + ax^2\bar{j} - 3xz^2\bar{k}$ is irrotational.

[**Hint:** $\nabla\times\bar{A} = \bar{i}\,(0-0) - \bar{j}\,(-3z^2+3z^2) + \bar{k}\,(4x-2ax)$. This is zero since $\bar{A}$ is irrotational. Hence $4x - 2ax = 0$ or $a = 2$]

66. Show that in the following vector point functions the first three are irrotational and the last one is both irrotational and solenoidal.

(*i*) $(4xy - z^3)\mathbf{i} + 2x^2\mathbf{j} - 3xz^2\mathbf{k}$.

(*ii*) $(3x^2+2y^2+1)\mathbf{i} + (4xy - 3y^2z - 3)\mathbf{j} + (2-y^3)\mathbf{k}$. **(1980)**

(*iii*) $(y^2 + 2xz^2 - 1)\mathbf{i} + 2xy\mathbf{j} + 2x^2z\mathbf{k}$. **(1981)**

(*iv*) $(y^2 - z^2 + 3yz - 2x)\mathbf{i} + (3xz + 2xy)\mathbf{j} + (3xy - 2xz + 2z)\mathbf{k}$. **(1985)**

[**Hint:** (*i*) $\nabla\times\bar{A} = \bar{i}\,(0-0) - \bar{j}\,(-3z^2+3z^2) + \bar{k}\,(4x-4x) = \bar{0}$. So $\bar{A}$ is irrotational.

(*iv*) $\nabla.\mathbf{A} = 0$ and $\nabla\times\mathbf{A} = \bar{0}$]

67. *Solved Sum.* Prove that

$$\bar{A} = (2x+yz)\bar{i} + (4y+xz)\bar{j} - (6z-xy)\bar{k},$$

is solenoidal as well as irrotational. Also find the scalar potential of $\bar{A}$.

$$\nabla.\bar{A} = \frac{\partial}{\partial x}(2x+yz) + \frac{\partial}{\partial y}(4y+xz) - \frac{\partial}{\partial z}(6z-xy)$$

$= 2+4-6=0.$

Since $\nabla.\overline{A} = 0$, $\overline{A}$ is solenoidal. Also

$$\nabla \times \overline{A} = \begin{vmatrix} \overline{i} & \overline{j} & \overline{k} \\ \frac{\partial}{\partial x} & \frac{\partial}{\partial y} & \frac{\partial}{\partial z} \\ 2x+yz & 4y+xz & -6z+xy \end{vmatrix}$$

$$= \overline{i}(x-x)-\overline{j}(y-y)+\overline{k}(z-z)=\overline{0}.$$

Since $\nabla \times \overline{A} = \overline{0}$, $\overline{A}$ is irrotational.

If ϕ is the scalar potential of $\overline{A}$, then $\overline{A} = \nabla\phi$.

i.e., $(2x+yz)\overline{i} + (4y+xz)\overline{j} - (6z-xy)\overline{k} = \overline{i}\frac{\partial\phi}{\partial x}+\overline{j}\frac{\partial\phi}{\partial y}+\overline{k}\frac{\partial\phi}{\partial z}.$

$$\therefore \quad \frac{\partial\phi}{\partial x}=2x+yz, \quad \frac{\partial\phi}{\partial y}=4y+xz, \quad \frac{\partial\phi}{\partial z}=-6z+xy.$$

$$\phi = x^2+xyz+f(y,z),$$
$$\phi = 2y^2+xyz+g(x,z)$$
$$\phi = -3z^2+xyz+h(x,y).$$

Taking the repeated terms only once, we get the scalar potential ϕ as $x^2+2y^2-3z^2+xyz+C$.

68. Show that the following vectors are irrotational. Find their scalar potential.

(*i*) $\overline{F} = (6xy+z^3)\overline{i} + (3x^2-z)\overline{j} + (3xz^2-y)\overline{k}$.

(*ii*) $\overline{F} = (y^2\cos x+z^3)\overline{i} + (2y\sin x-4)\overline{j} + 3xz^2\overline{k}$.

(*iii*) $\overline{F} = (3x^2+2y^2+1)\overline{i} + (4xy-3y^2z-3)\overline{j} + (2-y^3)\overline{k}$.

[**Hint:** Prove $\nabla \times \overline{F} = \overline{0}$. Then $\overline{F}$ is irrotational.

(*i*) $\phi = 3x^2y+xz^3-yz+C.$

(*ii*) $\phi = y^2\sin x+xz^3-4y+C.$

(*iii*) $\phi = x^3+2xy^2-y^3z+x-3y+2z+C$]

69. Find the value of m if $\overline{F} = (6xy+z^3)\overline{i} + (mx^2-z)\overline{j} + (3xz^2-y)\overline{k}$, is irrotational. Find also ϕ such that $\overline{F} = \nabla\phi$.

[**Hint:** $\nabla \times \overline{F} = \overline{0}$ implies $2m-6=0$ or $m=3$. Then

$$\overline{i}\frac{\partial\phi}{\partial x}+\overline{j}\frac{\partial\phi}{\partial y}+\overline{k}\frac{\partial\phi}{\partial z}=(6xy+z^3)\overline{i}+(3x^2-z)\overline{j}+(3xz^2-y)\overline{k}.$$

Thus $\phi = 3x^2y+xz^3-yz+C$]

70. *Solved sum.* If $\mathbf{r} = x\mathbf{i}+y\mathbf{j}+z\mathbf{k}$, for all $f(r)$, show that $\nabla \times \{f(r)\mathbf{r}\} = 0.$ **(1975)**

$$\nabla \times \{f(r)\mathbf{r}\} = \{\nabla f(r) \times \mathbf{r} + f(r)\{\nabla \times \mathbf{r}\} \quad \text{by (2.8.2)}$$
$$= \{f'(r)\hat{\mathbf{r}}\}\times\mathbf{r}+f(r)(0)$$
$$= \{f'(r)\hat{\mathbf{r}}\}\times(r\hat{\mathbf{r}}) = f'(r)r(\hat{\mathbf{r}}\times\hat{\mathbf{r}})=0.$$

71. Evaluate $\nabla \times r^3\mathbf{r}$, $\nabla \times r^2\mathbf{r}$, $\nabla \times \hat{\mathbf{r}}$, $\nabla \times (\mathbf{r}/r^2)$, curl ($\phi$ grad ϕ) **(Ans. 0, 0, 0, 0, 0)**

72. *Solved Sum.* If $\mathbf{r} = x\mathbf{i}+y\mathbf{j}+z\mathbf{k}$, show that $(\nabla \times \{r^n\mathbf{r}\} = 0.$ **(1980, 81)**

$$\nabla \times \{r^n\mathbf{r}\} = (\nabla r^n)\times\mathbf{r}+r^n.(\nabla\times\mathbf{r}) \quad \text{by (2.8.2)}$$
$$= (nr^{n-1}\hat{\mathbf{r}})\times\mathbf{r}+0 = nr^n(\hat{\mathbf{r}}\times\hat{\mathbf{r}})=0.$$

73. Show that $\nabla \times \hat{\mathbf{r}} = \mathbf{0}$.

$\left[\textbf{Hint: } \hat{\mathbf{r}} = \left(\frac{1}{r}\right)\mathbf{r}\right]$

74. *Solved sum.* If $\mathbf{v} = \mathbf{w} \times \mathbf{r}$, where $\mathbf{w}$ is a constant vector and $\mathbf{r} = x\mathbf{i} + y\mathbf{j} + z\mathbf{k}$, show that $\frac{1}{2}$ curl $\mathbf{v} = \mathbf{w}$. **(1979, 80, 85)**

$$\text{curl } \mathbf{v} = \Sigma \mathbf{i} \times \frac{\partial \mathbf{v}}{\partial x} = \Sigma \mathbf{i} \times \frac{\partial}{\partial x}(\mathbf{w} \times \mathbf{r})$$

$$= \Sigma \mathbf{i} \times \left\{\mathbf{w} \times \frac{\partial \mathbf{r}}{\partial x}\right\} \qquad (\because \mathbf{w} \text{ is constant})$$

$$= \Sigma \mathbf{i} \times \{\mathbf{w} \times \mathbf{i}\} \qquad \left(\because \frac{\partial \mathbf{r}}{\partial x} = \mathbf{i}\right)$$

$$= \Sigma[(\mathbf{i} \cdot \mathbf{i})\,\mathbf{w} - (\mathbf{i} \cdot \mathbf{w})\,\mathbf{i}] = \Sigma[\mathbf{w} - w_1\mathbf{i}]$$

$$= 3\mathbf{w} - \mathbf{w} = 2\mathbf{w}.$$

75. Show that $\nabla \cdot (\mathbf{A} \times \mathbf{r}) = \mathbf{r} \cdot (\nabla \times \mathbf{A}) = \mathbf{r} \cdot \text{curl } \mathbf{A}$. Evaluate $\nabla \cdot (\mathbf{A} \times \mathbf{r})$ if $\nabla \times \mathbf{A} = \mathbf{0}$.

76. *Solved sum.* If $\mathbf{a}$ is a constant vector, show that

(*i*) $\nabla \times \{(\mathbf{a} \cdot \mathbf{r})\mathbf{r}\} = \mathbf{a} \times \mathbf{r}$. (*ii*) $\nabla \cdot (\mathbf{a} \times \mathbf{r}) = 0$.

(*i*) If $\mathbf{a} = a_1\mathbf{i} + a_2\mathbf{j} + a_3\mathbf{k}$, then $\mathbf{a} \cdot \mathbf{r} = a_1x + a_2y + a_3z$ and

$$\nabla \times \{(\mathbf{a} \cdot \mathbf{r})\mathbf{r}\} = \nabla \times \{(a_1x + a_2y + a_3z)\mathbf{r}\}$$

$$= \{\nabla(a_1x + a_2y + a_3z)\} \times \mathbf{r} + (a_1x + a_2y + a_3z)(\nabla \times \mathbf{r}) \text{ by (2.8.2)}$$

$$= \mathbf{a} \times \mathbf{r} + (a_1x + a_2y + a_3z)\,0 = \mathbf{a} \times \mathbf{r}.$$

(*ii*) $\nabla \cdot (\mathbf{a} \times \mathbf{r}) = (\nabla \times \mathbf{a}) \cdot \mathbf{r} - (\nabla \times \mathbf{r}) \cdot \mathbf{a}$ by (2.8.6)

$= 0$ ($\because \nabla \times \mathbf{a} = \mathbf{0}$ and $\nabla \times \mathbf{r} = \mathbf{0}$).

77. If $\mathbf{A}$ and $\mathbf{B}$ are irrotational, show that $\mathbf{A} \times \mathbf{B}$ is solenoidal.

[**Hint:** Since irrotational, $\nabla \times \mathbf{A} = \mathbf{0}$, $\nabla \times \mathbf{B} = \mathbf{0}$; $\nabla \cdot (\mathbf{A} \times \mathbf{B}) = 0$ since

$\nabla \cdot (\mathbf{A} \times \mathbf{B}) = (\nabla \times \mathbf{A}) \cdot \mathbf{B} - (\nabla \times \mathbf{B}) \cdot \mathbf{A}$]

78. *Solved Sum:* If $\overline{\mathrm{F}} = xyz\,\bar{i} + xyz^2\,\bar{j} + x^2yz\,\bar{k}$, then find div curl $\overline{\mathrm{F}}$.

$$\nabla \times \overline{\mathrm{F}} = \begin{vmatrix} \bar{i} & \bar{j} & \bar{k} \\ \frac{\partial}{\partial x} & \frac{\partial}{\partial y} & \frac{\partial}{\partial z} \\ xyz & xyz^2 & x^2yz \end{vmatrix}$$

$$= \bar{i}\,(x^2z - 2xyz) - \bar{j}\,(2xyz - xy) + \bar{k}\,(yz^2 - xz).$$

Next, finding the divergence of this curl, we have

$$\nabla \cdot (\nabla \times \overline{\mathrm{F}}) = \frac{\partial}{\partial x}(x^2z - 2xyz) - \frac{\partial}{\partial y}(2xyz - xy) + \frac{\partial}{\partial z}(yz^2 - xz)$$

$$= (2xz - 2yz) - (2xz - x) + (2yz - x) = 0.$$

Note. We will prove later that, for all $\overline{\mathrm{F}}$, div curl $\overline{\mathrm{F}} = 0$.

79. Find div curl $\overline{\mathrm{F}}$, if $\overline{\mathrm{F}} = x^2y\,\bar{i} + xy\,\bar{j} + yz\,\bar{k}$.

[**Hint:** $\nabla \times \overline{\mathrm{F}} = z\,\bar{i} - (0)\,\bar{j} + (y - x^2)\,\bar{k}$; div curl $\overline{\mathrm{F}} = 0$]

80. *Solved Sum:* **Prove that curl $(\bar{r}\times\bar{a}) = -2\bar{a}$, where $\bar{a}$ is a constant vector.**

Now $\bar{r} = x\bar{i} + y\bar{j} + z\bar{k}$. Let $\bar{a} = a_1\bar{i} + a_2\bar{j} + a_3\bar{k}$.

$$\bar{r}\times\bar{a} = \begin{vmatrix} \bar{i} & \bar{j} & \bar{k} \\ x & y & z \\ a_1 & a_2 & a_3 \end{vmatrix} = (a_3y - a_2z)\bar{i} + (a_1z - a_3x)\bar{j} + (a_2x - a_1y)\bar{k}$$

$$= \begin{vmatrix} \bar{i} & \bar{j} & \bar{k} \\ \dfrac{\partial}{\partial x} & \dfrac{\partial}{\partial y} & \dfrac{\partial}{\partial z} \\ a_3y - a_2z & a_1z - a_3x & a_2x - a_1y \end{vmatrix}$$

$$= \bar{i}\,[-a_1 - a_1] + \bar{j}\,[-2a_2]\bar{k}\,[-2a_3] = -2\bar{a}$$

81. *Solved Sum:* **If $\bar{F} = x^2y\bar{i} + y^2z\bar{j} + z^2x\bar{k}$, then find curl curl $\bar{F}$.**

Now we have to find curl of $\bar{F}$ twice.

$$\nabla\times\bar{F} = \begin{vmatrix} \bar{i} & \bar{j} & \bar{k} \\ \dfrac{\partial}{\partial x} & \dfrac{\partial}{\partial y} & \dfrac{\partial}{\partial z} \\ x^2y & y^2z & z^2x \end{vmatrix} = \bar{i}(0-y^2) - \bar{j}(z^2-0) + \bar{k}(0-x^2) = -y^2\bar{i} - z^2\bar{j} - x^2\bar{k}.$$

Next, finding the curl of this curl,

$$\nabla\times(\nabla\times\bar{F}) = -\begin{vmatrix} \bar{i} & \bar{j} & \bar{k} \\ \dfrac{\partial}{\partial x} & \dfrac{\partial}{\partial y} & \dfrac{\partial}{\partial z} \\ y^2 & z^2 & x^2 \end{vmatrix}$$

$$= -[\bar{i}\,(0-2z) - \bar{j}\,(2x-0) + \bar{k}\,(0-2y)]$$

$$= 2(z\bar{i} + x\bar{j} + y\bar{k}).$$

82. **If $\bar{F} = x^3y\bar{i} + y^3z\bar{j} + z^3x\bar{k}$, find curl curl $\bar{F}$.**

[Hint: $\nabla\times\bar{F} = -(y^3\bar{i} + z^3\bar{j} + x^3\bar{k})$, $\nabla\times(\nabla\times\bar{F}) = 3(z^2\bar{i} + x^2\bar{j} + y^2\bar{k})$]

2.9. Laplacian differential operator

83. Prove that

$$\nabla^2(\phi\psi) = \phi\nabla^2\psi + 2(\nabla\phi)\,.\,(\nabla\psi) + \psi\nabla^2\phi.$$

84. *Solved Sum.* Show that $\nabla^2 f(r) = f''(r) + \dfrac{2}{r}f'(r)$. Also, show that if $\nabla^2 f(r) = 0$, then $f(r) = \dfrac{\alpha}{r} + \beta$, where α and β are arbitrary constants. **(1975)**

$$\nabla^2 f(r) = \frac{\partial^2}{\partial x^2}f(r) + \frac{\partial^2}{\partial y^2}f(r) + \frac{\partial^2}{\partial z^2}f(r).$$

Now $\dfrac{\partial}{\partial x}f(r) = f'(r)\dfrac{\partial r}{\partial x}$, where $\dfrac{\partial r}{\partial x} = \dfrac{x}{r}$ from $r^2 = x^2 + y^2 + z^2$.

$$\frac{\partial^2}{\partial x^2} f(r) = \frac{\partial}{\partial x}\left\{f'(r)\frac{x}{r}\right\}$$

$$= \frac{r\left\{f''(r)\frac{\partial r}{\partial x}+x+f'(r)\right\}-\{f'(r)x\}\frac{\partial r}{\partial x}}{r^2}$$

$$= \frac{r\left\{f''(r)\frac{x^2}{r}+f'(r)\right\}-f'(r)\frac{x^2}{r}}{r^2}$$

$$= \frac{x^2}{r^2}f''(r)+\frac{f'(r)}{r}\left\{1-\frac{x^2}{r^2}\right\}.$$

$$\nabla^2 f(r) = \frac{f''(r)}{r^2}(x^2+y^2+z^2) + \frac{f'(r)}{r}\left\{\left(1-\frac{x^2}{r^2}\right)+\left(1-\frac{y^2}{r^2}\right)+\left(1-\frac{z^2}{r^2}\right)\right\}$$

$$= f''(r) + \frac{2}{r}f'(r) \text{ since } r^2 = x^2+y^2+z^2.$$

If $\nabla^2 f(r) = 0$, then $f''(r) + \frac{2}{r}f'(r) = 0$ and hence

$$\frac{f''(r)}{f'(r)} = -\frac{2}{r}.$$

Integrating, $\log f'(r) = -2\log r + \log C = \log\frac{C}{r^2}$. So $f'(r) = \frac{C}{r^2}$ and $f(r) = -\frac{C}{r} + D$.

85. *Solved sum.* Show that $\nabla^2 r^n = n(n+1)\,r^{n-2}$ where n is a constant. **(1977, 79, 81)**

$$\frac{\partial r^n}{\partial x} = nr^{n-1}\frac{\partial r}{\partial x} = nr^{n-1}\frac{x}{r} = nr^{n-2}x.$$

$$\frac{\partial^2 r^n}{\partial x^2} = n\left\{(n-2)r^{n-3}\frac{x^2}{r}+r^{n-2}\right\}$$

$$= nr^{n-2}\left\{(n-2)\frac{x^2}{r^2}+1\right\}.$$

$$\therefore \quad \nabla^2 r^n = nr^{n-2}\{(n-2)+3\} = n\,r^{n-2}(n+1).$$

86. *Solved Sum:* Show that

$$\nabla^2(\phi\psi) = \phi\nabla^2\psi + 2\nabla\phi\,.\,\nabla\psi + \psi\nabla^2\phi.$$

$$\nabla^2(\phi\psi) = \Sigma\frac{\partial^2}{\partial x^2}(\phi\psi) = \Sigma\frac{\partial}{\partial x}(\phi_x\psi+\phi\psi_x)$$

$$= \Sigma(\phi_{xx}\psi + \phi_x\psi_x + \phi_x\psi_x + \phi\psi_{xx})$$

$$= \Sigma(\phi_{xx}\psi + 2\phi_x\psi_x + \phi\psi_{xx})$$

$$= \psi\Sigma\phi_{xx} + 2\,\Sigma\phi_x\psi_x + \phi\Sigma\psi_{xx}$$

$$= \psi\nabla^2\phi + 2(\nabla\phi)\,.\,(\nabla\psi) + \phi\nabla^2\psi$$

since $\quad 2(\nabla\phi)\,.\,(\nabla\psi) = 2(\phi_x\bar{i} + \phi_y\bar{j} + \phi_z\bar{k})\,.\,(\psi_x\bar{i} + \psi_y\bar{j} + \psi_z\bar{k})$

$$= 2\Sigma\phi_x\psi_x.$$

87. *Solved Sum*. Show that $\nabla^2 \log r = \frac{1}{r^2}$.

Differentiating $r^2 = x^2 + y^2 + z^2$, partially w.r.t. x,

$$2r\frac{\partial r}{\partial x} = 2x \text{ or } \frac{\partial r}{\partial x} = \frac{x}{r}.$$

Differentiating $\log r$ twice partially w.r.t. x,

$$\frac{\partial}{\partial x}\log r = \frac{1}{r}\frac{\partial r}{\partial x} = \frac{1}{r}\frac{x}{r} = \frac{x}{r^2},$$

$$\frac{\partial^2 \log r}{\partial x^2} = \frac{\partial}{\partial x}\frac{x}{r^2} = \frac{r^2 - x\cdot 2r(x/r)}{r^4} = \frac{r^2 - 2x^2}{r^4}.$$

Similarly, we have

$$\frac{\partial^2 \log r}{\partial y^2} = \frac{r^2-2y^2}{r^4}, \frac{\partial^2 \log r}{\partial z^2} = \frac{r^2-2z^2}{r^4}.$$

$$\therefore \quad \nabla^2 \log r = \left(\frac{\partial^2}{\partial x^2}+\frac{\partial^2}{\partial y^2}+\frac{\partial^2}{\partial z^2}\right)\log r$$

$$= \frac{(r^2-2x^2)+(r^2-2y^2)+(r^2-2z^2)}{r^4}$$

$$= \frac{3r^2-2(x^2+y^2+z^2)}{r^4} = \frac{3r^2-2r^2}{r^4} = \frac{1}{r^2}.$$

88. Show that $\nabla^2\left(\frac{x}{r^3}\right) = 0$.

[Hint: $\frac{\partial^2}{\partial x^2}\frac{x}{r^3} = -\frac{3x}{r^5} - 3x\left[\frac{2r^5-5x^2r^3}{r^{10}}\right]$, and also

$$\frac{\partial^2}{\partial y^2}\frac{x}{r^3} = -3x\left[\frac{r^5-5y^2r^3}{r^{10}}\right], \frac{\partial^2}{\partial z^2}\frac{x}{r^3} = -3x\left[\frac{r^5-5z^2r^3}{r^{10}}\right]$$]

2.11. Divergence and Curl of a Gradient

89. *Solved sum*. If $\phi = x^2y^3z^4$, find (*i*) $\nabla . (\nabla\phi)$ (*ii*) $\nabla \times (\nabla\phi)$ **(1983)**

(*i*) $\nabla . (\nabla\phi) = \nabla^2\phi = \left(\frac{\partial^2}{\partial x^2}+\frac{\partial^2}{\partial y^2}+\frac{\partial^2}{\partial z^2}\right)(x^2y^3z^4) = 2y^3z^4 + 6x^2yz^4 + 12x^2y^3z^2.$

(*ii*) Since $\nabla \times (\nabla\phi) = 0$, $\nabla \times [\nabla(x^2y^3z^4)] = 0$.

90. Show that $\nabla \times (\phi\nabla\phi) = \bar{0}$.

[Hint: In $\nabla \times (\phi\bar{A}) = (\nabla\phi) \times \bar{A} + \phi[\nabla \times (\bar{A})]$, set $\bar{A} = \nabla\phi$ and use $\nabla \times (\nabla\phi) = \bar{0}$]

91. If ϕ is a harmonic function, then show that $\nabla\phi$ is solenoidal and irrotational.

[Hint: Since ϕ is harmonic, $\nabla^2\phi = 0$; $\nabla.(\nabla\phi) = \nabla^2\phi$; $\nabla \times \nabla\phi = \mathbf{0}$.]

92. *Solved sum*. Show that $\nabla.(\phi\nabla\Psi - \Psi\nabla\phi) = \phi\nabla^2\Psi - \Psi\nabla^2\phi$.

Now $\nabla\Psi$ is a vector, say **A**. So, by (2.8.1),

$$\nabla . (\phi\nabla\Psi) = \nabla . (\phi \mathbf{A}) = (\nabla\phi) . \mathbf{A} + \phi(\nabla . \mathbf{A})$$

$$= (\nabla\phi) . (\nabla\Psi) + \phi(\nabla . \nabla\Psi)$$

$$= (\nabla\phi) . (\nabla\Psi) + \phi\nabla^2\Psi \quad \text{by (2.8.1).} \quad (1)$$

Interchanging the roles of ϕ and Ψ,

$$\nabla . (\Psi \nabla \phi) = (\nabla\Psi) . (\nabla\phi) + \Psi\nabla^2\phi. \quad (2)$$

From (1) and (2) the result follows.

93. *Solved sum.* Show that $(\nabla\phi) \times (\nabla\Psi)$ is solenoidal

$$\text{Let } \nabla\phi = \mathbf{A}.\ \nabla\Psi = \mathbf{B}. \text{ Then, from}$$

$$\nabla.\{(\nabla\phi) \times (\nabla\Psi)\} = \nabla.(\mathbf{A} \times \mathbf{B}) = \mathbf{B}.(\nabla \times \mathbf{A}) - \mathbf{A}.(\nabla \times \mathbf{B}) = 0$$

since $\nabla \times \mathbf{A} = \nabla \times (\nabla\phi)$ and $\nabla \times \mathbf{B} = \nabla \times (\nabla\psi)$ vanish by (2.11.2). So $(\nabla\phi) \times (\nabla\Psi)$ is solenoidal.

94. Prove that

(*i*) $\nabla.(\phi\nabla\Psi) = (\nabla\phi).(\nabla\Psi) + \phi(\nabla^2\Psi)$ (*ii*) $\nabla.(\phi\nabla\phi) = (\nabla\phi)^2 + \phi(\nabla^2\phi)$

(*iii*) $\nabla \times (\phi\nabla\Psi) = (\nabla\phi) \times (\nabla\Psi)$ (*iv*) $\nabla \times (\phi\nabla\phi) = \mathbf{0}$.

95. If $\bar{A}$ and $\bar{B}$ are irrotational, then show that $\bar{A} \times \bar{B}$ is solenoidal.

[**Hint:** Since irrotational $\nabla \times \bar{A} = \bar{0}$, $\nabla \times \bar{B} = \bar{0}$. Show that $\nabla\cdot(\bar{A}\times\bar{B}) = 0$. Use $\nabla\cdot(\bar{A}\times\bar{B}) = (\nabla\times\bar{A}).\bar{B} - (\nabla\times\bar{B})\cdot\bar{A}$]

96. If $\bar{a}$ is a constant vector and $\bar{r} = x\bar{i} + y\bar{j} + z\bar{k}$, then show that $\nabla.(\bar{a}\times\bar{r}) = 0$.

[**Hint:** Use $\nabla\cdot(\bar{A}\times\bar{B}) = (\nabla\times\bar{A})\cdot\bar{B} - (\nabla\times\bar{B})\cdot\bar{A}$]

97. Prove that $\nabla\times(\bar{u}\times\bar{v}) = \bar{u}(\nabla\cdot\bar{v}) - (\bar{u}\cdot\nabla)\bar{v}$ if $\bar{u}$ is a constant vector.

[**Hint:** Using the formula for $\nabla\times(\bar{u}\times\bar{v})$, we have

$$\nabla\times(\bar{u}\times\bar{v}) = (\bar{v}\cdot\nabla)\bar{u} - (\nabla\cdot\bar{u})\bar{v} - [(\bar{u}\cdot\nabla)\bar{v} - (\nabla\cdot\bar{v})\bar{u}]$$

If $\bar{u}$ is a constant, then $(\bar{v}.\nabla)\bar{u} = 0$ and $(\nabla.\bar{u}) = 0$]

98. *Solved sum.* Show that

(*i*) $(\mathbf{V}.\nabla)\mathbf{r} = \mathbf{V}$. (*ii*) $(\mathbf{V} \times \nabla) \times \mathbf{r} = -2\mathbf{V}$. **(1976, 77, 82)**

(*i*) Suppose $\mathbf{V} = V_1\mathbf{i} + V_2\mathbf{j} + V_3\mathbf{k}$. Then

$$(\mathbf{V}.\nabla)\mathbf{r} = \left(\mathbf{V}.\Sigma\mathbf{i}\frac{\partial}{\partial x}\right)\mathbf{r} = \left(\Sigma V_1\frac{\partial}{\partial x}\right)(x\mathbf{i} + y\mathbf{j} + z\mathbf{k})$$

$$= \Sigma V_1\mathbf{i} = \mathbf{V}.$$

(*ii*)

$$(\mathbf{V} \times \nabla) \times \mathbf{r} = \left(\mathbf{V}\times\Sigma\mathbf{i}\frac{\partial}{\partial x}\right) \times (x\mathbf{i} + y\mathbf{j} + z\,\mathbf{k})$$

$$= \begin{vmatrix} \mathbf{i} & \mathbf{j} & \mathbf{k} \\ V_1 & V_2 & V_3 \\ \frac{\partial}{\partial x} & \frac{\partial}{\partial y} & \frac{\partial}{\partial z} \end{vmatrix} \times (x\mathbf{i} + y\mathbf{j} + z\,\mathbf{k})$$

$$= \begin{vmatrix} \mathbf{i} & \mathbf{j} & \mathbf{k} \\ V_2\frac{\partial}{\partial z} - V_3\frac{\partial}{\partial y} & V_3\frac{\partial}{\partial x} - V_1\frac{\partial}{\partial z} & V_1\frac{\partial}{\partial y} - V_2\frac{\partial}{\partial x} \\ x & y & z \end{vmatrix}$$

$$= -2(V_1\mathbf{i} + V_2\mathbf{j} + V_3\mathbf{k}) = -2\mathbf{V}.$$

99. Show that $(\mathbf{V}.\nabla)\mathbf{V} = \frac{1}{2}\nabla\,\mathbf{V}^2 - \mathbf{V}\times(\nabla\times\mathbf{V})$.

CHAPTER

3

Multiple Integrals – Line, Surface, Volume Integrals

SINGLE, DOUBLE AND TRIPLE INTEGRALS

Single integrals are lengths, double integrals are areas, and triple integrals are volumes. As such we have to define the region on which integration is to be done.

INFINITESIMAL VECTORS

INFINITESIMAL LINE VECTOR

Consider an infinitesimal displacement vector **dS** of a line between two points P_1 and P_2. This vector can be decomposed into $d\mathbf{S} = dx\mathbf{i} + dy\mathbf{j} + dz\mathbf{k}$

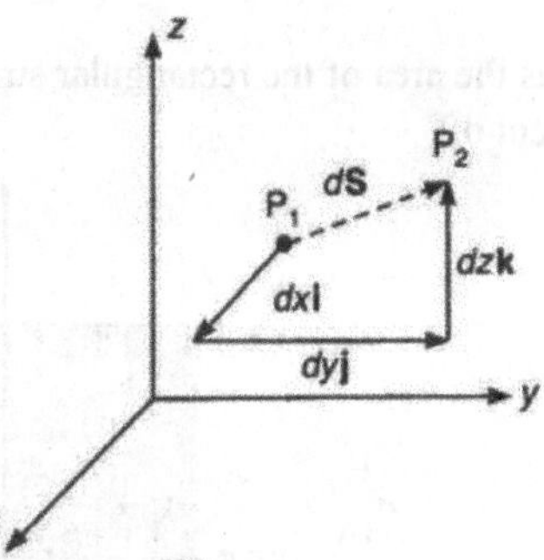

INFINITESIMAL AREA VECTOR

Consider an infinitesimal area element dA of a surface. The vector $d\mathbf{A}$ can be written as $d\mathbf{A} = dxdy\,\hat{\mathbf{n}}$ where $\hat{\mathbf{n}}$ is perpendicular to the surface defined by the area in the outward direction from a closed surface, defined by the right-hand rule.

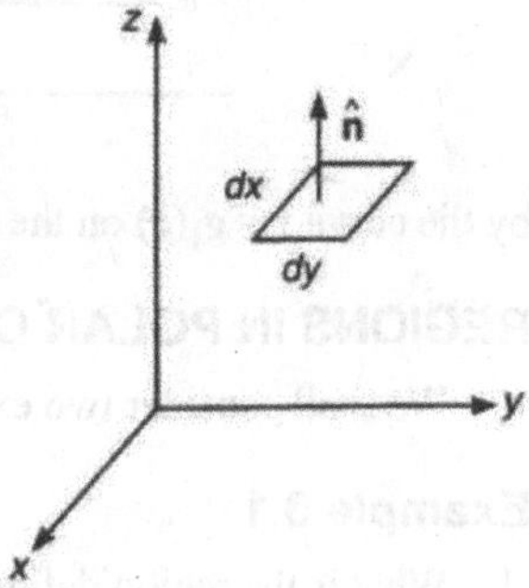

INFINITESIMAL VOLUME

Consider an infinitesimal volume element dV of an object. In Cartesian coordinates, scalar $dV = dxdydz$.

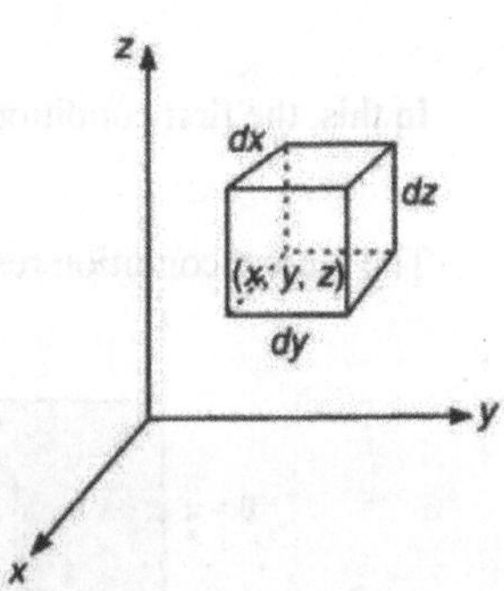

TWO DIMENSIONAL REGIONS

First consider the region, in the xOy plane, defined by

$$\{(x, y) \mid a \le x \le b, f_1(x) \le y \le f_2(x)\}.$$

Here $a \le x \le b$ specifies the area containing all the points whose

x coordinates are greater than or equal to a and less than or equal to b. This area is the infinite rectangular strip perpendicular to the x axis, bounded by the lines $x = a$, $x = b$. The region, in question, lies in this strip and it is bounded below by the curve $y = f_1(x)$ and above by the curve $y = f_2(x)$.

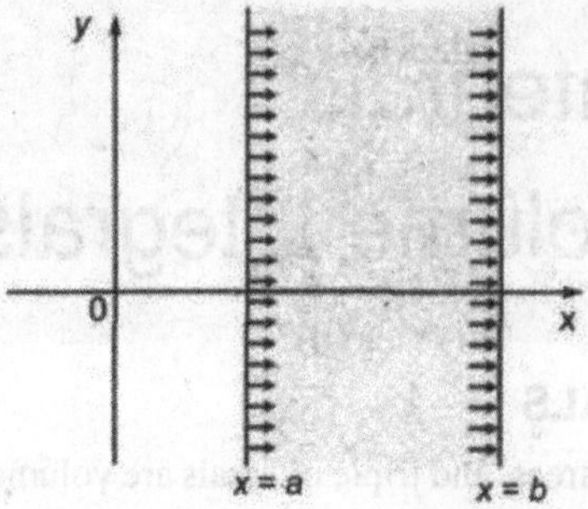

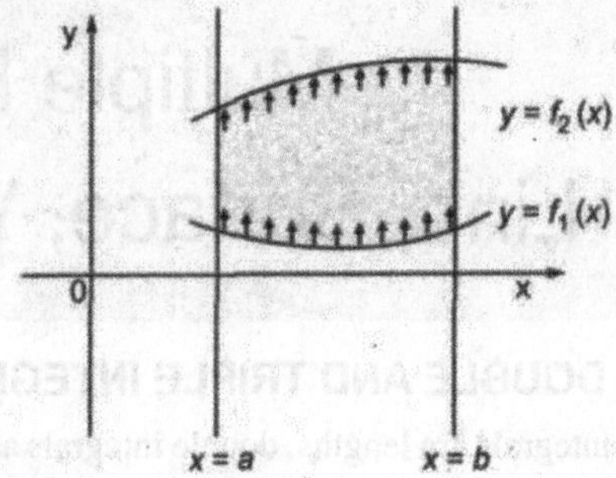

Similarly the region in the xOy plane defined by

$$\{(x, y) \mid c \le y \le d, g_1(y) \le x \le g_2(y)\}$$

is the area of the rectangular strip perpendicular to the y axis bounded by the lines $y = c$ and $y = d$, cut off

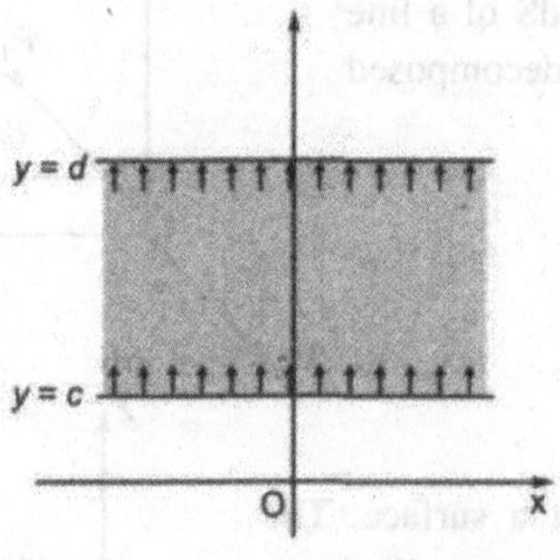

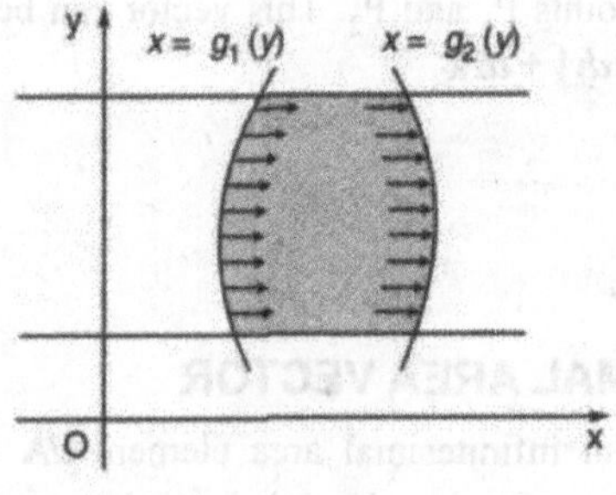

by the curve $x = g_1(y)$ on the left and by the curve $x = g_2(y)$ on the right.

REGIONS IN POLAR COORDINATES.

We shall consider two examples for the regions defined in polar coordinates.

Example 3.1

What is the region defined by

$$0 \le \theta \le \pi/2, \quad 0 \le r \le 2a \cos \theta\ ?$$

In this, the first condition specifies the infinite sector area between the rays

$$\theta = 0, \theta = \pi/2.$$

The second condition restricts the sector area to the area between the circles.

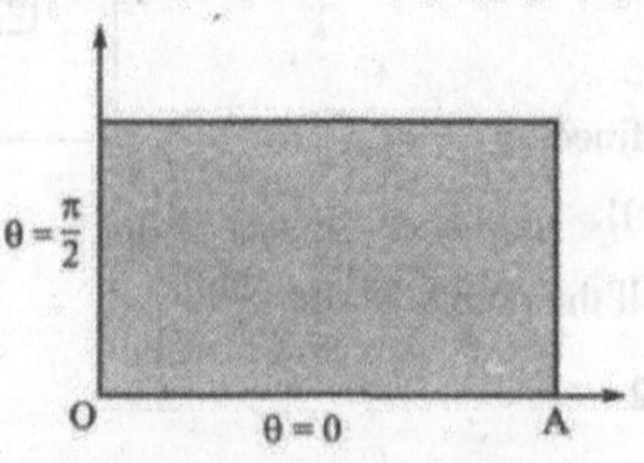

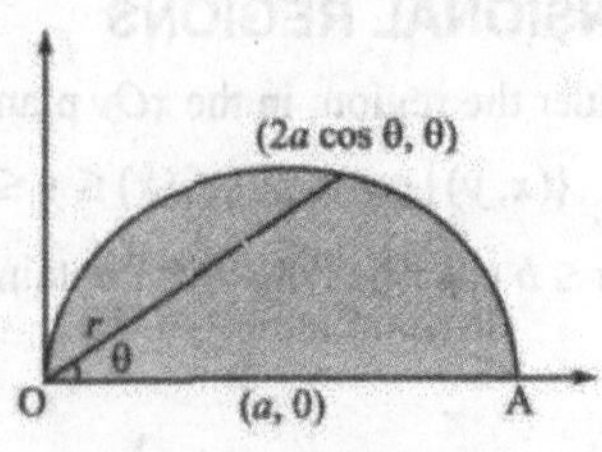

where $r = 0$ denotes the point pole and $r = 2a \cos \theta$ denotes the circle whose centre is $(a, 0)$ and radius is a.

Remark 1. It is evident that the restrictions $-\pi/2 \le \theta \le \pi/2$, $0 \le r \le 2a \cos \theta$ specify the full circle.

Remark 2. The equation $r = a \cos \theta$ represents the circle whose centre is at $(a/2, 0)$ and radius is $a/2$.

Example 3.2

What is the region specified by

$$0 \le \theta \le \pi, \qquad 0 \le r \le a((1 + \cos \theta)?$$

The region is the upper half of the cardioid

$$r = a(1 + \cos \theta)\ \lambda.$$

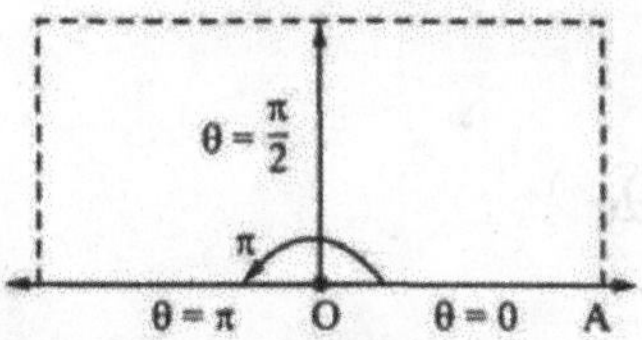

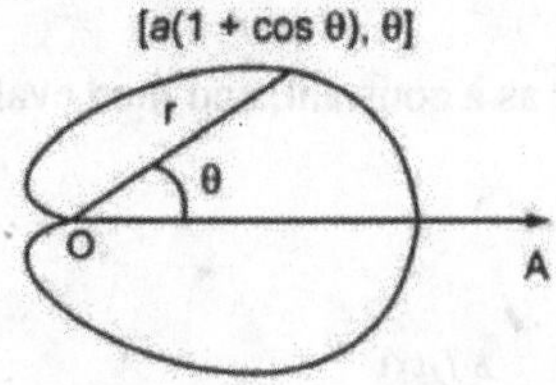

The full cardiodid is specified by

$0 \le \theta \le 2\pi$, $0 \le r \le a(1 + \cos \theta)$.

SINGLE INTEGRALS

First we shall recall the integral of $f(x)$ along the interval $[a, b]$. To get this integral, we divided $[a, b]$ into n subintervals and assuming Δx_i to be the length

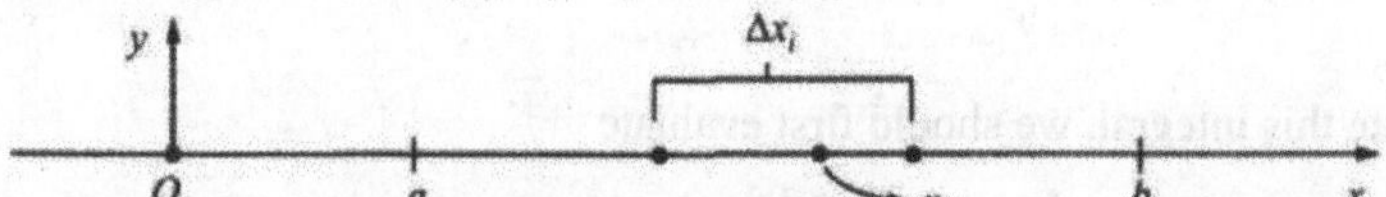

of the ith subinterval and x_i to be any point in that interval, obtained the limit of the sum of the products of $f(x_i)$ and Δx_i, namely,

$$\Sigma f(x_i)\, \Delta x_i,$$

as $n \to \infty$ and max $\Delta x_i \to 0$ as the integral

$$\int_a^b f(x)\, dx$$

if the limit existed.

DOUBLE INTEGRALS

In the same way we define now a double integral on an area. Suppose R is a region defined by

$$\{(x, y) \mid a \le x \le b, f_1(x) \le y \le f_2(x)\}$$

and $f(x, y)$ is a function defined in R. Divide R into subregions by lines parallel to x and y axes. The subregions are rectangles, full or partial. Let there be m rows and n columns of subregions. Let Δx_i and Δy_j be the length and breadth of the subregion in the ith row and j^{th} column and (x_i, y_j) be any point in that region. Then the limit of the sum of the products of $f(x_i, y_j)$ and the area $\Delta x_i\, \Delta y_j$, as m and $n \to \infty$ and maximum of $\Delta x_i\, \Delta y_j \to 0$, if the limit exists, is

$$\int_a^b \left(\int_{y=f_1(x)}^{y=f_2(x)} f(x,y)dy \right) dx.$$

This double integral can also be written as

$$\int_a^b \int_{f_1(x)}^{f_2(x)} f(x,y)dy\,dx.$$

To evaluate this integral we should first evaluate

$$\int_{f_1(x)}^{f_2(x)} f(x,y)\,dy$$

treating x as a constant, and then evaluate

$$\int_a^b (.....)dx.$$

Remark 1. $\int_a^b \int_{f_1(x)}^{f_2(x)} 1\,dy\,dx$ is the area of the region R.

Remark 2. The integral of $f(x, y)$ on the region defined by

$$\{(x, y) |\ c \le y \le d, g_1(y) \le x \le g_2(y)\}$$

is $\int_c^d \int_{g_1(y)}^{g_2(y)} f(x,y)\,dx\,dy.$

To evaluate this integral, we should first evaluate

$$\int_{g_1(y)}^{g_2(y)} f(x,y)\,dx,$$

treating y as a constant, and then evaluate

$$\int_c^d (.....)dy.$$

ORDER OF INTEGRATION WHEN LIMITS ARE CONSTANTS

If $f(x, y)$ is a function defined on the region

$$\{(x, y) |\ a \le x \le b, c \le y \le d\},$$

where a, b, c, d are constants, the integral of $f(x, y)$ on this region is

$$\int_a^b \left[\int_c^d f(x,y)\,dy \right] dx \text{ or } \int_a^b \int_c^d f(x,y)\,dydx.$$

This integral is same as the integral

$$\int_c^d \left[\int_a^b f(x,y)dx \right] dy \text{ or } \int_c^d \int_a^b f(x,y)\,dx\,dy$$

which means that the order of integration can be reversed taking care that the limits go along with the variable w.r.t. which the integration is performed. This is so because the region is a rectangle.

An important note. It is important to note that the integral.

$$\int_a^b\int_c^d f(x,y)\,dy\,dx$$

is, in general, not equal to the integral

$$\int_a^b\int_c^d f(x,y)\,dx\,dy$$

as illustrated below.

(i)
$$\int_0^1\int_0^2 6xy^2\,dx\,dy = \int_0^1\left[y^2\int_0^2 6x\,dx\right]dy = \int_0^1 y^2[3x^2]_0^2\,dy$$

$$= \int_0^1 12y^2 dy = 4\left[y^3\right]_0^1 = 4.$$

(ii)
$$\int_0^1\int_0^2 6xy^2\,dy\,dx = \int_0^1\left[x\int_0^2 6y^2 dy\right]dx$$

$$= \int_0^1 x[2y^3]_0^2\,dx = \int_0^1 16x\,dx = 8[x^2]_0^1 = 8.$$

so
$$\int_0^1\int_0^2 6xy^2\,dx\,dy \neq \int_0^1\int_0^2 6xy^2\,dy\,dx.$$

Example 3.3

Evaluate $\iint dx\,dy$ over the region bounded by $x = 0, x = 2, y = 0, y = 2$.

$$I = \int_0^2\int_0^2 1\,dx\,dy = \int_0^2\left[\int_0^2 1\,dx\right]dy = \int_0^2 [x]_0^2\,dy$$

$$= \int_0^2 2\,dy = 2[y]_0^2 = 4.$$

Example 3.4

Evaluate $\displaystyle\int_0^a \int_0^{\sqrt{a^2-x^2}} dy\,dx.$

$$I = \int_0^a [y]_0^{\sqrt{a^2-x^2}}\,dx = \int_0^a \sqrt{a^2-x^2}\,dx$$

$$= \left[\frac{x}{2}\sqrt{a^2-x^2} + \frac{a^2}{2}\sin^{-1}\frac{x}{a}\right]_0^a$$

$$= 0+\frac{a^2}{2}\sin^{-1}1-0=\frac{a^2}{2}\cdot\frac{\pi}{2}=\frac{\pi a^2}{4}.$$

Note. The region of integration is the first quadrant of the circle $x^2+y^2=a^2$. Its area is $\frac{1}{4}(\pi a^2)$

Example 3.5

Evaluate $\int_0^1\int_0^2 xy^2\,dy\,dx$.

First integrating w.r.t. y, treating x as a constant, etc., we get

$$I = \int_0^1 x\left[\frac{y^3}{3}\right]_0^2 dx = \int_0^1 x\cdot\frac{2^3}{3}dx = \frac{8}{3}\left[\frac{x^2}{2}\right]_0^1 = \frac{4}{3}.$$

Example 3.6

Evaluate $\int_0^a\int_x^a (x^2+y^2)\,dy\,dx$.

First integrating w.r.t y, treating x as a constant, etc., we get

$$I = \int_0^a\left[x^2y+\frac{y^3}{3}\right]_x^a dx = \int_0^a\left[ax^2+\frac{a^3}{3}-\left(x^3+\frac{x^3}{3}\right)\right]dx$$

$$= \left[a\frac{x^3}{3}+\frac{a^3x}{3}-\frac{4}{3}\frac{x^4}{4}\right]_0^a = \frac{a^4}{3}.$$

Example 3.7

Evaluate $\int_0^a\int_0^{\sqrt{a^2-x^2}} y^3\,dy\,dx.$

$$I = \int_0^a\left[\frac{y^4}{4}\right]_0^{\sqrt{a^2-x^2}} dx = \int_0^a\frac{(a^2-x^2)^2}{4}dx$$

$$= \frac{1}{4}\int_0^a (a^4-2a^2x^2+x^4)\,dx$$

$$= \frac{1}{4}\left[a^4x-\frac{2a^2x^3}{3}+\frac{x^5}{5}\right]_0^a = \frac{2a^5}{15}.$$

Example 3.8

Evaluate $\iint (x^2+y^2)\,dx\,dy$ over the region

$$\{(x,y)\,|\,x\geq 0, y\geq 0, x+y\leq 1\}.$$

The region is the area bounded by the lines

$$x=0,\ y=0,\ x+y=1.$$

It is triangular in shape having the vertices at (0, 0), (1, 0), (0, 1). Considering first the x coordinates of the points in the region, we see that the region lies between the lines $x = 0$ and $x = 1$ so that $0 \le x \le 1$.

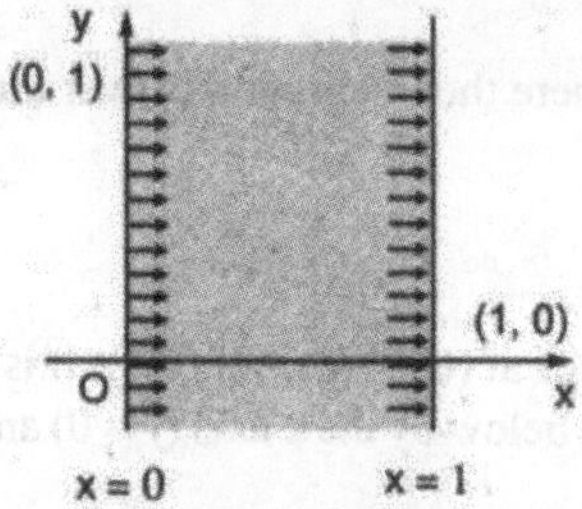

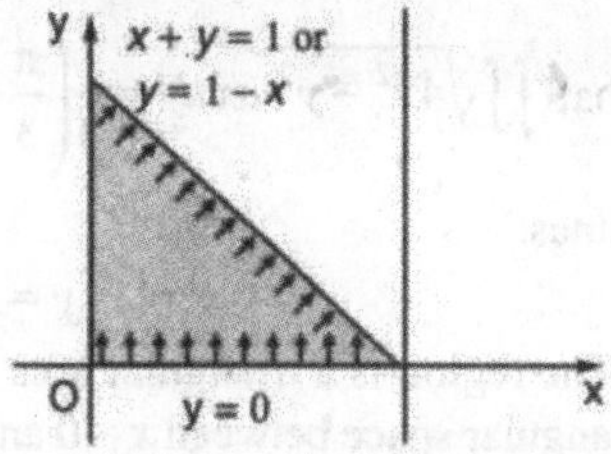

Considering the y coordinates, we see that the region lies above the x axis ($y = 0$) and below the line $x + y = 1$ or $y = 1 - x$ so that

$$0 \le y \le 1 - x.$$

Thus denoting the integral by I, we get

$$I = \int_0^1 \int_0^{1-x} (x^2 + y^2)\,dy\,dx = \int_0^1 \left[x^2 y + \frac{y^3}{3}\right]_0^{1-x} dx$$

$$= \int_0^1 \left(x^2(1-x) + \frac{(1-x)^3}{3}\right) dx = \left[\frac{x^3}{3} - \frac{x^4}{4} - \frac{(1-x)^4}{12}\right]_0^1$$

$$= \left(\frac{1}{3} - \frac{1}{4}\right) + \frac{1}{12} = \frac{1}{6}.$$

EXERCISES

1. Evaluate $\int_0^a \int_0^b (x+y)\,dx\,dy$. [Ans. $I = \int_0^a \left[\frac{x^2}{2} + xy\right]_0^b dy = \int_0^a \left[\frac{b^2}{2} + by\right] dy = \frac{ab^2 + a^2 b}{2}$]

2. Evaluate $\int_1^2 \int_2^3 xy^2\,dxdy$. [Ans. $I = \int_1^2 y^2 \left[\frac{x^2}{2}\right]_2^3 dy = \int_1^2 \frac{5y^2}{2} dy = \left[\frac{5y^3}{6}\right]_1^2 = \frac{35}{6}$]

3. Evaluate $\int_2^3 \int_1^2 \frac{1}{xy}\,dx\,dy$. [Ans. $I = \int_2^3 \frac{1}{y} [\log x]_1^2\, dx = \log 2 \int_2^3 \frac{1}{y} dy = \log 2 \cdot \log \frac{3}{2}$]

4. Evaluate $\int_0^3 \int_0^2 x\,dx\,dy$. [Ans. $I = \int_0^3 \left[\frac{x^2}{2}\right]_0^2 dy = \int_0^3 2\,dy = [2y]_0^3 = 6$]

5. Evaluate $\int_0^a \int_0^b xy(x-y)\,dx\,dy$. [Hint. $I = \int_0^a \left[\int_0^b xy(x-y)dx\right] dy = \int_0^a \left(y.\frac{x^3}{3} - y^2 \cdot \frac{x^2}{2}\right)_0^b dy$

$$= \int_0^a \left(\frac{b^3}{3} \cdot y - \frac{b^2}{2} \cdot y^2\right) dy = \left(\frac{b^3}{3} \cdot \frac{y^2}{2} - \frac{b^2}{2} \cdot \frac{y^3}{3}\right)_0^a$$

$$= \frac{a^2 b^3}{6} - \frac{a^3 b^2}{6} = \frac{a^2 b^2}{6}(b-a)]$$

6. Evaluate $\int_0^3\int_0^2 e^{x+y}\,dy\,dx$. [**Hint:** $I=\int_0^3\left[e^x\int_0^2 e^y dy\right]dx=(e^3-1)(e^2-1)$]

7. Show that $\iint\sqrt{4x^2-y^2}\,dx\,dy=\frac{1}{3}\left(\frac{\pi}{3}+\frac{\sqrt{3}}{2}\right)$, where the integration is over the area bounded by the lines.

$$y=0,\ y=x,\ x=1.$$

[**Hint:** The region is a triangular area with vertices at (0, 0), (1, 0), (0,1). It is obtained from the rectangular space between $x=0$ and $x=1$, cut below by the x axis ($y=0$) and above by the line $y=x$.

$$\therefore\quad I=\int_0^1\int_0^x\sqrt{4x^2-y^2}\,dy\,dx$$

$$=\int_0^1\left[\frac{y}{2}\sqrt{4x^2-y^2}+\frac{4x^2}{2}\sin^{-1}\frac{y}{2x}\right]_0^x dx$$

$$=\int_0^1\left[\frac{x}{2}\sqrt{3x^2}+2x^2\sin^{-1}\frac{1}{2}\right]dx=\frac{\sqrt{3}}{2}\cdot\frac{1}{3}+\frac{2}{3}\cdot\frac{\pi}{6}\,]$$

TRANSFORMATION OF COORDINATES

In certain cases the evaluation of a double integral which is in terms of x and y is made simpler by changing the coordinates to other suitable coordinates.

If the transformation is to polar coordinates, then

$$x=r\cos\theta,\ y=r\sin\theta.$$

$$dx\,dy=\left|\frac{\partial(x,y)}{\partial(r,\theta)}\right|dr\,d\theta=r\,dr\,d\theta,$$

where $\frac{\partial(x,y)}{\partial(r,\theta)}$ is the Jacobian of x, y with respect to r, θ.

Three dimensions. In three dimensions there are two important systems of polar coordinates, namely,

(*i*) cylindrical polar coordinates : r, θ, z. (*ii*) spherical polar coordinates: r, θ, ϕ.

CYLINDRICAL POLAR COORDINATES

If P is (x, y, z) and if PL is the perpendicular from P to xOy plane in the figure, then

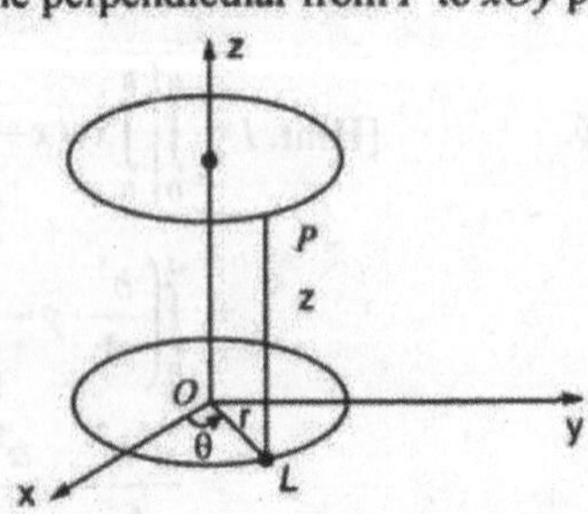

OL, angle *xOL*, *LP* are the cylindrical polar coordinates of *P*. They are denoted by r, θ, z, where $0 \le r < \infty,\ 0 \le \theta \le 2\pi,\ -\infty < z < \infty$. The relations between x, y, z and r, θ, z and that between $dx\ dy\ dz$ and $dr\ d\theta\ dz$ are

$$\left.\begin{aligned} x &= r\cos\theta \\ y &= r\sin\theta \\ z &= z \end{aligned}\right\},\ dx\,dy\,dz = \left|\frac{\partial(x,y,z)}{\partial(r,\theta,z)}\right| dr\,d\theta\,dz = r\,dr\,d\theta\,dz.$$

SPHERICAL POLAR CO-ODINATES

If P is (x, y, z) and if PL is the perpendicular from P to the xOy plane in the figure, then OP, angle zOP, angle xOL are the spherical polar coordinates of P. They are denoted by r, θ, ϕ, where $0 \le r < \infty$, $0 \le \theta \le \pi$, $0 \le \phi \le 2\pi$. The relations between x, y, z and r, θ, ϕ and that between $dx\ dy\ dz$ and $dr\ d\theta\ d\phi$ are

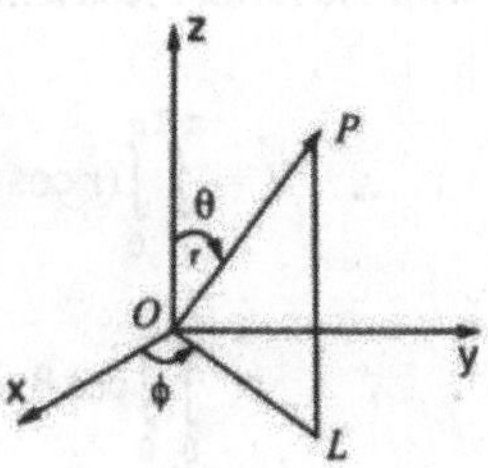

$$\left.\begin{aligned} x &= r\sin\theta\cos\phi \\ y &= r\sin\theta\sin\phi \\ z &= r\cos\theta \end{aligned}\right\},\ dx\,dy\,dz = \left|\frac{\partial(x,y,z)}{\partial(r,\theta,\phi)}\right| dr\,d\theta\,d\phi = (r^2\sin\theta)\,dr\,d\theta\,d\phi.$$

Example 3.9

Evaluate $\iint xy\,dx\,dy$ taken over the positive quadrant of the circle $x^2 + y^2 = a^2$.

Method using cartesian coordinates. The region of integration lies in the area between the lines $x = 0, x = a$.

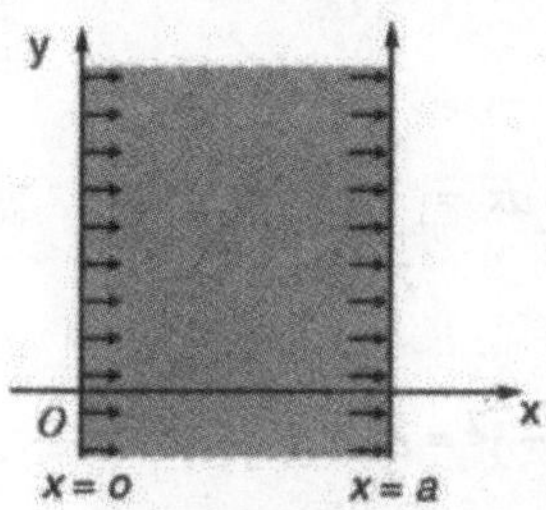

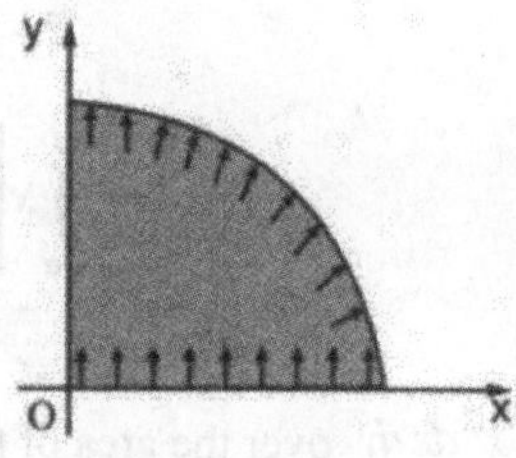

with the further restriction that it extends from the x axis to the circle $x^2 + y^2 = a^2$, that is, from

$$y = 0 \text{ to } y = \sqrt{a^2 - x^2}.$$

$$\therefore \quad I = \int_0^a \int_0^{\sqrt{a^2-x^2}} xy\,dy\,dx = \int_0^a x\left[\frac{y^2}{2}\right]_0^{\sqrt{a^2-x^2}} dx$$

$$= \int_0^a x\frac{a^2 - x^2}{2}\,dx = \frac{1}{2}\left[\frac{a^2x^2}{2} - \frac{x^4}{4}\right]_0^a = \frac{a^4}{8}.$$

Method using polar coordinates. The region of integration lies in the area between the rays.

$$\theta = 0, \theta = \pi/2$$

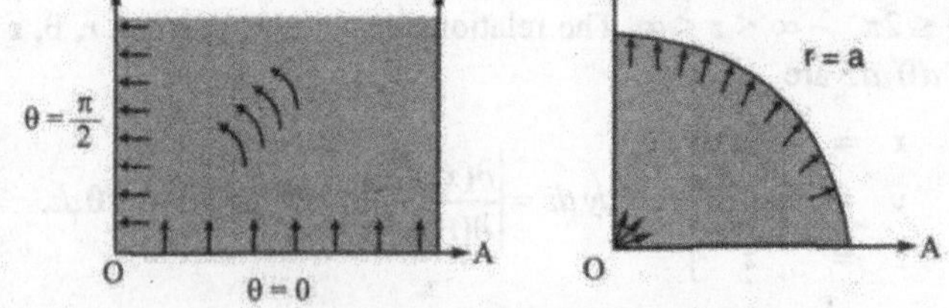

with the further restriction that it extends from the pole to the circle $r = a$, that is, from

$$r = 0 \text{ to } r = a.$$

$$\therefore \quad I = \int_0^{\pi/2}\int_0^a (r\cos\theta)(r\sin\theta)[r\,dr\,d\theta]$$

$$= \int_0^{\pi/2}\int_0^a \cos\theta\sin\theta\, r^3 dr\, d\theta = \int_0^{\pi/2} \cos\theta\sin\theta \left[\frac{r^4}{4}\right]_0^a d\theta$$

$$= \frac{a^4}{4}\int_0^{\pi/2} \cos\theta\,\sin\theta\, d\theta = \frac{a^4}{4} I_{1,1} = \frac{a^4}{4}\cdot\frac{1}{1+1} = \frac{a^4}{8}.$$

Example 3.10

Evaluate $\iint xy\,dx\,dy$ over the area of the circle $x^2 + y^2 = a^2$.

Proceeding as in the previous example,

$$I = \int_{-a}^{a}\int_{-\sqrt{a^2-x^2}}^{\sqrt{a^2-x^2}} xy\,dy\,dx$$

$$= \int_{-a}^{a} x\left[\frac{y^2}{2}\right]_{-\sqrt{a^2-x^2}}^{\sqrt{a^2-x^2}} dx = \int_{-a}^{a} x\times 0\,dx = 0.$$

Example 3.11

Evaluate $\iint x^2 dx\,dy$ over the area of the circle $x^2 + y^2 = a^2$.

Method 1. The circular area lies between the lines

$$x = -a, x = a$$

with the further restriction that it extends from the lower semicircle to the upper semicircle whose equations are

$$y = -\sqrt{a^2-x^2},\ y = \sqrt{a^2-x^2}.$$

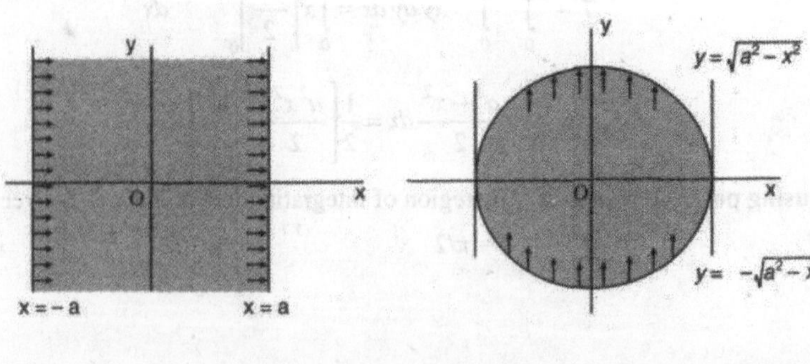

$$\therefore \quad I = \int_{-a}^{a} \int_{-\sqrt{a^2-x^2}}^{\sqrt{a^2-x^2}} x^2 dy\, dx = 2\int_{-a}^{a} \int_{0}^{\sqrt{a^2-x^2}} x^2 dy\, dx$$

$$= 2\int_{-a}^{a} x^2 [y]_0^{\sqrt{a^2-x^2}} dx = 2\int_{-a}^{a} x^2\sqrt{a^2-x^2}\, dx$$

Put $\begin{cases} x = a\sin t & \text{When } x = 0, t = 0 \\ dx = a\cos t\, dt & \text{When } x = a, t = \pi/2 \end{cases}$

$$= 2\times 2\int_{0}^{a} x^2\sqrt{a^2-x^2}\, dx$$

$$= 4\int_{0}^{\pi/2} (a^2\sin^2 t)(a\cos t)(a\cos t)\, dt$$

$$= 4a^4 I_{2,2} = 4a^4\left(\frac{1}{2+2} I_2\right)$$

$$= a^4 I_2 = a^4\left(\frac{1}{2}\frac{\pi}{2}\right) = \frac{a^4\pi}{4}.$$

Method 2. For the full circular area, θ varies from 0 to 2π and r varies from 0 to a.

$$\therefore \quad I = \int_{0}^{2\pi}\left[\int_{0}^{a} (r\cos\theta)^2 r\, dr\right] d\theta = \int_{0}^{2\pi} \cos^2\theta\left[\frac{r^4}{4}\right]_0^a d\theta$$

$$= \frac{a^4}{4}\int_{0}^{2\pi} \cos^2\theta\, d\theta = \frac{a^4}{4}\int_{0}^{2\pi} \frac{1+\cos 2\theta}{2} d\theta$$

$$= \frac{a^4}{8}\left[\theta + \frac{\sin 2\theta}{2}\right]_0^{2\pi} = \frac{a^4}{8}[2\pi] = \frac{a^4\pi}{4}.$$

Example 3.12

Evaluate $\iint r\sqrt{a^2-r^2}\, dr\, d\theta$ over the upper half of the circle $r = a\cos\theta$.

Here the region is defined by

$$\left\{(r,\theta) \mid 0 \le \theta \le \frac{\pi}{2}, 0 \le r \le a\cos\theta\right\}.$$

$$\therefore \quad I = \int_{0}^{\pi/2}\left[\int_{0}^{a\cos\theta} r\sqrt{a^2-r^2}\, dr\right] d\theta$$

Put $a^2 - r^2 = t^2$

$-2r\, dr = 2t\, dt$

When $r = 0, t = a$

$$= \int_{0}^{\pi/2}\left[\int_{a}^{a\sin\theta} \sqrt{t^2}(-t\, dt)\right] d\theta$$

When $r = a\cos\theta$, $t = a\sin\theta$

$$= \int_{0}^{\pi/2}\left[\int_{a\sin\theta}^{a} t^2 dt\right] d\theta = \int_{0}^{\pi/2}\left[\frac{t^3}{3}\right]_{a\sin\theta}^{a} d\theta$$

$$= \frac{1}{3}\int_0^{\pi/2} (a^3 - a^3 \sin^3 \theta)\, d\theta = \frac{a^3}{3}\cdot\frac{\pi}{2} - \frac{a^3}{3}\int_0^{\pi/2} \sin^3 \theta\, d\theta$$

$$= \frac{a^3\pi}{6} - \frac{a^3}{3} I_3 = \frac{a^3\pi}{6} - \frac{a^3}{3}\cdot\frac{2}{3}\cdot 1 = a^3\left(\frac{\pi}{6} - \frac{2}{9}\right)$$

Example 3.13

Evaluate $\iint xy\, dx\, dy$ over the first quadrant area of the ellipse $x^2/a^2 + y^2/b^2 = 1$, using the transformation

$$x = au \cos t,\ y = bu \sin t,\ 0 \le t \le 2\pi,\ 0 \le u \le 1.$$

$$\frac{\partial(x,y)}{\partial(u,t)} = \begin{vmatrix} a\cos t & -au\sin t \\ b\sin t & bu\cos t \end{vmatrix} = abu.$$

$$\therefore \quad I = \int_0^{\pi/2}\int_0^{1} (au\cos t)(bu\sin t)\, abu\, du\, dt$$

$$= \frac{a^2b^2}{4}\int_0^{\pi/2} \cos t \sin t\, dt = \frac{a^2b^2}{8}\int_0^{\pi/2} \sin 2t\, dt = \frac{a^2b^2}{8}.$$

Example 3.14

If R is the square region bounded by the line $x + y = \pm a, x - y = \pm a$, then show that

$$\iint_R e^{x+y}\, dx\, dy = 2a \sin h\, a.$$

The bounding lines of the region R are inclined to x and y axes at angles 45° and 135°. However, if we put

$$x + y = u,\ x - y = v,$$

then the bounding lines of R, namely

$$x + y = a,\ x + y = -a,\ x - y = a,\ x - y = -a$$

are transformed to the lines

$$u = a,\ u = -a;\ v = a,\ v = -a$$

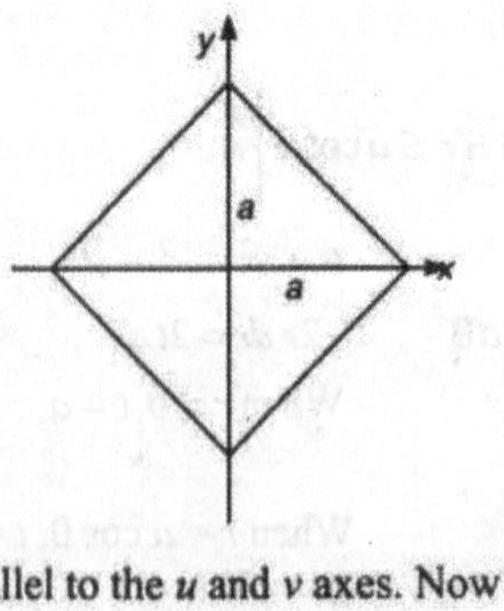

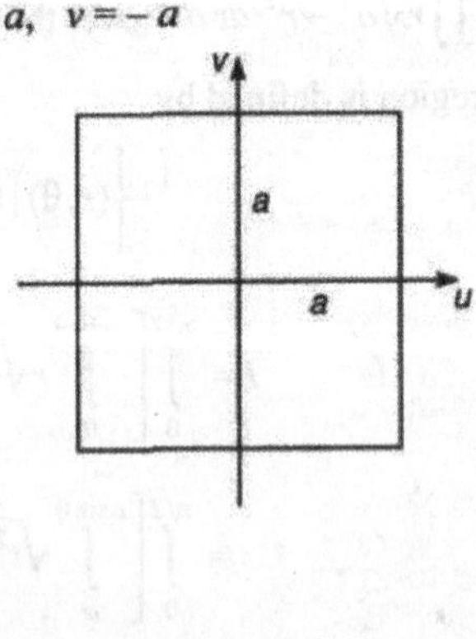

which are parallel to the u and v axes. Now

$$\frac{\partial(u,v)}{\partial(x,y)} = \begin{vmatrix} u_x & u_y \\ v_x & v_y \end{vmatrix} = \begin{vmatrix} 1 & 1 \\ 1 & -1 \end{vmatrix} = -2.$$

$$\therefore \qquad \frac{\partial(x,y)}{\partial(u,v)} = -\frac{1}{2}, \left|\frac{\partial(x,y)}{\partial(u,v)}\right| = \frac{1}{2}.$$

$$\therefore \qquad dx\,dy = \left|\frac{\partial(x,y)}{\partial(u,v)}\right| du\,dv = \frac{1}{2}du\,dv.$$

$$\therefore \qquad \iint_R e^{x+y}dx\,dy = \int_{-a}^{a}\int_{-a}^{a} e^u \frac{1}{2}dv\,du$$

$$= \frac{1}{2}\int_{-a}^{a} e^u [v]_{-a}^{a}\,du = a\int_{-a}^{a} e^u\,du$$

$$= a[e^a - e^{-a}] = 2a\frac{e^a - e^{-a}}{2} = 2a\sin h\,a.$$

Example 3.15

By changing to polar coordinates find the value of the integral

$$\int_0^\infty\int_0^\infty e^{-(x^2+y^2)}dx\,dy.$$

It is evident from the limits that the region of integration is the entire first quadrant which is specified, in polar coordinates, by

$\{(r,\theta)\,|\,0 \le \theta \le \pi/2,\ 0 \le r < \infty\}$.

$$\therefore \qquad I = \int_0^{\pi/2}\left[\int_0^\infty e^{-(r^2\cos^2\theta + r^2\sin^2\theta)}\,r dr\right]d\theta$$

$$= \int_0^{\pi/2}\left[\int_0^\infty e^{-r^2}\,r\,dr\right]d\theta$$

Put $r^2 = t$

$2r\,dr = dt$

When $r = 0,\ t = 0$

When $r = \infty,\ t = \infty$

$$= \frac{1}{2}\int_0^{\pi/2}\left[\int_0^\infty e^{-t}dt\right]d\theta = \frac{1}{2}\int_0^{\pi/2}\left[-\frac{1}{e^t}\right]_0^\infty d\theta$$

$$= \frac{1}{2}\int_0^{\pi/2} 1 d\theta = \frac{1}{2}\cdot\frac{\pi}{2} = \frac{\pi}{4}.$$

Remark. Since $e^{-(x^2+y^2)} = e^{-x^2}e^{-y^2}$, we have

$$\int_0^\infty\int_0^\infty e^{-(x^2+y^2)}dx\,dy = \int_0^\infty e^{-x^2}dx\int_0^\infty e^{-y^2}\,dy$$

$$= \int_0^\infty e^{-x^2}dx\int_0^\infty e^{-x^2}dx = \left[\int_0^\infty e^{-x^2}dx\right]^2$$

Thus from (1), $\int_0^\infty e^{-x^2}dx = \sqrt{\frac{\pi}{4}} = \frac{\sqrt{\pi}}{2}.$

Example 3.16

By changing to polar coordinates, find the value of the integral $\int_0^{2a}\int_0^{\sqrt{2ax-x^2}} (x^2+y^2)\,dy\,dx$.

Now the region lies between the lines $x = 0$, $x = 2a$, and also between the x axis $y = 0$ and the circle.

$$y = \sqrt{2ax-x^2} \quad \text{or} \quad x^2+y^2-2ax=0.$$

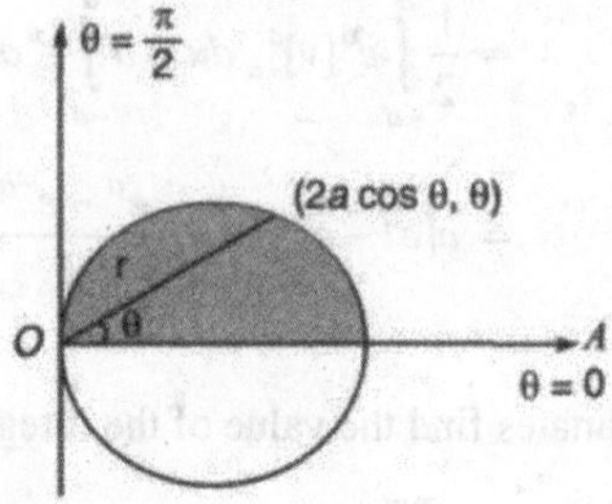

In this, the polar equation of the circle is

$$(r\cos\theta)^2+(r\sin\theta)^2-2a(r\cos\theta)=0 \text{ or } r=2a\cos\theta.$$

Not it is evident that the region of integration is

$$\{(r,\theta)| \quad 0\le\theta\le\pi/2, \quad 0\le r\le 2a\cos\theta\}.$$

$$\therefore \qquad I = \int_0^{\pi/2}\int_0^{2a\cos\theta} [(r\cos\theta)^2+(r\sin\theta)^2]\,r\,dr\,d\theta$$

$$= \int_0^{\pi/2}\int_0^{\cos\theta} r^3 dr\,d\theta = \int_0^{\pi/2}\left[\frac{r^4}{4}\right]_0^{2a\cos\theta} d\theta$$

$$= 4a^4\int_0^{\pi/2}\cos^4\theta\,d\theta = 4a^4 I_4 = 4a^4\frac{3}{4}\cdot\frac{1}{2}\frac{\pi}{2} = \frac{3\pi a^4}{4}.$$

Example 3.17

Changing the order of the integration evaluate

$$\int_{-a}^{a}\int_0^{\sqrt{a^2-y^2}} x\,dx\,dy.$$

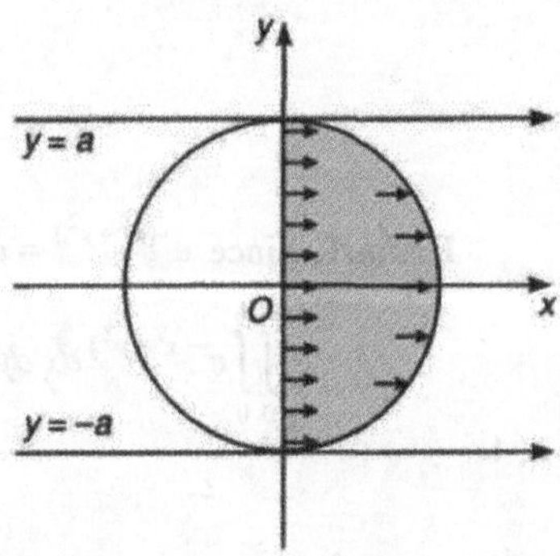

From the given integral, it is seen that the region of integration is inscribed in the strip bounded by

$$y=-a, y=a$$

are bounded, on the left and right by

$$x=0, x=\sqrt{a^2-y^2},$$

or

$$x=0, x^2+y^2=a^2.$$

So the region is the semicircular area lying in the 1st and 4th quadrants.

Considering the region as one inscribed in the strip bounded by

$$x = 0, x = a$$

and bounded, below and above, by the circular arcs

$$y = -\sqrt{a^2 - x^2},\ y = \sqrt{a^2 - x^2},$$

we get the integral as

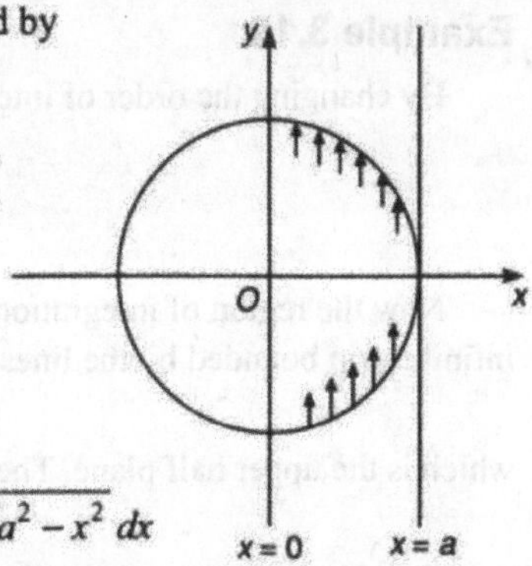

$$I = \int_0^a \int_{-\sqrt{a^2-x^2}}^{\sqrt{a^2-x^2}} x\,dy\,dx$$

$$= \int_0^a x[y]_{-\sqrt{a^2-x^2}}^{\sqrt{a^2-x^2}} dx = 2\int_0^a x\sqrt{a^2 - x^2}\,dx$$

Put $x = a\sin t$

$dx = a\cos t\,dt$

When $x = 0, t = 0$

When $x = a, t = \pi/2$

$$= 2\int_0^{\pi/2} a\sin t \cdot a\cos t \cdot a\cos t\,dt$$

$$= 2a^3 \int_0^{\pi/2} \cos^2 t \sin t\,dt$$

$$= 2a^3 I_{2,1} = 2a^3 \frac{1}{2+1} = \frac{2a^3}{3}.$$

Example 3.18

Evaluate, by changing the order of integration,

$$\int_0^a \int_0^{\sqrt{a^2-x^2}} \sqrt{a^2 - x^2 - y^2}\,dy\,dx. \qquad \textbf{(1997)}$$

The region is the area of the circle $x^2 + y^2 = a^2$ which lies in the first quadrant. In the given integral it is specified as

$$0 \le x \le a, 0 \le y \le \sqrt{a^2 - x^2}.$$

It can also be specified as

$$0 \le y \le a, 0 \le x \le \sqrt{a^2 - y^2}.$$

So, changing the order of integration,

$$I = \int_{y=0}^{y=a} \left[\int_{x=0}^{x=\sqrt{a^2-y^2}} \sqrt{a^2 - x^2 - y^2}\,dx \right] dy$$

$$= \int_0^a \left[\frac{x}{2}\sqrt{(a^2 - y^2) - x^2} + \frac{a^2 - y^2}{2} \sin^{-1} \frac{x}{\sqrt{a^2 - y^2}} \right]_0^{\sqrt{a^2-y^2}} dy$$

$$= \int_0^a \frac{a^2 - y^2}{2} \sin^{-1} 1 dy = \int_0^a \frac{a^2 - y^2}{2} . \frac{\pi}{2} dy = \frac{\pi}{4}\left[a^2 y - \frac{y^3}{3} \right]_0^a$$

$$= \frac{\pi}{4}\left[a^3 - \frac{a^3}{3} \right] = \frac{\pi}{4} . \frac{2a^3}{3} = \frac{\pi a^3}{6}.$$

Example 3.19

By changing the order of integration, show that

$$\int_0^\infty \int_0^y ye^{-y^2/x} dx\, dy = \frac{1}{2}.$$

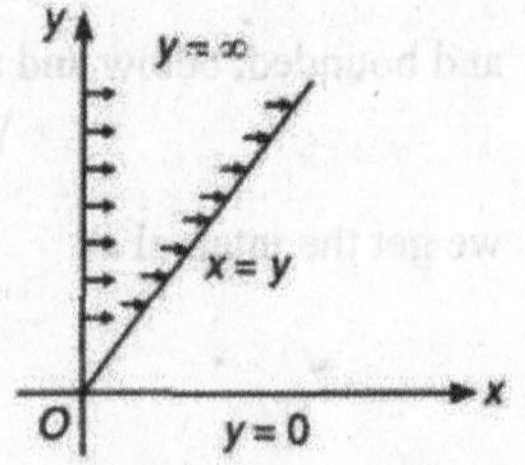

Now the region of integration is an area inscribed in the rectangular infinite strip bounded by the lines.

$$y = 0, y = \infty$$

which is the upper half plane. The area is bounded on the left and right by

$$x = 0, x = y.$$

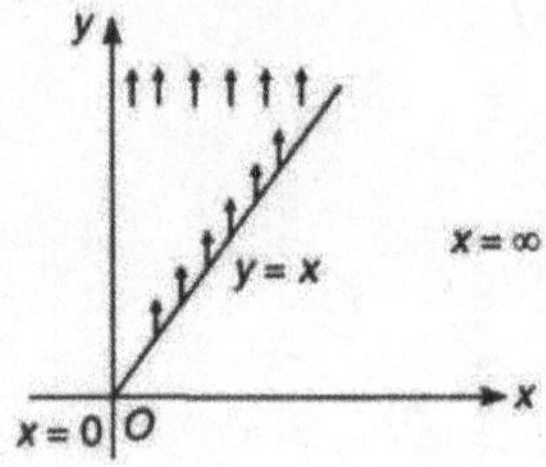

Considering the region as the area inscribed in the right half plane bounded by the lines.

$$x = 0, x = \infty$$

are bounded, below and above, by the lines

$$y = x, y = \infty,$$

we have

$$I = \int_0^\infty \int_x^\infty ye^{-y^2/x} dy\, dx.$$

Now $\frac{d}{dy} e^{-y^2/x} = e^{-y^2/x}\left(-\frac{2y}{x}\right)$ and using $e^{-\infty} = 0$,

$$I = \int_0^\infty \left\{-\frac{x}{2}\left[e^{-y^2/x}\right]_x^\infty\right\} dx$$

$$= \int_0^\infty \left(\frac{x}{2} e^{-x^2/x}\right) dx = \frac{1}{2}\int_0^\infty xe^{-x}\, dx$$

$$= \frac{1}{2}\left[-e^{-x}x + \int e^{-x} dx\right]_0^\infty$$

$$= \frac{1}{2}\left[-e^{-x}x - e^{-x}\right]_0^\infty$$

$$= \frac{1}{2}\left[0 + 0 + 0 + e^0\right] = \frac{1}{2}.$$

Example 3.20

Change the order of integration and evaluate:

$$\int_0^3 \int_1^{\sqrt{4-y}} (x + y)\, dx\, dy.$$

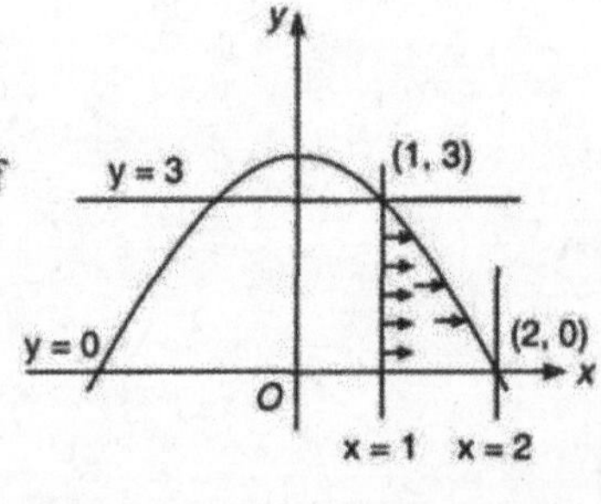

From the given integral, it is seen that the region of integration is in the strip bounded by

$$y = 0, y = 3$$

and bounded, on the left and right, by

$$x = 1, x = \sqrt{4 - y}.$$

or $$x = 1, x^2 = -(y-4).$$

The parabola, $x^2 = -(y-4)$, intersects the line $y = 3$ at (1, 3) and the x axis at (2, 0).

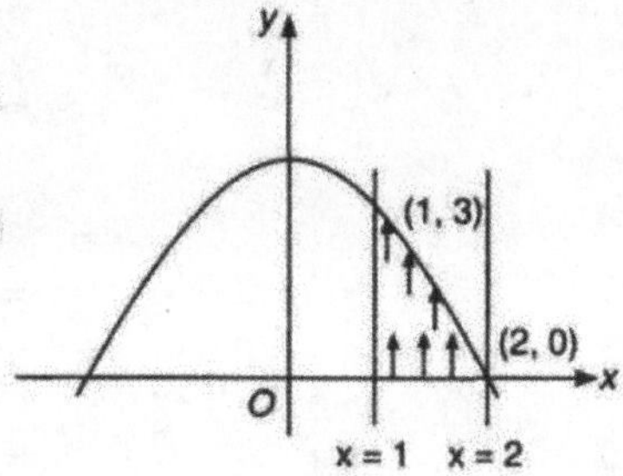

Considering the region as one inscribed in the strip bounded by

$$x = 1, x = 2$$

and bounded, below and above, by

$$y = 0, y = 4 - x^2,$$

we get the integral as

$$I = \int_1^2 \int_0^{4-x^2} (x+y)\,dy\,dx = \int_1^2 \left[xy + \frac{y^2}{2} \right]_0^{4-x^2} dx$$

$$= \int_1^2 \left[x(4-x^2) + \frac{(4-x^2)^2}{2} \right] dx$$

$$= \int_1^2 \left[(4x - x^3) + \frac{16 - 8x^2 + x^4}{2} \right] dx$$

$$= \left[2x^2 - \frac{x^4}{4} + 8x - \frac{4x^3}{3} + \frac{x^5}{10} \right]_1^2 = \frac{241}{60}.$$

Example 3.21

Change the order of integration and evaluate.

$$\int_0^{4a} \int_{x^2/4a}^{2\sqrt{ax}} xy\,dy\,dx. \qquad \textbf{(2005)}$$

Now the region of integration is enclosed by

$$\left. \begin{aligned} y &= 2\sqrt{ax} \\ y &= \frac{x^2}{4a} \end{aligned} \right\} \text{ or } \left\{ \begin{aligned} y^2 &= 4ax \\ x^2 &= 4ay \end{aligned} \right. \text{ or } \left\{ \begin{aligned} x &= \frac{y^2}{4a} \\ x &= \sqrt{4ay} \end{aligned} \right.$$

It is evident that they meet at (0, 0) and $(4a, 4a)$. Now considering the region of integration as the infinite rectangular area, between the lines parallel to x axis, namely,

$$y = 0, y = 4a,$$

cut off on the left and right by parabolas

$$x = \frac{y^2}{4a}, \; x = \sqrt{4ay},$$

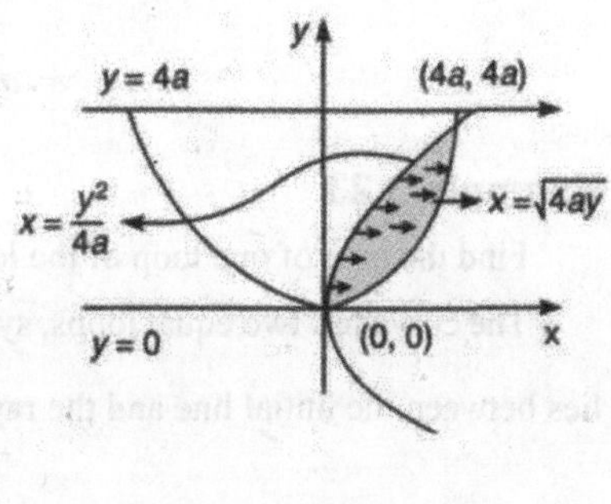

we get the integral as

$$I = \int_0^{4a} \int_{y^2/4a}^{\sqrt{4ay}} xy\,dx\,dy$$

$$= \int_0^{4a} y\left[\frac{x^2}{2}\right]_{y^2/4a}^{\sqrt{4ay}} dy = \int_0^{4a} y\left[\frac{4ay}{2} - \frac{y^4}{2\times4^2a^2}\right]dyt$$

$$= \left[\frac{4ay^3}{2\times3} - \frac{y^6}{2\times4^2a^2\times6}\right]_0^{4a} = \frac{4a(4a)^3}{2\times3} - \frac{4^6a^6}{2\times4^2a^2\times6}$$

$$= a^4\left[\frac{4^4}{2\times3} - \frac{4^4}{2\times6}\right] = a^4\left[\frac{128}{3} - \frac{64}{3}\right] = \frac{64}{3}a^4.$$

Example 3.22

Find the area inside the circle $r = 3a\cos\theta$ and outside the cardioid $r = a(1+\cos\theta)$.

Solving the equations, we have

$$3a\cos\theta = a(1+\cos\theta)$$

or $$\theta = -\frac{\pi}{3}, \frac{\pi}{3}.$$

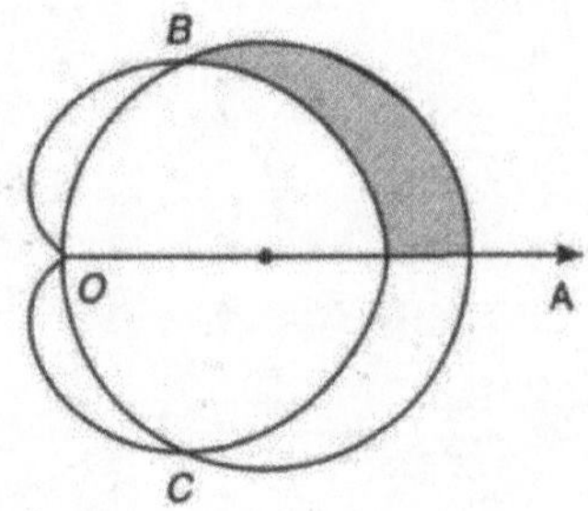

They are the vectorial angles of the common points. B, C, say. The curves are symmetrical about the initial line and the required area is

$$= 2\int_0^{\pi/3}\int_{a(1+\cos\theta)}^{3a\cos\theta} r\,dr\,d\theta$$

$$= 2\int_0^{\pi/3}\left[\frac{r^2}{2}\right]_{a(1+\cos\theta)}^{3a\cos\theta} d\theta$$

$$= \int_0^{\pi/3}[(3a\cos\theta)^2 - a^2(1+\cos\theta)^2]d\theta$$

$$= a^2\int_0^{\pi/3}(8\cos^2\theta - 2\cos\theta - 1)d\theta$$

$$= a^2\int_0^{\pi/3} 4(1+\cos2\theta) - 2\cos\theta - 1]d\theta$$

$$= a^2[3\theta + 2\sin2\theta - 2\sin\theta]_0^{\pi/3}$$

$$= a^2\left[\pi + 2\times\frac{\sqrt{3}}{2} - 2\frac{\sqrt{3}}{2}\right] = \pi a^2.$$

Example 3.23

Find the area of one loop of the lemniscate $r^2 = a^2\cos2\theta$.

The curve has two equal loops, symmetrical about the initial line. Half the area of the right loop lies between the initial line and the ray $\theta = \frac{\pi}{4}$. If the area of this loop is A, then

$$A = 2\int_0^{\pi/4}\int_0^{a\sqrt{\cos 2\theta}} r\,dr\,d\theta$$

$$= 2\int_0^{\pi/4}\left[\frac{r^2}{2}\right]_0^{a\sqrt{\cos 2\theta}} d\theta$$

$$= a^2\int_0^{\pi/4} \cos 2\theta\, d\theta = \frac{a^2}{2}$$

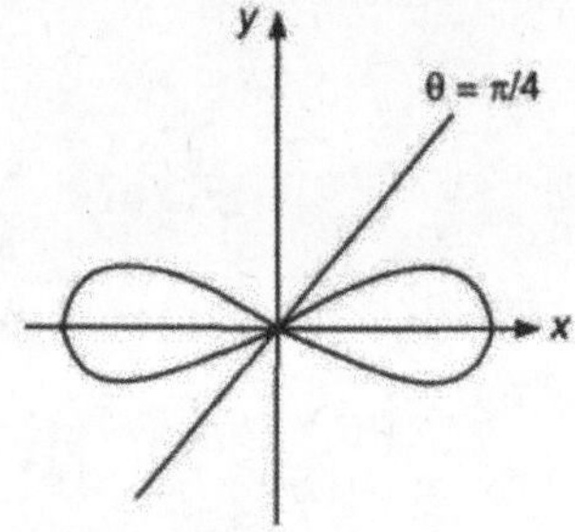

3.4.2. Mass centre

Mass centre is also called the centroid and the centre of gravity. We shall consider a plane area as a uniform thin plate. Then the formula for mass centre $(\bar{x}, \bar{y})$ of an area of region R is

$$\bar{x} = \frac{\iint_R x\,dy\,dx}{\iint_R dy\,dx},\ \bar{y} = \frac{\iint_R x\,dy\,dx}{\iint_R dy\,dx}$$

The double integral in the denominator is the area of R.
If $\rho(x, y)$ is the density at (x, y), then

$$\bar{x} = \frac{\iint_R \rho x\,dy\,dx}{\iint_R \rho\,dy\,dx},\ \bar{y} = \frac{\iint_R \rho y\,dy\,dx}{\iint_R \rho\,dy\,dx}.$$

The double integral in the denominator is the mass of the area.
In the polar coordinates we have

$$\bar{x} = r\cos\theta,\ y = r\sin\theta$$

and hence the centroid $(\bar{x}, \bar{y})$ is such that

$$\bar{x} = \frac{\iint_R x(r\,dr\,d\theta)}{\iint_R r\,dr\,d\theta} = \frac{\iint_R r^2\cos\theta\,dr\,d\theta}{\iint_R r\,dr\,d\theta}$$

$$\bar{y} = \frac{\iint_R y(r\,dr\,d\theta)}{\iint_R r\,dr\,d\theta} = \frac{\iint_R r^2\sin\theta\,dr\,d\theta}{\iint_R r\,dr\,d\theta}.$$

For a solid of volume V, the mass centre $(\bar{x}, \bar{y}, \bar{z})$ is such that

$$\bar{x} = \frac{\iiint_V x\,dV}{\iiint_V dV},\ \bar{y} = \frac{\iiint_V y\,dV}{\iiint_V dV},\ \bar{z} = \frac{\iiint_V z\,dV}{\iiint_V dV}$$

Example 3.24

Find the mass centre of the first quadrant area of the circle $x^2 + y^2 = a^2$, using double integral.

$$\bar{x} = \frac{\int_0^a \int_0^{\sqrt{a^2-x^2}} x\,dy\,dx}{(1/4)\pi a^2} = \frac{4}{\pi a^2}\int_0^a x[y]_0^{\sqrt{a^2-x^2}}\,dx$$

$$= \frac{4}{\pi a^2}\int_0^a x\sqrt{a^2-x^2}\,dx$$

$$= \frac{4}{\pi a^2}\int_0^a t^2\,dt = \frac{4}{\pi a^2}\cdot\frac{a^3}{3}$$

Put $a^2 - x^2 = t^2$

$x\,dx = -t\,dt$

When $x = 0, t = a$

When $x = 0, t = 0$

$$= \frac{4a}{3\pi}.$$

So also, by symmetry,

$$\bar{y} = \frac{4a}{3\pi}$$

Thus the centroid is $\left(\frac{4a}{3\pi}, \frac{4a}{3\pi}\right)$ which is on the middle radius at a distance $\frac{4a}{3\pi}\sqrt{2}$ from the centre.

Example 3.25

Find the mass centre of the first quadrant area of the ellipse $\frac{x^2}{a^2}+\frac{y^2}{b^2}=1$, using double integral.

$$\bar{x} = \frac{\int_0^a \int_0^{(b/a)\sqrt{a^2-x^2}} x\,dy\,dx}{(1/4)\pi ab} \quad \text{since area of ellipse is } \pi \text{ ab}$$

$$= \frac{4}{\pi ab}\int_0^a x[y]_0^{(b/a)\sqrt{a^2-x^2}}\,dx$$

$$= \frac{4}{\pi ab}\cdot\frac{b}{a}\int_0^a x\sqrt{a^2-x^2}\,dx = \frac{4}{\pi a^2}\cdot\frac{a^3}{3}.$$

$$= \frac{4a}{3\pi}.$$

By symmetry, we also have

$$\bar{y} = \frac{4b}{3\pi}.$$

Thus the mass centre is $\left(\frac{4a}{3\pi}, \frac{4b}{3\pi}\right)$.

Example 3.26

Using double integral, find the centroid of the area enclosed by one arch and the base of the cycloid

$$x = a(\theta - \sin\theta),\ y = a(1 - \cos\theta).$$

For one branch from the origin, θ varies from 0 to 2π. This branch is symmetrical about the line $x = a\pi$. So $\bar{x} = a\pi$.

$$\bar{y} = \frac{\int_0^{2a\pi}\int_0^{y} y\,dy\,dx}{\int_0^{2a\pi}\int_0^{y} dy\,dx} = \frac{\int_0^{2a\pi}\int_0^{a(1-\cos\theta)} y\,dy\,dx}{\int_0^{2a\pi}\int_0^{a(1-\cos\theta)} dy\,dx}$$

$$= \frac{\int_0^{2a\pi}\left[\frac{y^2}{2}\right]_0^{a(1-\cos\theta)} dx}{\int_0^{2a\pi}[y]_0^{a(1-\cos\theta)}\,dx}$$

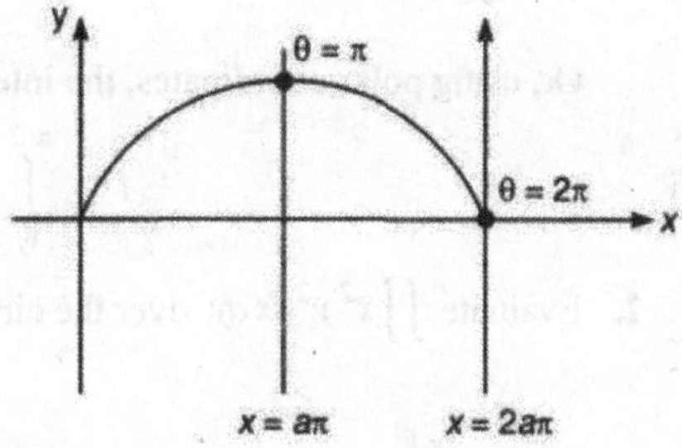

$$= \frac{\frac{1}{2}\int_0^{2a\pi} a^2(1-\cos\theta)^2\,dx}{\int_0^{2a\pi} a(1-\cos\theta)\,dx}$$

$$= \frac{a}{2}\,\frac{\int_0^{2\pi}(1-\cos\theta)^3\,d\theta}{\int_0^{2x}(1-\cos\theta)^2\,d\theta} \qquad \left(\begin{array}{l} x = a(\theta - \sin\theta) \\ dx = a(1-\cos\theta)\,d\theta \\ \theta \text{ varies from } 0 \text{ to } 2\pi \end{array}\right)$$

We know that $\int_0^{2a} f(x)\,dx = 2\int_0^{a} f(x)\,dx$ if $f(2a - x) = f(x)$

$$\therefore \qquad \bar{y} = \frac{a}{2}\,\frac{2\int_0^{\pi}(1-\cos\theta)^3\,d\theta}{2\int_0^{\pi}(1-\cos\theta)^2\,d\theta} = \frac{a}{2}\,\frac{\int_0^{\pi} 8\sin^6\frac{\theta}{2}\,d\theta}{\int_0^{\pi} 4\sin^4\frac{\theta}{2}\,d\theta}$$

$$= a\,\frac{\int_0^{\pi/2}\sin^6 t\,(2\,dt)}{\int_0^{\pi/2}\sin^4 t\,(2\,dt)}, \text{ where } t = \frac{\theta}{2}$$

$$= a\,\frac{I_6}{I_4} = a\left(\frac{5}{6}\,\frac{3}{4}\,\frac{1}{2}\,\frac{\pi}{2}\right)\left(\frac{4}{3}\cdot\frac{2}{1}\cdot\frac{2}{\pi}\right) = \frac{5a}{6}.$$

Thus the centroid is $\left(a\pi, \frac{5a}{6}\right)$.

Exercise

1. Evaluate $\iint x^2 y\,dx\,dy$ taken over the positive quadrant of the circle $x^2+y^2=a^2$.

   [**Hint:** In cartesian coordinates the integral is

$$I=\int_0^a\int_0^{\sqrt{a^2-x^2}} x^2 y\,dy\,dx=\int_0^a x^2\left[\frac{y^2}{2}\right]_0^{\sqrt{a^2-x^2}} dx$$

$$=\int_0^a x^2\left[\frac{a^2-x^2}{2}\right]dx=\frac{a^5}{15}.$$

   Or, using polar coordinates, the integral is

$$I=\int_0^{\pi/2}\int_0^a (r\cos\theta)^2(r\sin\theta)r\,dr\,d\theta=\frac{a^5}{5}I_{2,1}=\frac{a^5}{5}\frac{1}{2+1}]$$

2. Evaluate $\iint x^2y^2\,dx\,dy$ over the circular area

$$x^2+y^2\le 1.$$

   [**Hint:** $\int_0^{2\pi}\int_0^1 (r\cos\theta)^2(r\sin\theta)^2\,r\,dr\,d\theta=\frac{\pi}{24}$]

3. Show that $\iint\sqrt{4x^2-y^2}\,dx\,dy=\frac{1}{3}\left(\frac{\pi}{3}+\frac{\sqrt{3}}{2}\right)$ where the integration is over the area bounded by the lines $y=0, y=x, x=1$. **(1983, 90)**

   [**Hint:** The region is a triangular area with vertices at (0, 0), (1, 0), (0, 1). It is obtained from the rectangular space between $x=0$ and $x=1$, cut below by the x axis ($y=0$) and above by the line $y=x$.

$$\therefore\quad I=\int_0^1\int_0^x\sqrt{4x^2-y^2}\,dy\,dx$$

$$=\int_0^1\left[\frac{y}{2}\sqrt{4x^2-y^2}+\frac{4x^2}{2}\sin^{-1}\frac{y}{2x}\right]_0^x dx$$

$$=\int_0^1\left[\frac{x}{2}\sqrt{3x^2}+2x^2\sin^{-1}\frac{1}{2}\right]dx=\frac{\sqrt{3}}{2}\cdot\frac{1}{3}+\frac{2}{3}\cdot\frac{\pi}{6}]$$

4. Evaluate $\iint x^2 y\,dx\,dy$ taken over the positive quadrant of the circle $x^2+y^2=a^2$.

   [**Hint:** In the cartesian coordinates the integral is

$$I=\int_0^a\int_0^{\sqrt{a^2-x^2}} x^2 y\,dy\,dx=\int_0^a x^2\left[\frac{y^2}{2}\right]_0^{\sqrt{a^2-x^2}} dx$$

$$=\int_0^a x^2\left[\frac{a^2-x^2}{2}\right]dx=\frac{a^5}{15}.$$

Or, using polar coordinates, the integral is

$$I=\int_0^{\pi/2}\int_0^a (r\cos\theta)^2(r\sin\theta)r\,dr\,d\theta=\frac{a^5}{5}I_{2,1}=\frac{a^5}{5}\frac{1}{2+1}]$$

5. Evaluate $\iint x^2y^2\,dx\,dy$ over the circular area

$$x^2+y^2\le 1. \qquad \textbf{(1986)}$$

[**Hint:** $I=\int_0^{2\pi}\int_0^1 (r\cos\theta)^2(r\sin\theta)^2\,r\,dr\,d\theta=\frac{\pi}{24}$]

6. Evaluate $\iint y^2\,dx\,dy$ over the area of the ellipse

$$\frac{x^2}{a^2}+\frac{y^2}{b^2}=1. \qquad \textbf{(1985)}$$

[**Hint:** The area lies between $x=a$ and $x=-a$ and further between the lower and upper halves of the ellipse whose equations are

$$y=-\frac{b}{a}\sqrt{a^2-x^2},\quad y=\frac{b}{a}\sqrt{a^2-x^2}.$$

$$\therefore \qquad I=\int_{-a}^{a}\int_{-(b/a)\sqrt{a^2-x^2}}^{(b/a)\sqrt{a^2-x^2}} y^2\,dy\,dx=\frac{\pi ab^3}{4}$$

or, putting $x=au\cos t$, $y=bu\sin t$, $0\le t\le 2\pi$, $0\le u\le 1$,

$$I=\int_0^{2\pi}\int_0^1 (bu\sin t)^2\,abu\,du\,dt=\frac{\pi ab^3}{4}]$$

TRIPLE INTEGRALS

The triple integral over a volume is defined in the same way as a double integral is defined over an area.

In general the triple integral of $f(x, y, z)$ in cartesian coordinates is of the form

$$\int_a^b\int_{g_1(x)}^{g_2(x)}\int_{h_1(x,y)}^{h_2(x,y)} f(x,y,z)\,dz\,dy\,dx,$$

where the region of integration is defined by

$$a\le x\le b,\; g_1(x)\le y\le g_2(x),\; h_1(x,y)\le z\le h_2(x,y).$$

In the evaluation of the integral, first the integration is to be done with respect to z, treating x and y as constants, and taking limits from $h_1(x, y)$ to $h_2(x, y)$ and so on.

Volume. The volume of the above region is given by

$$\int_a^b\int_{g_1(x)}^{g_2(x)}\int_{h_1(x,y)}^{h_2(x,y)} 1\,dz\,dy\,dx.$$

IMPORTANT SURFACES

The following are some important surfaces:

(*i*) $\boxed{x^2+y^2=a^2}$

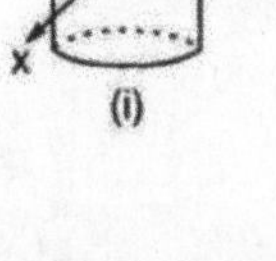

(i)

A cylinder whose axis is the z axis and radius is a. It extends above and below the xy plane infinitely.

(ii) $$\boxed{x^2+y^2 = z^2}$$

A cone whose axis is the z axis, the semivertical angle is 45° and vertex is the origin.

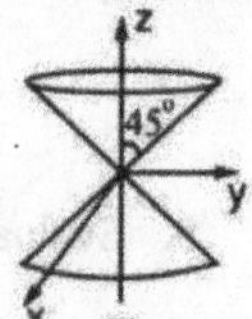

(ii)

(iii) $$\boxed{x^2+y^2 = 4az}$$

A paraboloid; the section of it by any plane through the z axis is a parabola whose axis is the z axis and the vertex is the origin and l.r. = $4a$.

(iii)

(iv) $$\boxed{x^2+y^2 = 2ax}$$

A cylinder whose radius is a, its axis is parallel to the z axis and passes through $(a, 0, 0)$. The equations of its circle of intersection by $z = 0$ are

$$r = 2a\cos\theta, z = 0.$$

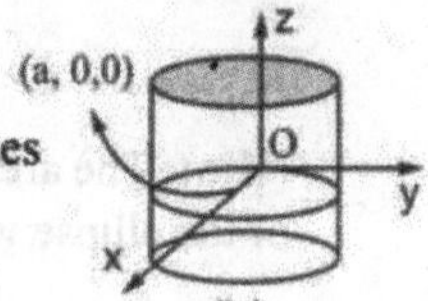

(iv)

(v) $$\boxed{x^2+y^2 = ax}$$

A cylinder whose section by $z = 0$ is the circle $r = a\cos\theta$, $z = 0$.

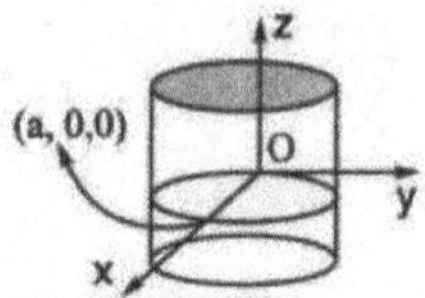

(v)

COORDINATES OF POINTS OF REGIONS

(i) For the points of the region, bounded by the planes

$$x = 0, \quad y = 0, \quad z = 0, \quad x + y + z = 1,$$

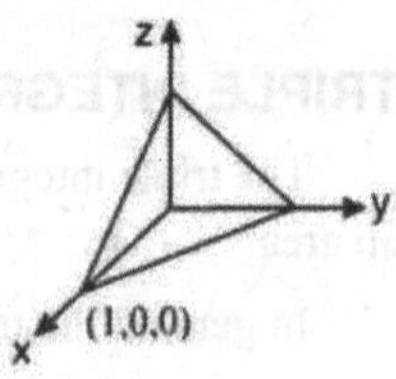

x varies between the planes

$$x = 0, x = 1,$$

y varies between the planes

$$y = 0, y = 1 - x,$$

z varies between the planes

$$z = 0, z = 1 - x - y.$$

(ii) For the points of the region, bounded by the cylinder $x^2 + y^2 = a^2$, the plane $z = 0$ and the plane $y + z = 1$, the cylindrical polar coordinates r, θ, z are such that

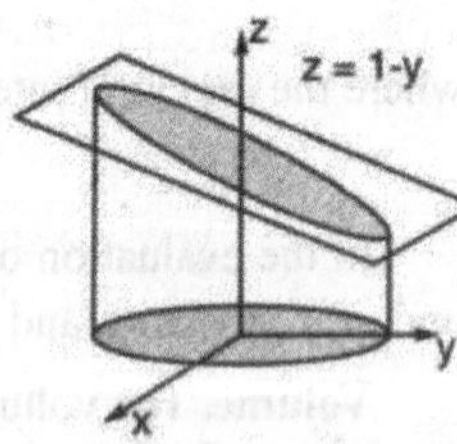

r varies from 0 to a,
θ varies from 0 to 2π,
z varies from 0 to $1 - y$

i.e., z varies from 0 to $1 - r\sin\theta$.

(iii) For the points of the region bounded by the hemisphere $x^2 + y^2 + z^2 = a^2$ and the plane $z = 0$, the spherical polar coordinates r, θ, ϕ are such that

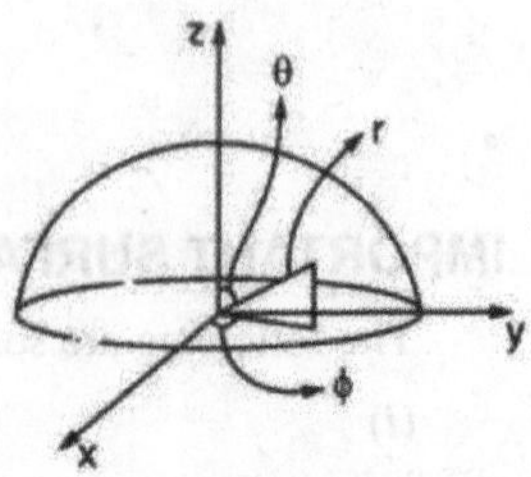

r varies from 0 to a,
θ varies from 0 to $\pi/2$,
ϕ varies from 0 to 2π.

(iv) For the points of the region bounded by the sphere $x^2+y^2+z^2=a^2$, the spherical coordinates r, θ, ϕ are such that

$$0 \le r \le a,\ \ 0 \le \theta \le \pi,\ \ 0 \le \phi \le 2\pi.$$

(v) For the points of the region of the sphere $x^2+y^2+z^2=a^2$, lying in the positive octant,

$$0 \le r \le a,\ \ 0 \le \theta \le \pi/2,\ \ 0 \le \phi \le \pi/2.$$

Example 3.22

Evaluate $\int_0^1\int_0^2\int_0^3 (x+y+z)\,dx\,dy\,dz.$

Here the order of integration is as follows:

(i) w.r.t. x from 0 to 3

(ii) w.r.t. y from 0 to 2

(iii) w.r.t. z from 0 to 1.

$$\therefore \quad I = \int_0^1\int_0^2\left[\frac{x^2}{2}+yx+zx\right]_0^3 dy\,dz = \int_0^1\int_0^2\left[\frac{9}{2}+3y+3z\right]dy\,dz$$

$$= \int_0^1\left[\frac{9y}{2}+\frac{3y^2}{2}+3zy\right]_0^2 dz = \int_0^1 (9+6+6z)dz$$

$$= \left[15z+3z^2\right]_0^1 = 18.$$

Example 3.23

Evaluate $\int_0^1\int_0^{\sqrt{1-x^2}}\int_0^{\sqrt{1-x^2-y^2}} \frac{1}{\sqrt{1-x^2-y^2-z^2}}\,dz\,dy\,dx.$

$$I = \int_0^1\int_0^{\sqrt{1-x^2}}\int_0^{\sqrt{1-x^2-y^2}} \frac{1}{\sqrt{(1-x^2-y^2)-z^2}}\,dz\,dy\,dx$$

$$= \int_0^1\int_0^{\sqrt{1-x^2}}\left[\sin^{-1}\frac{z}{\sqrt{1-x^2-y^2}}\right]_0^{\sqrt{1-x^2-y^2}} dy\,dx$$

$$= \int_0^1\int_0^{\sqrt{1-x^2}} \sin^{-1}\frac{\sqrt{1-x^2-y^2}}{\sqrt{1-x^2-y^2}}\,dy\,dx$$

$$= \int_0^1\int_0^{\sqrt{1-x^2}} \frac{\pi}{2}\,dy\,dx = \int_0^1 \frac{\pi}{2}[y]_0^{\sqrt{1-x^2}}\,dx$$

$$= \frac{\pi}{2}\int_0^1 \sqrt{1-x^2}\,dx = \frac{\pi}{2}\left[\frac{x}{2}\sqrt{1-x^2}+\frac{1}{2}\sin^{-1}x\right]_0^1$$

$$= \frac{\pi}{2}\left[\frac{1}{2}\cdot\frac{\pi}{2}\right] = \frac{\pi^2}{8}.$$

Remark. The region of integration is the volume of the sphere $x^2+y^2+z^2=1$ in the positive octant.

Example 3.24

Evaluate $\int_0^1\int_0^{1-x}\int_0^{x+y} e^z\,dz\,dy\,dx$.

$$I=\int_0^1\int_0^{1-x}\int_0^{x+y} e^z\,dz\,dy\,dx=\int_0^1\int_0^{1-x}\left[e^z\right]_0^{x+y}dy\,dx$$

$$=\int_0^1\int_0^{1-x}(e^{x+y}-1)\,dy\,dx$$

$$=\int_0^1\int_0^{1-x}(e^xe^y-1)\,dy\,dx=\int_0^1\left[e^xe^y-y\right]_0^{1-x}dx$$

$$=\int_0^1[e^xe^{1-x}-(1-x)-e^x]\,dx$$

$$=\int_0^1(e-1+x-e^x)\,dx=\left[ex-x+\frac{x^2}{2}-e^x\right]_0^1=\frac{1}{2}.$$

Example 3.25

Evaluate $\int_0^1\int_0^{1-x}\int_0^{1-x-y} dx\,dy\,dz$.

Writing the order of integration properly,

$$I=\int_0^1\int_0^{1-x}\int_0^{1-x-y} dz\,dy\,dx=\int_0^1\int_0^{1-x}[z]_0^{1-x-y}dy\,dx$$

$$=\int_0^1\int_0^{1-x}(1-x-y)\,dy\,dx=\int_0^1\left[y-xy-\frac{y^2}{2}\right]_0^{1-x}dx$$

$$=\int_0^1\left[(1-x)-x(1-x)-\frac{(1-x)^2}{2}\right]dx$$

$$=\int_0^1\left[(1-x)(1-x)-\frac{(1-x)^2}{2}\right]dx$$

$$=\int_0^1\frac{(1-x)^2}{2}dx=\left[-\frac{(1-x)^3}{2\times 3}\right]_0^1=\frac{1}{6}.$$

Remark. The region of integration is the volume enclosed by the planes

$$x=0,\ y=0,\ z=0,\ x+y+z=1$$

Example 3.26

Evaluate $\iiint \frac{1}{(x+y+z+1)^3} dx\,dy\,dz$ over the region bounded by the planes

$$x = 0,\; y=0,\; z=0,\; x+y+z=1.$$

$$I = \int_0^1 \int_0^{1-x} \int_0^{1-x-y} (x+y+z+1)^{-3}\, dz\, dy\, dx$$

$$= \int_0^1 \int_0^{1-x} \left[\frac{(x+y+z+1)}{-2}\right]_0^{1-x-y} dy\, dx$$

$$= \int_0^1 \int_0^{1-x} \left[-\frac{1}{8} + \frac{(x+y+1)^{-2}}{2}\right] dy\, dx$$

$$= \int_0^1 \left[-\frac{y}{8} - \frac{(x+y+1)^{-1}}{2}\right]_0^{1-x} dx$$

$$= \int_0^1 \left(-\frac{1-x}{8} - \frac{1}{4} + \frac{1}{2(1+x)}\right) dx$$

$$= \int_0^1 \left(-\frac{3}{8} + \frac{x}{8} + \frac{1}{2(1+x)}\right) dx$$

$$= \left[-\frac{3x}{8} + \frac{x^2}{16} + \frac{\log(1+x)}{2}\right]_0^1 = -\frac{5}{16} + \frac{1}{2}\log 2.$$

Example 3.27

Evaluate $\iiint xyz\, dx\, dy\, dz$ over the positive octant of the sphere $x^2 + y^2 + z^2 = a^2$.

In cartesian coordinates the integral is

$$\int_0^a \int_0^{\sqrt{a^2-x^2}} \int_0^{\sqrt{a^2-x^2-y^2}} xyz\, dz\, dy\, dx = \frac{a^6}{48}$$

on simplification. Alternatively, using the spherical polar coordinates,

$$x = r\sin\theta\cos\phi,\; y = r\sin\theta\sin\phi,\; z = r\cos\theta,$$

$$dx\,dy\,dz = (r^2\sin\theta)\, dr\, d\theta\, d\phi,$$

and the integral is

$$\int_0^{\pi/2} \int_0^{\pi/2} \int_0^a r^5 \sin^3\theta\cos\theta\sin\phi\cos\phi\; d\theta\, d\phi$$

$$= \int_0^{\pi/2} \int_0^{\pi/2} \left[\frac{r^6}{6}\right]_0^a \sin^3\theta\cos\theta\sin\phi\cos\phi\, d\theta\, d\phi$$

$$= \frac{a^6}{6} \int_0^{\pi/2} \left[I_{3,1}\right] \sin\phi\cos\phi\, d\phi$$

$$= \frac{a^6}{6} \int_0^{\pi/2} \frac{1}{3+1} \sin\phi\cos\phi\, d\phi$$

$$= \frac{a^6}{24} I_{1,1} = \frac{a^6}{24} \cdot \frac{1}{1+1} = \frac{a^6}{48}.$$

Example 3.28

Evaluate $\iiint \frac{1}{\sqrt{1-x^2-y^2-z^2}}\,dx\,dy\,dz$ over the region bounded by the sphere $x^2+y^2+z^2=1$.

Using spherical polar coordinates, as in the previous example, we have

$$\int_0^{2\pi}\int_0^{\pi}\int_0^{1}\frac{1}{\sqrt{1-r^2}}r^2\sin\theta\,dr\,d\theta\,d\phi$$

$$=\left(\int_0^{2\pi}d\phi\right)\left(\int_0^{\pi}\sin\theta\,d\theta\right)\left(\int_0^{1}\frac{r^2}{\sqrt{1-r^2}}dr\right)$$

$$=[2\pi]\,[-\cos\theta]_0^{\pi}\times\int_0^1\frac{r^2}{\sqrt{1-r^2}}dr$$

Put $r=\sin t$

$dr=\cos t\,dt$

When $r=0,\ t=0$

When $r=1,\ t=\pi/2$

$$=2\pi[2]\int_0^1\frac{r^2}{\sqrt{1-r^2}}dr=4\pi\int_0^{\pi/2}\frac{\sin^2 t}{\sqrt{\cos^2 t}}\cos t\,dt$$

$$=4\pi I_2=4\pi\left(\frac{1}{2}\frac{\pi}{2}\right)=\pi^2.$$

Example 3.29

Using cylindrical polar coordinates, show that

$$I=\iiint_R z^2\cdot dz\,dy\,dx=\frac{\pi a^8}{12},$$

where R is the region bounded by the cylinder $x^2+y^2=a^2$ and the paraboloid $x^2+y^2=z$ and the xy plane.

The region R is vertically over the circular area

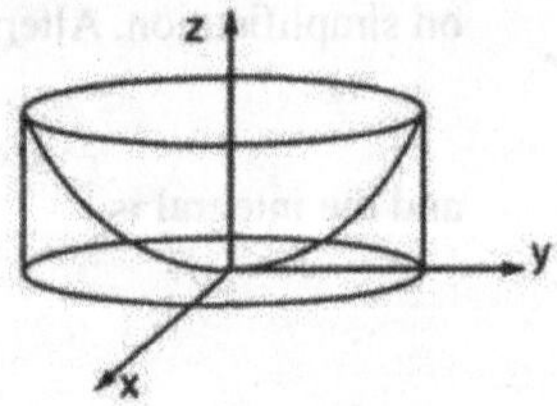

$$x^2+y^2=a^2,\ z=0$$

upto the paraboloid

$$z=x^2+y^2.$$

So, in cylindrical polar coordinates,

$$x=r\cos\theta,\quad y=r\sin\theta,\quad z=z$$

and R is defined by the limits $0\le r\le a,\ \ 0\le\theta\le 2\pi,\ \ 0\le z\le x^2+y^2=r^2$.

$$\therefore\quad I=\int_0^a\int_0^{2\pi}\int_0^{r^2}z^2dz(r\,d\theta\,dr)$$

$$=\int_0^a\int_0^{2\pi}\left[\frac{z^3}{3}\right]_0^{r^2}(r\,d\theta\,dr)=\int_0^a\int_0^{2\pi}\frac{r^7}{3}d\theta\,dr$$

$$=\int_0^a\frac{r^7}{3}2\pi\,dr=\frac{\pi a^8}{12}.$$

Example 3.30

Show that the volume common to the sphere $x^2 + y^2 + z^2 = a^2$ and one branch of the cone $x^2 + y^2 = z^2$ is $(\pi/3)(2 - \sqrt{2})a^3$.

We shall use spherical polar coordinates. Now the semivertical angle of the cone is $\pi/4$. So θ varies from 0 to $\pi/4$ and

$$0 \le r \le a, 0 \le \phi \le 2\pi.$$

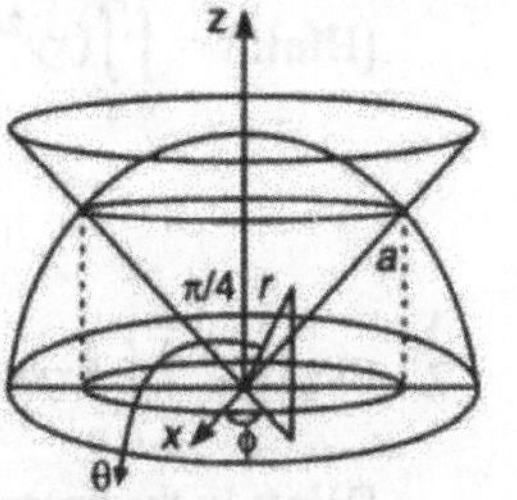

$$\text{Volume} = \iiint_V dx\,dy\,dz = \int_0^{2\pi}\int_0^{a}\int_0^{\pi/4} (r^2 \sin\theta)\,d\theta\,dr\,d\phi$$

$$= \int_0^{2\pi}\int_0^{a} r^2 [1 - \cos\theta]_0^{\pi/4}\,dr\,d\phi$$

$$= \int_0^{2\pi}\int_0^{a} r^2\left(1 - \frac{1}{\sqrt{2}}\right)dr\,d\phi = \int_0^{2\pi} \frac{a^3}{3}\left(1 - \frac{1}{\sqrt{2}}\right)d\phi$$

$$= \frac{2\pi a^3}{3}\left(1 - \frac{1}{\sqrt{2}}\right) = \frac{\pi}{3}(2 - \sqrt{2})a^3.$$

Example 3.31

Evaluate the triple integral $I = \int_0^1 \int_0^{\sqrt{1-x^2}} \int_{\sqrt{x^2+y^2}}^{1} \frac{1}{\sqrt{x^2+y^2+z^2}}\,dz\,dy\,dx,$

using spherical polar coordinates.

The region of integration is the portion of a cone which lies in the first octant and whose vertex is the origin, semivertical angle is $\pi/4$, axis is the z axis, cut off by the plane $z = 1$. So, of the spherical polar coordinates, θ is such that

$$0 \le \theta \le \pi/4.$$

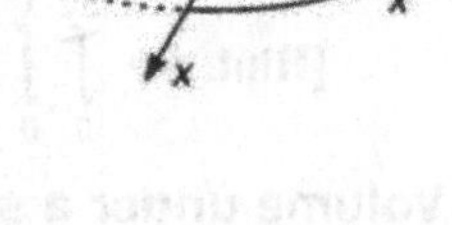

Let P be any point within the region. Let OP produced meet the plane $z = 1$ at Q. Then $OQ = \sec\theta$. So

$$0 \le r \le \sec\theta.$$

It is evident that ϕ is such that

$$0 \le \phi \le \pi/2.$$

$$\therefore \quad I = \int_0^{\pi/4}\int_0^{\sec\theta}\int_0^{\pi/2} \frac{1}{\sqrt{r^2}} r^2 \sin\theta\,d\phi\,dr\,d\theta$$

$$= \int_0^{\pi/4}\int_0^{\sec\theta} \left[\frac{\pi}{2}\right] r \sin\theta\,dr\,d\theta$$

$$= \frac{\pi}{2}\int_0^{\pi/4} \frac{\sec^2\theta}{2}\sin\theta\,d\theta = \frac{\pi}{4}\int_0^{\pi/4} \sec\theta\tan\theta\,d\theta$$

$$= \frac{\pi}{4}[\sec\theta]_0^{\pi/4} = \frac{\pi}{4}(\sqrt{2} - 1).$$

EXERCISE

1. Evaluate $\int_0^2\int_1^3\int_1^2 xy^2z\,dz\,dy\,dx$

[**Hint:** $I=\int_0^2\int_1^3 (xy^2)\left[\frac{z^2}{2}\right]_1^2 dy\,dx=\int_0^2\int_1^3 xy^2(3/2)\,dy\,dx$

$=\frac{3}{2}\int_0^2 x\left[\frac{y^3}{3}\right]_1^3 dx=\frac{3}{2}\cdot\frac{26}{3}\int_0^2 x\,dx=26$]

2. Evaluate $\int_0^a\int_0^x\int_0^y xyz\,dx\,dy\,dz$

[**Hint:** In the proper order of integration,

$$I=\int_0^a\int_0^x\int_0^y xyz\,dz\,dy\,dx=\int_0^a\int_0^x xy\left[\frac{z^2}{2}\right]_0^y dy\,dx$$

$$=\frac{1}{2}\int_0^a x\left[\frac{y^4}{4}\right]_0^x dx=\frac{1}{8}\int_0^a x^5\,dx=\frac{a^6}{48}\text{]}$$

3. Show that the volume of the region common to the cylinders $x^2+y^2=a^2$ and $x^2+z^2=a^2$ is $16a^3/3$.

[**Hint:** $I=8\int_{x=0}^{a}\int_{y=0}^{\sqrt{a^2-x^2}}\int_{z=0}^{\sqrt{a^2-x^2}} dz\,dy\,dx$ for 8 octants]

4. Using spherical polar coordinates, show that $I=\iiint\frac{1}{(x^2+y^2+z^2)^{3/2}}dz\,dy\,dx=4\pi\log\frac{a}{b}$,

where the integral is over the region between the spheres.

$$x^2+y^2+z^2=a^2,\ x^2+y^2+z^2=b^2,\ a>b.$$

[**Hint:** $I=\int_0^{2\pi}\int_0^{\pi}\int_b^a\frac{1}{(r^2)^{3/2}}(r^2\sin\theta)\,dr\,d\theta\,d\phi$]

5. Using spherical polar coordinates, show that

$$\int_0^1\int_0^{\sqrt{1-x^2}}\int_0^{\sqrt{1-x^2-y^2}}\frac{1}{\sqrt{1-x^2-y^2-z^2}}dz\,dy\,dx=\frac{\pi^2}{8}.$$

[**Hint:** $I=\int_0^{\pi/2}\int_0^1\int_0^{\pi/2}\frac{1}{\sqrt{1-r^2}}(r^2\sin\theta)\,d\phi\,dr\,d\theta$]

Volume under a surface and on the xy-plane.

Suppose a solid is in the form of a cylinder whose generators are vertical such that its top portion is a surface S whose equation is

$[z=\phi(x,y)]$.

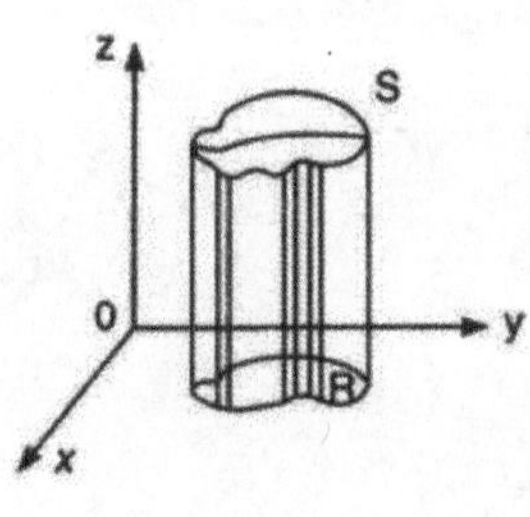

Let its bottom be the projection R of S on the xy plane, $z=0$. Then the volume of the solid is the double integral

$$\iint_R z\,dx\,dy \quad\text{or}\quad \iint_R \phi(x,y)\,dx\,dy.$$

EXAMPLES

Example 1. Find the volume of the tetrahedron bounded by

$$x = 0, y = 0, z = 0, (x/a) + (y/b) + (z/c) = 1.$$

The volume V is over the triangle OAB. So

$$V = \iint_{\Delta OAB} z\,dy\,dx$$

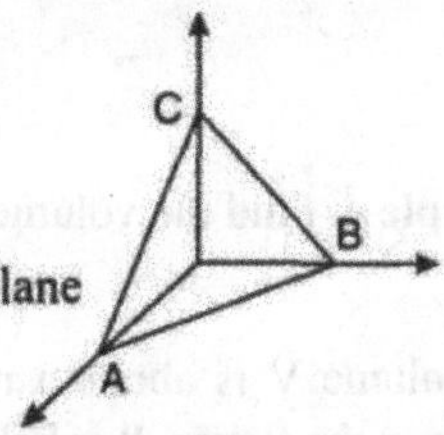

Where $z = c(1 - (x/a) - (y/b))$ since the volume is below the plane

$$(x/a) + (y/b) + (z/c) = 1.$$

$$\therefore \quad V = \int_0^a \int_0^{b(1-(x/a))} c(1-(x/a)-(y/b))\,dy\,dx$$

$$= c\int_0^a \left[\left(1-\frac{x}{a}\right)y - \frac{y^2}{2b}\right]_0^{b1-(x/a)} dx$$

$$= c\int_0^a \left[b\left(1-\frac{x}{a}\right)^2 - \frac{b}{2}\left(1-\frac{x}{a}\right)^2\right] dx$$

$$= \frac{bc}{2}\int_0^a \left(1-\frac{x}{a}\right)^2 dx = \frac{abc}{-6}\left[\left(1-\frac{x}{a}\right)^3\right]_0^a = \frac{abc}{6}.$$

Example 2. Find the volume of the region bounded by the planes $x = -a, x = a, y = -a, y = a,$ $z = 0$ and the paraboloid $x^2 + y^2 = z$.

$$V = \int_{-a}^{a}\int_{-a}^{a} z\,dy\,dx$$

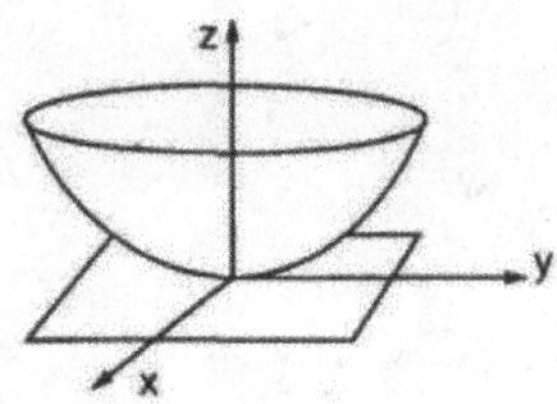

$$= \int_{-a}^{a}\int_{-a}^{a} (x^2+y^2)\,dy\,dx$$

$$= \int_{-a}^{a}\left[x^2 y + \frac{y^3}{3}\right]_{-a}^{a} dx$$

$$= \int_{-a}^{a}\left[2ax^2 + \frac{2a^3}{3}\right] dx = 2a\left[\frac{x^3}{3} + \frac{a^2 x}{3}\right]_{-a}^{a} = \frac{8a^4}{3}.$$

Example 3. Find the volume common to the cylinders $x^2 + y^2 = a^2, x^2 + z^2 = a^2$.

The required volume is symmetrical about the xy plane and half of it is above the xy plane. This half is over the circular area.

$$R : x^2 + y^2 = a^2, z = 0$$

$$\therefore \quad I = 2\iint_R z\,dx\,dy,\ z = \sqrt{a^2 - x^2}$$

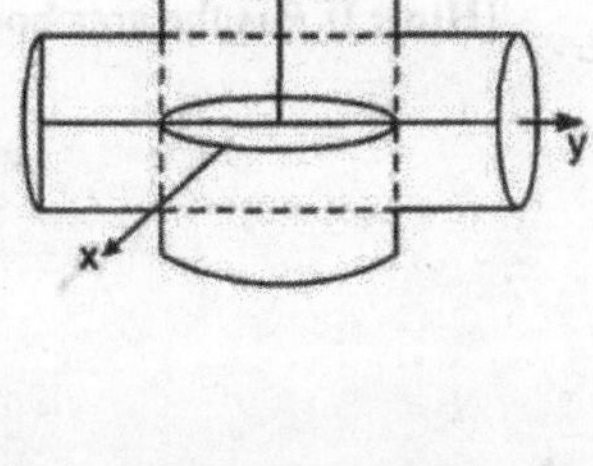

$$= 2\int_{-a}^{a}\int_{-\sqrt{a^2-x^2}}^{\sqrt{a^2-x^2}} \sqrt{a^2-x^2}\,dy\,dx$$

$$= 2\int_{-a}^{a} \sqrt{a^2-x^2}\,[y]_{-\sqrt{a^2-x^2}}^{\sqrt{a^2-x^2}}\,dx$$

$$= 2\int_{-a}^{a} 2(a^2-x^2)\,dx = \frac{16a^3}{3}.$$

Example 4. Find the volume bounded by the paraboloid $x^2 = a^2 - z$ and the planes.

$$x = 0, y = 0, z = 0, y = b.$$

The volume V is above a rectangle of sides a and b in the first quadrant as in the figure. It is below the surface of the paraboloid

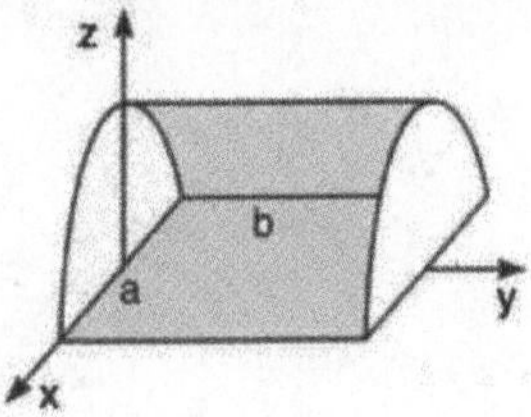

$$z = a^2 - x^2$$

Its volume is

$$V = \int_0^a\int_0^b z\,dy\,dx = \int_0^a\int_0^b (a^2-x^2)\,dy\,dx$$

$$= \int_0^a (a^2-x^2)\,b\,dx = b\left(a^3 - \frac{a^3}{3}\right) = \frac{2a^3b}{3}.$$

Example 5. Show that the volume of the region between the paraboloid $z = x^2 + y^2$ and the cone $z^2 = x^2 + y^2$ is $\pi/6$.

Subtracting the first equation from the second, we get that the region R lies between the planes $z = 0$ and $z = 1$. Now the semivertical angle of the cone is 45° and R is the vertically above the unit circle and bounded above and below by the cone z_1 and the paraboloid z_2.

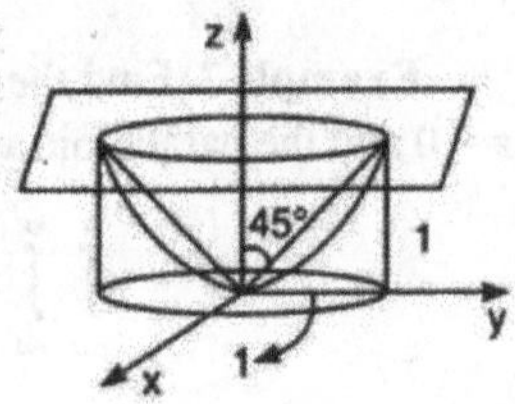

$$z_1 = \sqrt{x^2+y^2},\ z_2 = x^2 + y^2.$$

$$\therefore \quad V = \int_{-1}^{1}\int_{-\sqrt{1-x^2}}^{\sqrt{1-x^2}} \left[\sqrt{x^2+y^2} - (x^2+y^2)\right] dy\,dx$$

$$= \int_0^{2\pi}\int_0^1 (r - r^2)\,r\,dr\,d\theta = 2\pi\left[\frac{r^3}{3} - \frac{r^4}{4}\right]_0^1 = \frac{\pi}{6}.$$

EXERCISES

1. Show that the volume of the tetrahedron bounded by the planes $x = 0, y = 0, z = 0, x + y + z = 1$ is 1/6.

[**Hint:** $V = \int_0^1\int_0^{1-x} z\,dy\,dx,\ z = 1 - x - y$]

2. Show that the volume of the ellipsoid $(x^2/a^2) + (y^2/b^2) + (z^2/c^2) = 1$ is $4/3\ \pi abc$.

[**Hint:** If R is the area bounded by $(x^2/a^2) + (y^2/b^2) = 1, z = 0$, $V = 2\iint_R z\,dy\,dx$; set $x = au\cos t$,

$y = bu\sin t$, where $0 \le t \le 2\pi$, $0 \le u \le 1$]

3. Show that the volume bounded by the cylinder $x^2 + y^2 = a^2$ and the planes $y + z = c$ and $z = 0$ is $\pi a^2 c$.

[**Hint:** $V = \iint_R z\,dy\,dx = \int_0^{2\pi}\int_0^a (c - r\sin\theta)\,r\,dr\,d\theta$]

4. Show that the volume bounded by the cylinder $x^2 + y^2 = 4$, the paraboloid $x^2 + y^2 = 2z$ and the plane $z = 0$ is 4π.

[**Hint:** $V = \iint_R z\,dy\,dx = \int_0^{2\pi}\int_0^2 \frac{r^2}{2}\,r\,dr\,d\theta$]

5. Show that the volume common to the sphere $x^2 + y^2 + z^2 = a^2$ and the cylinder $x^2 + y^2 = ax$ is $(2/9)\,a^3(3\pi - 4)$.

[**Hint:** The section of the cylinder by $z = 0$ is the circle $r = a\cos\theta$, $z = 0$. Volume V is above and below this circular area symmetrically.

$$\therefore \quad V = 2\int_0^{2\pi}\int_{r=0}^{r=a\cos\theta} z(r\,dr\,d\theta),\ z = \sqrt{a^2 - (x^2 + y^2)} = \sqrt{a^2 - r^2}\ ]$$

6. Show that the volume bounded by the paraboloid $x^2 + y^2 = az$ and the cylinder $x^2 + y^2 = 2ax$ and the plane $z = 0$ is $(3/2)\,\pi a^3$.

[**Hint:** The section of the cylinder by $z = 0$ is the circle $r = 2a\cos\theta$, $z = 0$. The volume is above this circle and below the paraboloid. So

$$V = \int_0^{2\pi}\int_0^{2a\cos\theta} z(r\,dr\,d\theta),\ z = \frac{x^2 + y^2}{a} = \frac{r^2}{a}\ ]$$

7. Show that the volume bounded by the paraboloid $x^2 + y^2 = az$ and the cylinder $x^2 + y^2 = 2ay$ and the plane $z = 0$ is $(3/2)\,\pi a^3$.

INTEGRATION OF POINT FUNCTIONS

3.1. Line integrals

Suppose that C is an arc of a curve oriented from the end point A to the end point B. Let ϕ be a scalar point function defined at all points on C. Divide C arbitrarily into m parts. Let P_0P_1, P_1P_2, ... $P_{m-1}P_m$ be the m arcs. (Fig. 3.1.1). Also let Δs_i be the arcul distance of P_i

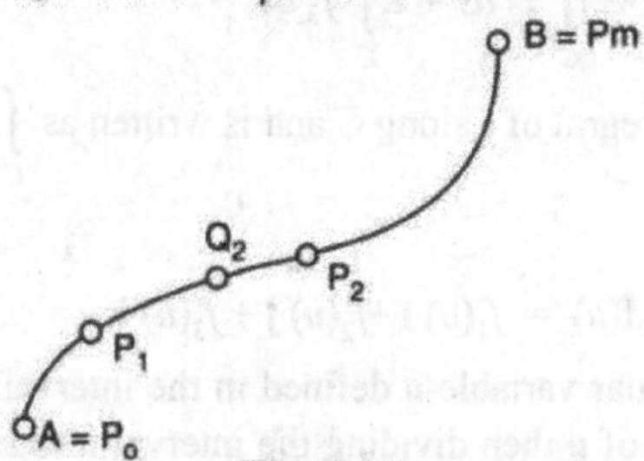

Fig. 3.1.1

from P_{i-1} and Q_i, a point on the arc $P_{i-1}P_i$. It is to be noted that Δs_i, $i = 1, 2....$ m, are all positive. Now the limit of the sum

$$\sum_{i=1}^{m} \phi(Q_i)\,\Delta s_i = \phi(Q_1)\Delta s_1 + \phi(Q_2)\Delta s_2 + ... + \phi(Q_m)\Delta s_m \tag{3.1.1}$$

as $m \to \infty$ and max $\Delta s_i \to 0$, if it exists, is called the *line* or *curvilinear integral* of ϕ along C and is denoted by

$$\int_C \phi ds \text{ or } \int_{C(AB)} \phi ds. \quad \text{... (3.1.2)}$$

C is called the *path of integration*. Now it is evident that

$$\int_{C(AB)} \phi ds = -\int_{(BA)} \phi ds.$$

When the equation of C is given in intrinsic form $r = x(s)\mathbf{i} + y(s)\mathbf{j} + z(s)\mathbf{k}$, $0 \le s \le l$, A and B correspond respectively to $s = 0$ and $s = l$. Now, for points on C, ϕ is a function of s only and so (3.1.1) is

$$\int_{C(AB)} \phi ds = \int_0^l \phi\{x(s), y(s), z(s)\} ds. \quad \text{(3.1.3)}$$

As such the integral (3.1.3) is a Riemann integral and so it can be evaluated by the methods learnt in Calculus. Further, if the equation of C is given in terms of a parameter u in the form $r = x(u)\mathbf{i} + y(u)\mathbf{j} + z(u)\mathbf{k}$, $u_1 \le u \le u_2$, where u_1 and u_2 correspond to A to B, then

$$\int_{C(AB)} \phi ds = \int_{u_1}^{u_2} \phi\{x(u), y(u), z(u)\} \frac{ds}{du} du$$

$$= \int_{u_1}^{u_2} \phi\{x(u), y(u), z(u)\} \left|\frac{dr}{du}\right| du$$

since $\frac{ds}{du} = \left|\frac{dr}{du}\right|$ from $\frac{dr}{du} = \frac{dr}{ds}\frac{ds}{du}$ and $\left|\frac{dr}{ds}\right| = 1$.

Next we consider the line integrals of vector functions along a curve C. For this we replace ϕ in (3.1.1) with a vector point function, say $\mathbf{f}(P) = f_1(P)\mathbf{i} + f_2(P)\mathbf{j} + f_3(P)\mathbf{k}$, defined at every point P on C. Then the sum reduces to

$$\mathbf{i}\left\{\sum_{i=1}^{m} f_1(Q_i)\,\Delta s_i\right\} + \mathbf{j}\left\{\sum_{i=1}^{m} f_2(Q_i)\Delta s_i\right\} + \mathbf{k}\left\{\sum_{i=1}^{m} f_3(Q_i)\,\Delta s_i\right\}$$

Its limit, if it exists, is $\mathbf{i}\int_C f_1 ds + \mathbf{j}\int_C f_2\, ds + \mathbf{k}\int_C f_3\, ds.$

This limit is called the line integral of $\mathbf{f}$ along C and is written as $\int_C \mathbf{f}\, ds$.

Further, in particular, if

$$\mathbf{f}(u) = f_1(u)\,\mathbf{i} + f_2(u)\,\mathbf{j} + f_3(u)\,\mathbf{k}$$

is a vector function of the scalar variable u defined in the interval $u_1 \le u \le u_2$, and $f_1(u)$, $f_2(u)$, $f_3(u)$ be continuous functions of u then dividing the interval into sub-intervals, we get the limit of the sum as

$$\mathbf{i}\int_{u=u_1}^{u_2} f_1(u)du + \mathbf{j}\int_{u=u_1}^{u_2} f_2(u)du + \mathbf{k}\int_{u=u_1}^{u_2} f_3(u)du.$$

This integral is called the *definite integral of f(u) from the limit $u = u_1$ to the limit $u = u_2$* and is written as $\int_{u=u_1}^{u_2} f(u)\,du$.

From Riemann theory of integration we can establish the following results:-

$$(i)\quad \frac{d}{du}\int_{u_1}^{u} f(u)\,du = f(u) \qquad \text{... (3.1.5)}$$

$$(ii)\quad \int_{u=u_1}^{u_2} f(u)\,du = \int_{t=\alpha}^{\beta} f[\phi(t)]\phi'(t)\,dt \qquad \text{... (3.1.6)}$$

under suitable conditions pertaining to the functions f_1, f_2, f_3 and ϕ. In (3.1.6) u is a function of t, namely, $\phi(t)$.

In what follows, while integrating functions, we assume the existence of the conditions of integrability. The study of such conditions forms mainly a part of Calculus.

It is evident that the line integrals (3.1.2) and (3.1.4) are respectively a scalar and a vector. The line integrals of the form $\int_C (\text{a scalar})\,ds$ and $\int_C (\text{a vector})\,ds$ are called *scalar line integrals and vector line integrals* respectively.

Particular forms of line integrals. We know that the unit tangent vector **T** at a general point on C defines a vector point function on C. So, if ϕ and **A** are point functions, defined on C, then some more point functions can be derived such as

$$\phi\mathbf{T},\ \mathbf{A}\times\mathbf{T},\ \mathbf{A}.\mathbf{T}.$$

Correspondingly, we have the line integrals

$$\int_C (\phi\mathbf{T})\,ds,\ \int_C (\mathbf{A}\times\mathbf{T})\,ds,\ \int_C (\mathbf{A}\cdot\mathbf{T})\,ds. \qquad (3.1.7)$$

If **r** is the position vector of a general point on C, $\mathbf{T}ds = d\mathbf{r}$ and so the integrals in (3.1.7) can be written as

$$\int_C \phi\,d\mathbf{r},\ \int_C \mathbf{A}\times d\mathbf{r},\ \int_C \mathbf{A}\cdot d\mathbf{r}.$$

Furthermore, if $\mathbf{r} = \mathbf{f}(u)$, $u_1 \le u \le u_2$, is the parametric equation of C, then, from $d\mathbf{r} = \frac{d\mathbf{r}}{du}du$, $\frac{ds}{du} = \left|\frac{d\mathbf{r}}{du}\right|$,

$$\int_C A\cdot d\mathbf{r} = \int_{u_1}^{u_2} A\cdot\frac{d\mathbf{r}}{du}\,du, \qquad (3.1.8)$$

$$\int_C \mathbf{A}\,ds = \int_{u_1}^{u_2} \mathbf{A}\left|\frac{d\mathbf{r}}{du}\right|du. \qquad (3.1.9)$$

Illustration. If a vector function $\mathbf{A}(x, y, z)$ is defined at all points on the curve whose parametric equation is

$\mathbf{r} = f_1(u)\mathbf{i} + f_2(u)\mathbf{j} + f_3(u)\mathbf{k}$, $u_1 \le u \le u_2$, then the integral (3.1. 8) is

$$\int_{u_1}^{u_2} \mathbf{A}\{f_1(u), f_2(u), f_3(u)\}\cdot\frac{d}{du}\{f_1(u)\mathbf{i} + f_2(u)\mathbf{j} + f_3(u)\mathbf{k}\}\,du.$$

Integral along a closed curve. If C is a *simple* closed curve, that is, a closed curve which does not intersect itself anywhere, the line integral along C is denoted by $\oint_C (\ldots)\, ds$ or simply by $\oint (\ldots)\, ds$.

In the next chapter some line integrals are evaluated in verifying Stokes' theorem. Therefore the reader has to refer to both the chapters for evaluation of line integrals.

LINE INTEGRAL ALONG AN ARC C.

Let (x, y, z) be any point on an arc C whose parametric equations are

$$x = x(t),\ \ y = y(t),\ \ z = z(t)$$

and the vector equation is

$$\overline{r} = x(t)\overline{i} + y(t)\overline{j} + z(t)\overline{k}.$$

Let the end points of C be A, B given by $t = t_1$, $t = t_2$.

Let a vector point function

$$\overline{f}(x, y, z) = f_1\overline{i} + f_2\overline{j} + f_3\overline{k}$$

be defined on C. Then the integral

$$\int_{t_1}^{t_2} \overline{f}\cdot d\overline{r} = \int_C \overline{f}\cdot d\overline{r} = \int_C \overline{f}\cdot(dx\overline{i} + dy\overline{j} + dz\overline{k})$$

$$= \int_C f_1 dx + f_2 dy + f_3 dz$$

is a line integral, called the integral of the tangential component of $\overline{f}$ along C because, if $\hat{t}$ is the unit vector tangential to C, then

$$d\overline{r} = dr\,\hat{t},\ \ \overline{f}\cdot d\overline{r} = (\overline{f}\cdot\hat{t})\,dr,$$

where $\overline{f}\cdot\hat{t}$ is the component of $\overline{f}$ in the direction of $\hat{t}$.

Working rule to evaluate $\int_C \overline{f}\cdot d\overline{r}$.

Step 1. Obtain the parameteric equations in t of C and the parametric values, say t_1 and t_2, for the end points of C.

Step 2. Get $\overline{f}\cdot d\overline{r}$ in the form $\phi(t)\, dt$, where $d\overline{r} = dx\overline{i} + dy\overline{j} + dz\overline{k}$.

Step 3. Evaluate $\int_{t_1}^{t_2} \phi(t)\, dt$.

Now we define a physical quantity belonging to Fluid Dynamics.

Circulation

If $\overline{v}$ is a vector point function, representing the velocity of a particle in a fluid motion and if C is a closed curve, then the line integral.

$$\int_C \overline{v}\cdot d\overline{r}$$

is called the circulation of $\bar{v}$ around the curve C. When the circulation of $\bar{v}$ around every closed curve in a region vanishes, $\bar{v}$ is said to be irrotational in that region.

Work done by a force

When a particle moves along an arc C in a force field $\bar{F}$, the integral of the tangential component of $\bar{F}$ along C is the work done by the force.

Work done in moving a particle in a field of force **F** along the curve C from one point to another point is $\int_C \mathbf{F}\cdot dr$

3.2 Independence of path of integration

Let f be a function defined on a domain D and A_1 and A_2, two fixed points in it. From the definition of the line integral, it is evident that the value of a line integral from A_1 to A_2 may vary with the configuration of the path joining A_1 and A_2. However, there are certain class of point functions whose line integrals of the form, $\int_{C(A_1A_2)} \mathbf{f}\cdot dr$ depends only on the end points A_1 and A_2 and not on the configuration of C.

Book work 3.1. The necessary and sufficient condition for the line integral $\int_{C(A_1A_2)} \mathbf{f}.dr$ to be independent of the path of integration is the existence of a scalar point function ϕ such that $\mathbf{f} = \nabla\phi$.

Necessity part: Given that the line integral depends on the end points alone, we have to establish the existence of a ϕ such that $\mathbf{f} = \nabla\phi$.

Suppose that **f** is defined in D and that the symbol (A_1P) denotes any curve in D joining A_1 and P. If $P(x, y, z)$ is a variable point in D, then the line integral.

$$\int_{(A_1P)} \mathbf{f}.dr \qquad \text{.... (3.2.1)}$$

depends on P and not on the curve (A_1P). Hence the integral (3.2.1) defines a scalar point function in D. Let this function be denoted by $\phi(P)$, that is, $\phi(x, y, z)$. Then

$$\phi(x, y, z) = \int_{(A_1P)} \mathbf{f}\cdot dr = \int_{(x_1,y_1,z_1)}^{(x,y,z)} \mathbf{f}\cdot dr.$$

Further, let $P_1(x + \Delta x, y, z)$ be a point in D in the neighbourhood of P. Then

$$\phi(x+\Delta x, y, z) = \int_{(A_1P_1)} \mathbf{f}\cdot dr = \int_{(A_1PP_1)} f\cdot dr$$

$$= \int_{(A_1P)} \mathbf{f}\cdot dr + \int_{(PP_1)} \mathbf{f}\cdot dr$$

$$= \phi(x, y, z) + \int_{(PP_1)} \mathbf{f}\cdot dr;$$

i.e., $$\phi(x+\Delta x, y, z) - \phi(x, y, z) = \int_{(PP_1)} \mathbf{f}\cdot dr \qquad (3.2.2)$$

Since the line integral is independent of the path of integration in D, we shall evaluate the integral in (3.2.2) by choosing (PP_1) as the straight line joining P and P_1 (Fig. 3.2.1).

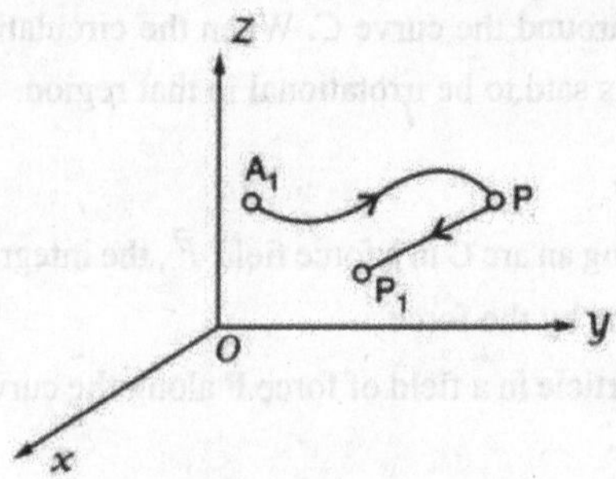

Fig. 3.2.1

This path is evidently parallel to the x-axis and along it $dy = 0$ and $dz = 0$. So, if $\mathbf{f} = f_1\mathbf{i} + f_2\mathbf{i} + f_3\mathbf{k}$, then (3.2.2) becomes

$$\phi(x+\Delta x, y, z) - \phi(x, y, z) = \int_P^{P_1} (f_1\mathbf{i} + f_2\mathbf{j} + f_3\mathbf{k}) \cdot (dx\mathbf{i})$$

$$= \int_P^{P_1} f_1 dx = \int_{(x,y,z)}^{(x+\Delta x,y,z)} f_1 dx\,;$$

$$i.e., \quad \frac{\phi(x+\Delta x, y, z) - \phi(x, y, z)}{\Delta x} = \frac{1}{\Delta x}\int_{(x,y,z)}^{(x+\Delta x,y,z)} f_1 dx.$$

Taking limit as $P_1 \to P$, that is, as $\Delta x \to 0$, we get $\dfrac{\partial\phi}{\partial x} = f_1$ and similarly $\dfrac{\partial\phi}{\partial y} = f_2, \dfrac{\partial\phi}{\partial z} = f_3$.

Thus

$$\mathbf{f} = f_1\mathbf{i} + f_2\mathbf{j} + f_3\mathbf{k} = \mathbf{i}\frac{\partial\phi}{\partial x} + \mathbf{j}\frac{\partial\phi}{\partial y} + \mathbf{k}\frac{\partial\phi}{\partial z} = \nabla\phi.$$

Sufficiency part: Given that there exists a ϕ such that $\nabla\phi = \mathbf{f}$, we have to show that the line integral is independent of the path of integration. Let C be an arbitrary curve with end points (x_1, y_1, z_1) and (x_2, y_2, z_2). Now

$$\int_C \mathbf{f}\cdot d\mathbf{r} = \int_C (\nabla\phi)\cdot d\mathbf{r}$$

$$= \int_C \left(\mathbf{i}\frac{\partial\phi}{\partial x} + \mathbf{j}\frac{\partial\phi}{\partial y} + \mathbf{k}\frac{\partial\phi}{\partial z}\right)\cdot(\mathbf{i}dx + \mathbf{j}dy + \mathbf{k}dz)$$

$$= \int_C \left(\frac{\partial\phi}{\partial x}dx + \frac{\partial\phi}{\partial y}dy + \frac{\partial\phi}{\partial z}dz\right) = \int_C d\phi \qquad (3.2.3)$$

$$= \phi\Big|_{(x_1,y_1,z_1)}^{(x_2,y_2,z_2)} = \phi(x_2, y_2, z_2) - \phi(x_1, y_1, z_1)$$

which is independent of the configuration of C.

Remark. The necessity part can also be proved as follows :

$$\phi(x, y, z) = \int_{(x_1,y_1,z_1)}^{(x,y,z)} \mathbf{f}\cdot d\mathbf{r} = \int_{(x_1,x_1,z_1)}^{(x,y,z)} \mathbf{f}.\mathbf{T}\,ds,$$

where **T** is the unit vector along an arbitrarily chosen curve through (x, y, z) and (x_1, y_1, z_1). Now, finding the directional derivatives of both sides along that curve, we get

$$\frac{d\phi}{ds} = \mathbf{f}.\mathbf{T} \quad \text{or} \quad (\nabla\phi).\mathbf{T} = \mathbf{f}.\mathbf{T}.$$

So $(\nabla\phi - \mathbf{f}).\mathbf{T} = 0$. But **T** is arbitrary depending on the choice of the curve.

So $\nabla\phi - \mathbf{f} = 0$ or $\mathbf{f} = \nabla\phi$.

3.3. Conservative field and scalar potential

If a vector field **f** is such that there exists a scalar point function ϕ such that $f = \nabla\phi$, then **f** is said to be a *conservative field* and ϕ is said to be its *scalar potential*.

Book Work 3.2. In a conservative field **f**, $\int_C \mathbf{f}\cdot d\mathbf{r} = 0$, where C is any simple closed curve.

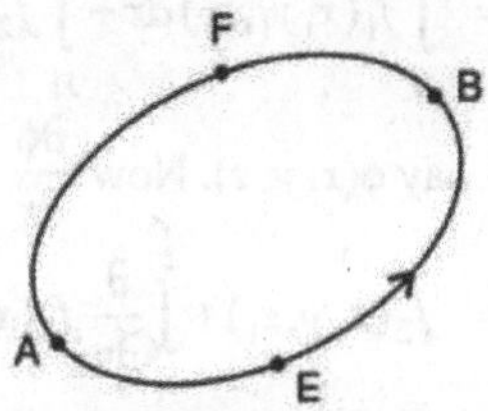

Fig. 3.3.1

Proof: Let A, E, B, F be points on C taken in the sense in which C is oriented as shown in Fig. 3.3.1. Now

$$\int_C = \int_{AEB} + \int_{BFA} = \int_{AEB} - \int_{AFB} = \int_{AEB} - \int_{AEB} = 0.$$

Book Work 3.3. The necessary and sufficient condition for a vector field **f** to be conservative is that $\nabla \times \mathbf{f} = \mathbf{0}$.

Necessity part: The field is conservative. Let ϕ be the scalar potential. Then

$$\nabla\times\mathbf{f} = \nabla\times(\nabla\phi) = \mathbf{0} \qquad \text{by (2.8.2).}$$

Sufficiency part: Now $\nabla\times\mathbf{f} = \mathbf{0}$

i.e.,

$$\begin{vmatrix} \mathbf{i} & \mathbf{j} & \mathbf{k} \\ \dfrac{\partial}{\partial x} & \dfrac{\partial}{\partial y} & \dfrac{\partial}{\partial z} \\ f_1 & f_2 & f_3 \end{vmatrix} = 0.$$

So

$$\mathbf{i}\left(\frac{\partial f_3}{\partial y} - \frac{\partial f_2}{\partial z}\right) + \mathbf{j}\left(\frac{\partial f_1}{\partial z} - \frac{\partial f_3}{\partial x}\right) + \mathbf{k}\left(\frac{\partial f_2}{\partial x} - \frac{\partial f_1}{\partial y}\right) = 0$$

or

$$\frac{\partial f_2}{\partial z} - \frac{\partial f_3}{\partial y} = 0, \quad \frac{\partial f_3}{\partial x} - \frac{\partial f_1}{\partial z} = 0, \quad \frac{\partial f_1}{\partial y} - \frac{\partial f_2}{\partial x} = 0.$$

Let $A(x_1, y_1, z_1)$ be a fixed point and $P(x, y, z)$, a variable point. Let (x, y_1, z_1) and (x, y, z_1)be B and C (Fig. 3.3.2). Then

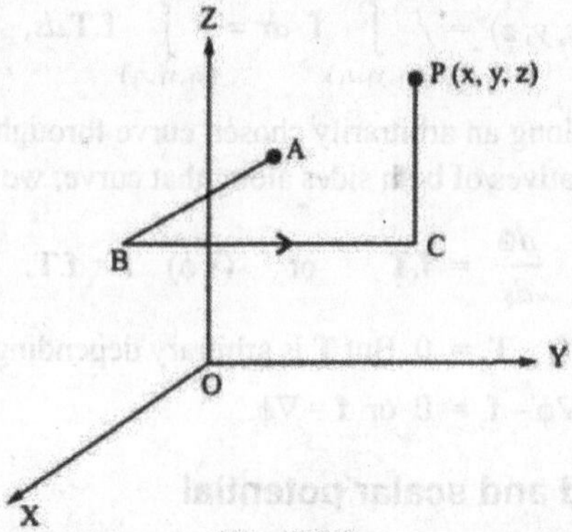

Fig. 3.3.2

$$\int_{ABCP} \mathbf{f}\cdot d\mathbf{r} = \int_{ABCP} f_1(x,y,z)dx + f_2(x,y,z)dy + f_3(x,y,z)dz$$

$$= \int_{x_1}^{x} f_1(x,y_1,z_1)dx + \int_{y_1}^{y} f_2(x,y,z_1)dy + \int_{z_1}^{z} f_3(x,y,z)dz.$$

it is a scalar point function of P, say $\phi(x, y, z)$. Now $\frac{\partial\phi}{\partial z} = f_3(x, y, z)$;

$$\frac{\partial\phi}{\partial y} = f_2(x,y,z_1) + \int_{z_1}^{z} \frac{\partial}{\partial y} f_3(x,y,z)dz$$

$$= f_2(x,y,z_1) + \int_{z_1}^{z} \frac{\partial}{\partial z} f_2(x,y,z)dz \text{ since } \frac{\partial f_3}{\partial y} = \frac{\partial f_2}{\partial z}$$

$$= f_2(x,y,z_1) + [f_2(x,y,z)]_{z_1}^{z} = f_2(x,y,z);$$

$$\frac{\partial\phi}{\partial x} = f_1(x,y_1,z_1) + \int_{y_1}^{y} \frac{\partial}{\partial x} f_2(x,y,z_1)dy + \int_{z_1}^{z} \frac{\partial}{\partial x} f_3(x,y,z)dz$$

$$= f_1(x,y_1,z_1) + \int_{y_1}^{y} \frac{\partial}{\partial y} f_1(x,y,z_1)dy + \int_{z_1}^{z} \frac{\partial}{\partial z} f_1(x,y,z)dz$$

$$= f_1(x,y_1,z_1) + [f_1(x,y,z_1)]_{y_1}^{y} + [f_1(x,y,z)]_{z_1}^{z}$$

$$= f_1(x,y,z).$$

Therefore a scalar ϕ exists such that

$$\mathbf{f} = \frac{\partial\phi}{\partial x}\mathbf{i} + \frac{\partial\phi}{\partial y}\mathbf{j} + \frac{\partial\phi}{\partial z}\mathbf{k} = \nabla\phi.$$

Hence **f** is a conservative vector field.

3.4 Line integral of a conservative vector

When **f** is conservative, the integral $\int_{(x_1,y_1,z_1)}^{(x_2,y_2,z_2)} \mathbf{f}\cdot d\mathbf{r}$ can also be evaluated by using the scalar potential since

$$\mathbf{f}.d\mathbf{r} = (\nabla\phi).d\mathbf{r} = \left(\mathbf{i}\frac{\partial\phi}{\partial x}+\mathbf{j}\frac{\partial\phi}{\partial y}+\mathbf{k}\frac{\partial\phi}{\partial z}\right).(dx\mathbf{i}+dy\mathbf{j}+dz\mathbf{k})$$

$$= \frac{\partial\phi}{\partial x}dx+\frac{\partial\phi}{\partial y}dy+\frac{\partial\phi}{\partial z}dz = d\phi$$

i.e.,
$$\int_{(x_1,y_1,z_1)}^{(x_2,y_2,z_2)} \mathbf{f}\cdot d\mathbf{r} = [\phi]_{(x_1,y_1,z_1)}^{(x_2,y_2,z_2)} = \phi(x_2,y_2,z_2)-\phi(x_1,y_1,z_1).$$

3.5 Surface integrals

Suppose that a surface is bounded by a simple closed curve C as in Fig. 3.5.1. Now the surface can be thought of as to have two sides separated by *C* of which one is arbitrarily chosen as the

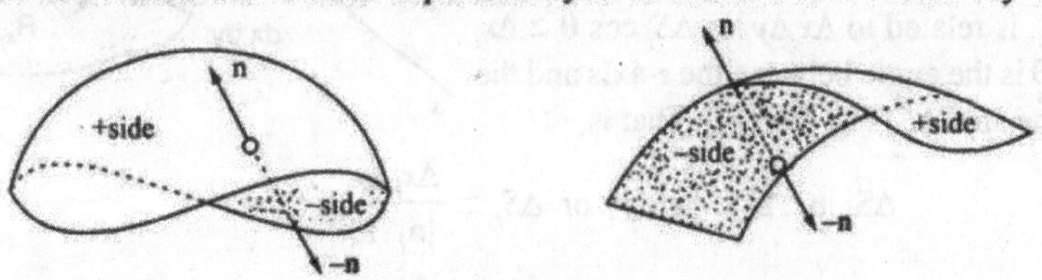

Fig. 3.5.1

positive side and the other, the negative side. In what follows we denote the positive side by *S*. In the case of a closed surface usually the outer side is chosen as the positive side. The unit normal at any point to the surface drawn in the positive side is denoted by **n**.

Consider a plane surface whose area is *A*. When one side of this surface is chosen as the positive side *S* with **n** as the unit normal drawn to this side, the area of this positive side can be specified vectorially as *A***n**. Also the area dS of one side of an infinitely small bit of a general surface can be specified as *dS***n** or simply as *d***S**, where **n** is the unit normal drawn to this side. Thus **n***dS* = *d***S**.

Surface integral. As in the case of a line integral we proceed now to define a surface integral. Given a scalar point function ϕ defined at all points of a surface *S*, the limit of the sum $\phi(Q_1)\,\Delta S_1+\phi(Q_2)\,\Delta S_2+\ldots+\phi(Q_m)\,\Delta S_m$ as $m\to\infty$ and max $\Delta S_i\to 0$, where $\Delta S_1, \Delta S_2, \ldots, \Delta S_m$ are the areas of the *m* small arbitrary bits of surfaces into which *S* is subdivided and Q_1 is a point in ΔS_i, is called the *surface integral* of ϕ on *S* and is denoted by $\iint_S \phi\, dS$. This integral is a scalar.

Similarly if we consider **f**, a vector point function defined on S, instead of ϕ, we get the surface integral of **f** on *S* to be

$$\iint_S \mathbf{f}\,ds = \mathbf{i}\iint_S f_1 dS+\mathbf{j}\iint_S f_2 dS+\mathbf{k}\iint_S f_3 dS, \text{ where}$$

$\mathbf{f}=f_1\mathbf{i}+f_2\mathbf{j}+f_3\mathbf{k}$. It is evident that the unit normal at every point of the surface *S* defines a vector field on *S*. So, if ϕ and **A** are two point functions defined on S, then some more point functions can be derived on *S* such as $\phi\mathbf{n}$, $\mathbf{A}.\mathbf{n}$, $\mathbf{A}\times\mathbf{n}$. Correspondingly, we have the surface integrals.

$$\iint_S (\phi\mathbf{n})\,dS, \iint_S (\mathbf{A}\cdot\mathbf{n})\,dS, \iint_S (\mathbf{A}\times\mathbf{n})\,dS. \qquad \ldots (3.5.1)$$

But, by using $d\mathbf{S} = \mathbf{n}dS$, these integrals can also be written as

$$\iint_S \phi\,d\mathbf{S}, \iint_S \mathbf{A}\cdot d\mathbf{S}, \iint_S \mathbf{A}\times d\mathbf{S}. \qquad \ldots (3.5.2)$$

Flux. In physical applications the integral $\iint_S \mathbf{A}\cdot d\mathbf{S}$ is called the *flux* of **A** through *S*.

Evaluation of surface integral. Evaluation of a surface integral can be made by reducing it to a double integral. Let *S* be a surface whose projection R_{xy} on the *xOy* plane is such that the points on *S* have a one-to-one correspondence with the points on R_{xy}. Then divide R_{xy} into infinitesimal rectangular bits of areas by drawing lines parallel to the *x* and *y* axes. If the area of a typical bit is $\Delta x_i\, \Delta y_i$, then the corresponding surface area on *S* pertaining to $\Delta x_i\, \Delta y_i$, say ΔS_i, is related to $\Delta x_i \Delta y_i$ as $\Delta S_i \cos\theta \simeq \Delta x_i\, \Delta y_i$, where θ is the angle between the *z*-axis and the normal to the area ΔS_i[Fig. (3.5.2)]. That is,

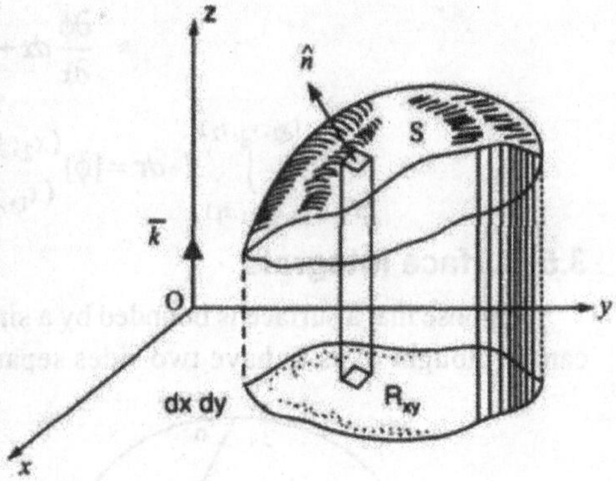

$$\Delta S_i\, |\mathbf{n}_i \cdot \mathbf{k}| \simeq \Delta x_i\, \Delta y_i \quad \text{or} \quad \Delta S_i \simeq \frac{\Delta x_i \Delta y_i}{|\mathbf{n}_i \cdot \mathbf{k}|}$$

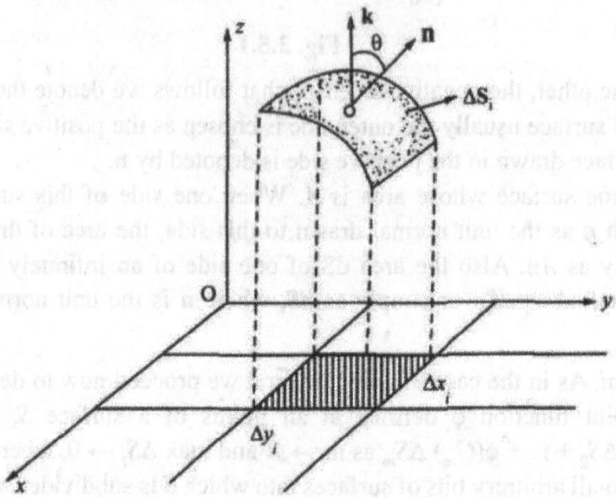

Fig. 3.5.2

where n_i is the unit vector normal to ΔS_i. Hence, if φ is a scalar or vector point function defined on *S*, then

$$\sum_{i=1}^{m} \phi(Q_i)\Delta S_i = \sum_{i=1}^{m} \phi(Q_i)\frac{\Delta x_i \Delta y_i}{|\mathbf{n}_i \cdot \mathbf{k}|}.$$

When $m \to \infty$ and $\Delta x_i, \Delta y_i \to 0$, this takes the form

$$\iint_S \phi\, dS = \iint_{R_{xy}} \phi \frac{dx\, dy}{|\mathbf{n}\cdot\mathbf{k}|} \qquad \text{... (3.5.3)}$$

of which the right integral can be evaluated after eliminating *z* from the integrand by using the equation of *S*. In a similar manner it can be shown that

$$\iint_S \phi\, dS = \iint_{R_{yz}} \phi \frac{dy\, dz}{|\mathbf{n}\cdot\mathbf{i}|}, \qquad (3.5.4)$$

$$\iint_S \phi dS = \iint_{R_{zx}} \phi \frac{dz\,dx}{|\mathbf{n}\cdot\mathbf{j}|}. \tag{3.5.5}$$

where R_{yz} and R_{zx} are resepctively the projections of the surface S on the yOz and zOx planes.

Remark. In the next chapter we learn a theorem called divergence theorem which converts a surface integral along a closed surface into a volume integral. Evaluation of the volume integral is easier than evaluation of the respective surface integral. Thus, in the next chapter, many surface integrals are evaluated as volume integrals by using divergence theorem. However, in the verification of divergence theorem, the surface integrals are evaluated directly. Therefore the reader has to refer to both the chapters for evaluation of surface integrals.

3.6 Volume integrals

Like the line and surface integrals, volume integrals over volumes are defined. The volume integral is denoted by $\iiint_V (\text{function})\, dV$.

If the vector $\mathbf{f} = f_1\mathbf{i} + f_2\mathbf{j} + f_3\mathbf{k}$, then the volume integral

$$\iiint_V \mathbf{f} dV = \mathbf{i}\iiint_V f_1 dV + \mathbf{j}\iiint_V f_2 dV + \mathbf{k}\iiint_V f_3 dV.$$

An important result. From the definition of the mass centre, it may be seen that, when the integrands are simply x, y, z, it is advantageous to remember and use the particular cases

$$\iiint_V x\,dV = V\bar{x}, \quad \iiint_V y\,dV = V\bar{y}, \quad \iiint_V z\,dV = V\bar{z},$$

where $\bar{x}, \bar{y}, \bar{z}$ are the x, y, z coordinates of the mass centre of the volume V.

For example, if $(\bar{x}, \bar{y}, \bar{z})$ is the mass centre of the volume enclosed by the planes,

$x = 0, x = 1, y = 0, y = 1, z = 0, z = 1,$

then the mass centre is (1/2, 1/2, 1/2) and $\bar{x} = \bar{y} = \bar{z} = 1/2$ and the volume is

$$V = \text{length} \times \text{breadth} \times \text{height} = 1 \times 1 \times 1 = 1.$$

$$\therefore \quad \iiint_V x\,dx\,dy\,dz = \bar{x} \times V = \frac{1}{2} \times 1 = \frac{1}{2}.$$

3.7. Cylindrical and spherical polar coordinates

In evaluating surface and volume integrals, in certain cases, it will be advantageous to change the variables x, y, z into cylindrical or spherical polar coordinates. So it is better to recall to memory the relations between these coordinates and the respective Jacobian of transformation.

Plane Polar coordinates. We know that in two dimensions the relations between x, y and the polar coordinates r, θ are $x = r\cos\theta, y = r\sin\theta$ and $0 \le r < \infty, 0 \le \theta \le 2\pi$. And in the tranformation from x, y to r, θ the Jacobian of transformation is r so that $dx\,dy = r\,dr\,d\theta$.

Cylindrical coordinates. If P is (x, y, z) (Fig. 3.7.1) and if PL is the perpendicular from P to the xOy plane, then OL, angle xOL, LP are the cylindrical polar coordinates which are denoted by r, θ, z, where $0 \le r < \infty, 0 \le \theta < 2\pi, -\infty < z < \infty$, where θ is the angle from the x axis (*i.e.*, reference ray) to OL in the (anti-clockwise direction) counterclockwise direction.

The relations between x, y, z and r, θ, z and that between $dx\,dy\,dz$ and $dr\,d\theta\,dz$ are

$$\left.\begin{aligned} x &= r\cos\theta \\ y &= r\sin\theta \\ z &= z \end{aligned}\right\}, \quad dx\,dy\,dz = (r\,dr\,d\theta)\,dz.$$

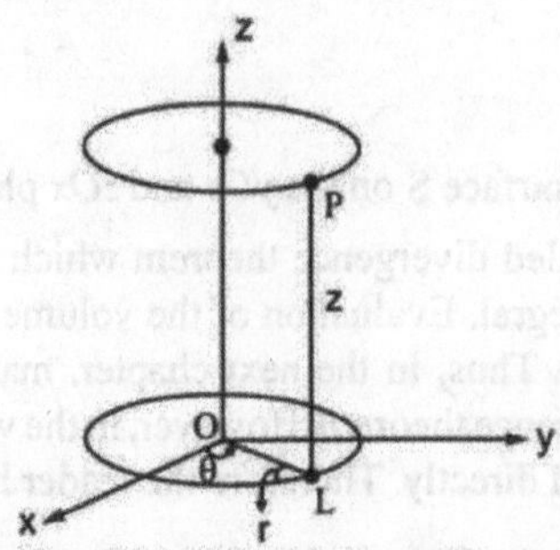

Fig. 3.7.1 Cylindrical coordinates

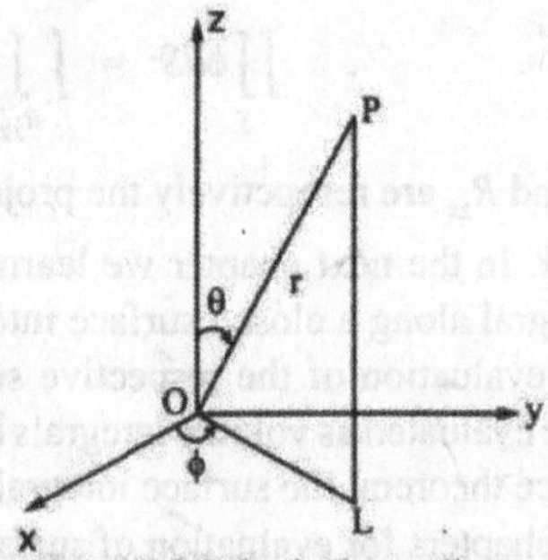

Fig. 3.7.2 Spherical coordinates

Spherical coordinates. If P is (x, y, z) (Fig. 3.7.2) and if PL is the perpendicular from P to the xOy plane, then OP, angle zOP, angle xOL are the spherical polar coordinates which are denoted by r, θ, ϕ, where $0 \le r < \infty$, $0 \le \theta < \pi$, $0 \le \phi < 2\pi$. The relations between x, y, z and r, θ, ϕ and that between $dx\,dy\,dz$ and $dr\,d\theta\,d\phi$ are

$$\left.\begin{aligned} x &= r\sin\theta\cos\phi \\ y &= r\sin\theta\sin\phi \\ z &= r\cos\theta \end{aligned}\right\}, \quad dx\,dy\,dz = (r^2\sin\theta)\,dr\,d\theta\,d\phi,$$

Conversely, if the cartesian coordinates (x, y, z) of a sphere is given, the spherical polar coordinates (r, θ, ϕ) can be determined from the coordinate transformations

$$r = +(x^2+y^2+z^2)^{1/2}$$

$$\theta = \cos^{-1}\left(\frac{z}{(x^2+y^2+z^2)^{\frac{1}{2}}}\right)$$

$$\phi = \tan^{-1}(y/x)$$

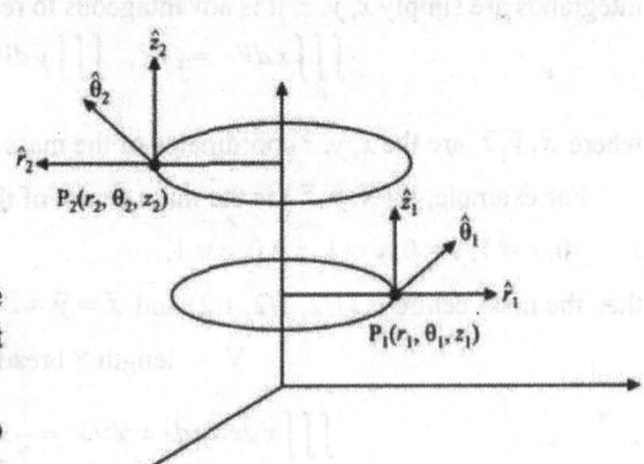

Cartesian coordinate system is a unique coordinate system in which the unit vector at different points in space are equal.

In polar coordinates, the unit vectors at two different points are not equal since they point in different directions.

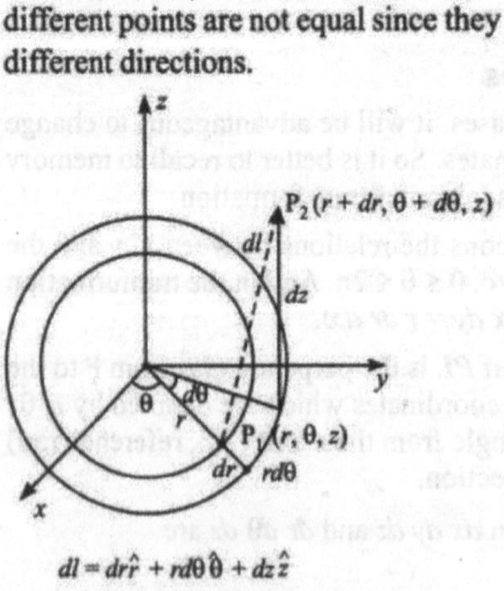

$dl = dr\,\hat{r} + r d\theta\,\hat{\theta} + dz\,\hat{z}$

Infinitesimal displacement

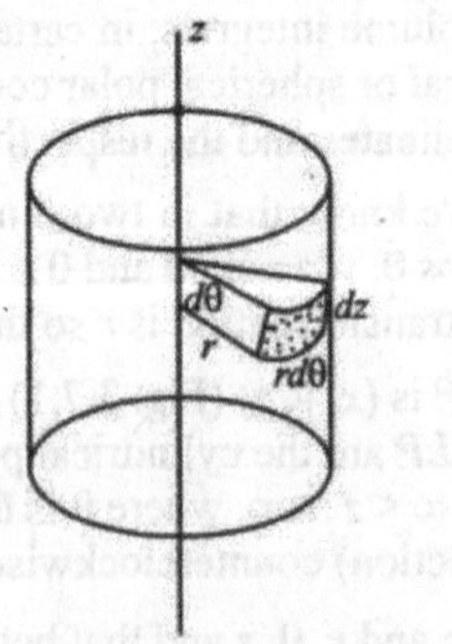

$dA = (rd\theta)\,dz$

Infinitesimal area

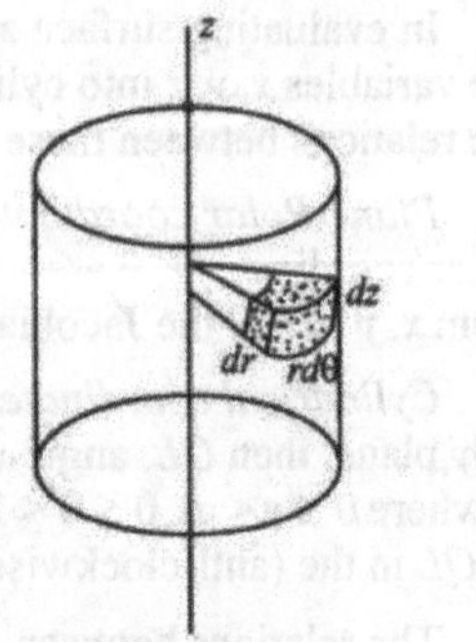

$dV = r dr d\theta dz$

Infinitesimal volume

$ds = rd\theta dz\,\hat{r} + dr dz\,\hat{\theta} + r dr\,d\theta\,\hat{z}$

In the case of a sphere

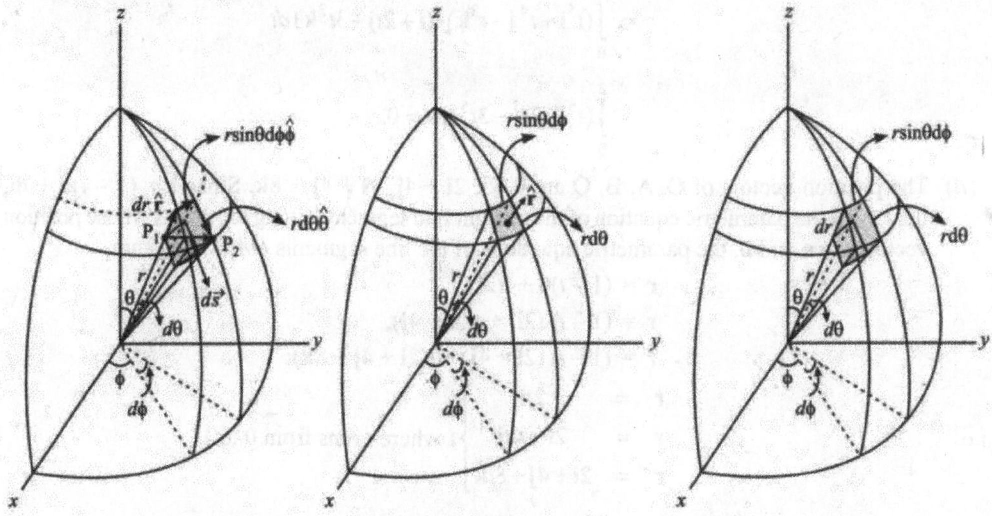

$dl = dr\hat{r} + rd\theta\,\hat{\theta} + r\sin\theta d\phi\hat{\phi}$ Infinitesimal displacement

$dA = (rd\theta)\,(r\sin\theta d\phi)$ Infinitesimal area

$dV = (dr)\,(rd\theta)\,(r\sin\theta d\phi)$ Infinitesimal volume

$dS = r^2\sin\theta d\theta d\phi\hat{r} + r\sin\theta drd\phi\hat{\theta} + rdr\,d\theta\hat{\phi}$

3.8 Examples

3.1 Line integrals.

1. *Solved sum.* Find the value of the integral $\int_C \mathbf{A}\cdot d\mathbf{r}$, where $A = yz\mathbf{i} + zx\mathbf{j} - xy\mathbf{k}$ in the following cases (Fig. 3.8.1) :

(*i*) C is the curve whose parametric equations are $x = t$, $y = t^2$, $z = t^3$ drawn from $O(0, 0, 0)$ to $Q(2, 4, 8)$;

(*ii*) C is the curve obtained by joining O to $A(2, 0, 0)$, then A to $B(2, 4, 0)$ and then B to Q by straight lines;

(*iii*) C is the straight line joining O to Q. **(1979)**

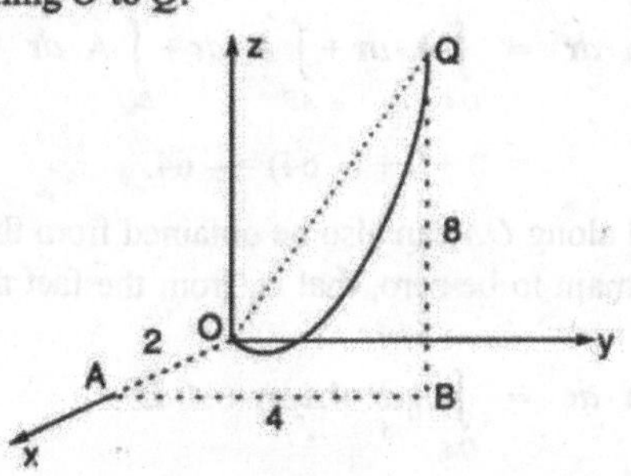

Fig. 3.8.1

(*i*) If $\mathbf{r}$ is the position vector of a general point P on C, then the equation of C is $\mathbf{r} = t\mathbf{i} + t^2\mathbf{j} + t^3\mathbf{k}$, where t takes the values from 0 to 2. The value of A at P is $(t^2)(t^3)\mathbf{i} + (t^3)(t)\mathbf{j} - (t)(t^2)\mathbf{k}$ or $t^5\mathbf{i} + t^4\mathbf{j} - t^3\mathbf{k}$. So, as in (3.1.8),

$$\int_C \mathbf{A}\cdot d\mathbf{r} = \int_0^2 (t^5\mathbf{i}+t^4\mathbf{j}-t^3\mathbf{k})\cdot\frac{d}{dt}(t\mathbf{i}+t^2\mathbf{j}+t^3\mathbf{k})\,dt$$

$$= \int_0^2 (t^5\mathbf{i}+t^4\mathbf{j}-t^3\mathbf{k})\cdot(\mathbf{i}+2t\mathbf{j}+3t^2\mathbf{k})\,dt$$

$$= \int_0^2 (t^5+2t^5-3t^5)\,dt = 0.$$

(*ii*) The position vectors of O, A, B, Q are 0, 2**i**, 2**i** + 4**j**, 2**i** + 4**j** + 8**k**. Since **r** = (1 − *t*)**a** + *t***b**, $0 \le t \le 1$ is the parametric equation of the straight line segment joining the points whose position vectors are **a** and **b**, the parametric equations of the line segments *OA*, *AB*, *BQ* are

$$\mathbf{r} = (1-t)0 + t(2\mathbf{i}),$$
$$\mathbf{r} = (1-t)(2\mathbf{i}) + t(2\mathbf{i}+4\mathbf{j}),$$
$$\mathbf{r} = (1-t)(2\mathbf{i}+4\mathbf{j}) + t(2\mathbf{i}+4\mathbf{j}+8\mathbf{k});$$

i.e.
$$\left.\begin{array}{l} \mathbf{r} = 2t\mathbf{i} \\ \mathbf{r} = 2\mathbf{i}+4t\mathbf{j} \\ \mathbf{r} = 2\mathbf{i}+4\mathbf{j}+8t\mathbf{k} \end{array}\right\}; \text{ where } t \text{ runs from 0 to 1.}$$

$$\int_{OA} \mathbf{A}\cdot d\mathbf{r} = \int_0^1 \{(0)(0)\mathbf{i}+(0)(2t)\mathbf{j}-(2t)(0)\mathbf{k}\}\cdot\frac{d(2t\mathbf{i})}{dt}dt = 0.$$

$$\int_{AB} \mathbf{A}\cdot d\mathbf{r} = \int_0^1 \{(4t)(0)\mathbf{i}+(0)(2)\mathbf{j}-(2)(4t)\mathbf{k}\}\cdot\frac{d(2\mathbf{i}+4t\mathbf{j})}{dt}dt$$

$$= -\int_0^1 (8t\mathbf{k})\cdot(4\mathbf{j})\,dt = 0.$$

$$\int_{BQ} \mathbf{A}\cdot d\mathbf{r} = \int_0^1 \{(4)(8t)\mathbf{i}+(8t)(2)\mathbf{j}-(2)(4)\mathbf{k}\}\cdot\frac{d(2\mathbf{i}+4\mathbf{j}+8t\mathbf{k})}{dt}dt$$

$$= \int_0^1 (32t\mathbf{i}+16t\mathbf{j}-8\mathbf{k})\cdot(8\mathbf{k})\,dt = \int_0^1 (-64)\,dt = -64.$$

So
$$\int_C \mathbf{A}\cdot d\mathbf{r} = \int_{OA} \mathbf{A}\cdot d\mathbf{r} + \int_{AB} \mathbf{A}\cdot d\mathbf{r} + \int_{BQ} \mathbf{A}\cdot d\mathbf{r}$$

$$= 0 + 0 + (-64) = -64.$$

The value of the integral along *OA* can also be obtained from the fact that, along *OA*, *x* varies from 0 to 2 while *y* and *z* remain to be zero, that is, from the fact that $0 \le x \le 2$, $y = 0$, $z = 0$, $dy = 0$, $dz = 0$.

$$\int_{OA} \mathbf{A}\cdot d\mathbf{r} = \int_{OA} A_1dx + A_2dy + A_3dz$$

$$= \int_{OA} A_1dx = \int_0^2 yz\,dx = \int_0^2 (0)(0)dx = 0.$$

Further, along *AB*, $x = 2$, y varies from 0 to 4, $z = 0$. So $x = 2$, $0 \le y \le 4$, $z = 0$, $dx = 0$, $dz = 0$. Thus

$$\int_{AB} \mathbf{A}\cdot d\mathbf{r} = \int_{AB} A_2dy = \int_{AB} (zx)\,dy = \int_0^4 (0)(2)\,dy = 0.$$

Similarly $$\int_{BQ} \mathbf{A}\cdot d\mathbf{r} = \int_{BQ} A_3 dz = \int_{BQ} (-xy)\,dz = -\int_0^8 (2)(4)\,dz = -64.$$

(*iii*) The equation of the straight line OQ is

$$\mathbf{r} = (1-t)\,\mathbf{0} + t(2\mathbf{i}+4\mathbf{j}+8\mathbf{k})$$

and that of the segment OQ is

$$\mathbf{r} = (2\mathbf{i}+4\mathbf{j}+8\mathbf{k})t,\ 0 \le t \le 1.$$

$$\int_C \mathbf{A}\cdot d\mathbf{r} = \int_0^1 \{(4t)(8t)\mathbf{i}+(8t)(2t)\mathbf{j}-(2t)(4t)\mathbf{k}\}\cdot(2\mathbf{i}+4\mathbf{j}+8\mathbf{k})dt = \frac{64}{3}.$$

2. *Solved sum.* In the vector field $\mathbf{F} = z(x\mathbf{i}+y\mathbf{j}+z\mathbf{k})$ evaluate $\int_C \mathbf{F}\cdot d\mathbf{r}$ along the following curves:-

(*i*) Curve $x = t, y = t^2, z = t^3$ from (0, 0, 0) to (1, 1, 1) :

(*ii*) Rectilinear curve obtained by joining $O(0, 0, 0)$, A(1, 0, 0), B(1, 1, 0), C (1, 1,1) by straight lines;

(*iii*) Straight line joining (0, 0, 0), (1, 1, 1). **(1978)**

(*i*) $$\mathbf{r} = x\mathbf{i}+y\mathbf{j}+z\mathbf{k} = t\mathbf{i}+t^2\mathbf{j}+t^3\mathbf{k} \text{ where } t \text{ runs from 0 to 1.}$$

$$\mathbf{F} = z(x\mathbf{i}+y\mathbf{j}+z\mathbf{k}) = t^3(t\mathbf{i}+t^2\mathbf{j}+t^3\mathbf{k}).$$

$$\therefore \quad \int \mathbf{F}\cdot d\mathbf{r} = \int_0^1 t^3(t\mathbf{i}+t^2\mathbf{j}+t^3\mathbf{k})\cdot(\mathbf{i}+2t\mathbf{j}+3t^2\mathbf{k})\,dt.$$

$$= \int_0^1 (t^4+2t^6+3t^8)\,dt = \frac{86}{105}.$$

(*ii*) $$OA : \mathbf{r} = t\mathbf{i},\ t \text{ runs from 0 to 1,}$$

$$AB : \mathbf{r} = \mathbf{i}+t\mathbf{j},\ t \text{ runs from 0 to 1,}$$

$$BC : \mathbf{r} = \mathbf{i}+\mathbf{j}+t\mathbf{k},\ t \text{ runs from 0 to 1.}$$

$$\therefore \int_{OA} 0(t\mathbf{i}+0\mathbf{j}+0\mathbf{k})\cdot d\mathbf{r} = 0,$$

$$\int_{AB} 0(\mathbf{i}+t\mathbf{j}+0\mathbf{k})\cdot d\mathbf{r} = 0.$$

$$\int_{BC} t(\mathbf{i}+\mathbf{j}+t\mathbf{k})\cdot(dt\,\mathbf{k}) = \int_0^1 t^2 dt = \frac{1}{3}.$$

Therefore the integral is $0+0+\left(\frac{1}{3}\right) = \frac{1}{3}$.

(*iii*) $OC : \mathbf{r} = \mathbf{0} + t(\mathbf{i}+\mathbf{j}+\mathbf{k})$, where t runs from 0 to 1.

$$\int_{OC} t(t\mathbf{i}+t\mathbf{j}+t\mathbf{k})\cdot\{(\mathbf{i}+\mathbf{j}+\mathbf{k})dt\} = \int_0^1 3t^2 dt = 1.$$

3. If $\mathbf{A} = y\mathbf{i} - z\mathbf{j} + x\mathbf{k}$ and C is the arc of the curve $\mathbf{r} = \cos u\mathbf{i} + \sin u\mathbf{j} + u\mathbf{k}$, where the parameter u takes the values from 0 to 2π, show that

$$\int_C \mathbf{A}\,ds = -2\sqrt{2}\pi^2\mathbf{j},\quad \int_C \mathbf{A}\cdot d\mathbf{r} = -\pi,$$

$$\int_C \mathbf{A}\times d\mathbf{r} = -(2\pi^2+\pi)\mathbf{i}+(2\pi)\mathbf{k}. \quad \textbf{(1976)}$$

4. *Solved sum.* Evaluate the following integrals if C is the arc of the parabola $y^2 = 2x$, $z = 0$ from $(0, 0, 0)$ to $(\frac{1}{2}, 1, 0)$ and $\phi = 3y - z$, $\mathbf{A} = 2xy\mathbf{i} + y\mathbf{j} + xz\mathbf{k}$:

$$\int_C \phi ds, \int_C \mathbf{A}\, ds, \int_C \mathbf{A}\cdot d\mathbf{r}, \int_C \mathbf{A}\times d\mathbf{r}.$$

Writing t for y, we get the parametric equations of C as

$$x = \frac{1}{2}t^2, y = t, z = 0, 0 \le t \le 1.$$

That is, the parametric vector equation of C is $\mathbf{r} = \frac{1}{2}t^2\mathbf{i} + t\mathbf{j}$, ($t$ runs from 0 to 1).

Since $\phi = 3y - z$ on C, it is $3(t) - 0$.

$$\int_C \phi ds = \int_0^1 \phi\left|\frac{dr}{dt}\right| dt = \int_0^1 \{3(t)-0\}\,|t\mathbf{i}+\mathbf{j}|\,dt$$

$$= \int_0^1 (3t)(1+t^2)^{1/2} dt = (1+t^2)^{3/2}\Big|_0^1 = 2\sqrt{2}-1.$$

$$\int_C \mathbf{A}\, ds = \int_0^1 \{2(\tfrac{1}{2}t^2)(t)\mathbf{i}+(t)\mathbf{j}+(\tfrac{1}{2}t^2)(0)\mathbf{k}\}|t\mathbf{i}+\mathbf{j}|\,dt$$

$$= \int_0^1 (t^3\mathbf{i}+t\mathbf{j})(1+t^2)^{1/2} dt$$

$$= \mathbf{i}\int_0^1 t^3(1+t^2)^{1/2} dt + \mathbf{j}\int_0^1 t(1+t^2)^{1/2} dt$$

$$= \frac{2\sqrt{2}+2}{15}i + \frac{2\sqrt{2}-1}{3}\mathbf{j}.$$

$$\int_C \mathbf{A}\cdot d\mathbf{r} = \int_0^1 (t^3\mathbf{i}+t\mathbf{j})\cdot\frac{d}{dt}\left(\frac{1}{2}t^2\mathbf{i}+t\mathbf{j}\right)dt = \int_0^1 (t^4+t)\,dt = \frac{7}{10}.$$

$$\int_C \mathbf{A}\times d\mathbf{r} = \int_0^1 (t^3\mathbf{i}+t\mathbf{j})\times\frac{d}{dt}(\frac{1}{2}t^2\mathbf{i}+t\mathbf{j})\,dt$$

$$= \mathbf{k}\int_0^1 (t^3-t^2)\,dt = -\frac{\mathbf{k}}{12}.$$

5. Find $\int_C \mathbf{F}\cdot d\mathbf{r}$, (*i*) if $\mathbf{F} = 3xy\mathbf{i} - y^2\mathbf{j}$ and C is the curve $x = t, y = 2t^2$ from (0, 0) to (1, 2), **(1979)**

(*ii*) if $\mathbf{F} = (3x^2 + 6y)\mathbf{i} - 14yz\,\mathbf{j} + 20\,xz^2\,\mathbf{k}$ and C is the curve $x = t, y = t^2, z = t^3$, from (0, 0, 0) to (1, 1, 1). **(1981)**

[**Hint:** (*i*) C is $\mathbf{r} = t\mathbf{i} + 2t^2\mathbf{j}$, where t runs from 0 to 1, $\mathbf{F}.d\mathbf{r} = 6t^3\,dt - 16t^5 dt$. Integral $= -7/6$.

(*ii*) C is $\mathbf{r} = t\mathbf{i} + t^2\mathbf{j} + t^3\mathbf{k}$, where t runs from 0 to 1.

$$\mathbf{F}.d\mathbf{r} = (9t^2 - 28t^6 + 60t^9)dt. \text{ Integral} = 5.]$$

6. Show that, if C is the circle $x = 3\cos t,\ \ y = 3\sin t,\ \ z = 0$, then

$$\oint_C \{(2x-y+z)\mathbf{i}+(x+y-z^2)\mathbf{j}+(3x-2y+4z)\mathbf{k}\}\cdot d\mathbf{r} = 18\pi.$$

7. Show that, if C is the ellipse $\frac{x^2}{a^2}+\frac{y^2}{b^2}=1, z=0$, then $\oint_C \left(\frac{b^2x^2}{a^2}+\frac{a^2y^2}{b^2}\right)^{1/2} ds = (a^2+b^2)\pi.$

[**Hint:** $\mathbf{r} = a\cos t\,\mathbf{i} + b\sin t\,\mathbf{j}, 0 \le t \le 2\pi.$]

8. Show that, if C is the semi-circle $x^2 + y^2 = 1, z = 0$ drawn from (1, 0, 0) to (–1, 0, 0) through (0, 1, 0), then

$$\int_C (x^3 - y^3)\,dy = \frac{3\pi}{8}.$$

[**Hint:** $\mathbf{r} = \cos t\,\mathbf{i} + \sin t\,\mathbf{j},\ 0 \le t \le \pi,\ \mathbf{A} = (x^3 - y^3)\mathbf{j},\ \mathbf{A}\,.\,d\mathbf{r} = (x^3 - y^3)\,dy.$]

9. Show that if C is the triangle with vertices (0, 0, 0), (1, 0, 0), (1, 1, 0) then

$$\oint_C y^2dx + x^2dy = \frac{1}{3}.$$

[**Hint:** $\mathbf{A} = y^2\mathbf{i} + x^2\mathbf{j}$. Use $\mathbf{r} = (1-t)\,a + t\,\mathbf{b}$ to get the equations of the sides.]

10. Show that the work done in moving a particle in a field of force **F**, where $\mathbf{F} = 3xy\,\mathbf{i} + (x+y)\,\mathbf{j} - z\mathbf{k}$ along the curve $x = t+1,\ y = t-1,\ z = t^2$ from the point (2, 0, 1) to (4, 2, 9) is –12.

[**Hint:** Work done $= \int_C \mathbf{F}\cdot d\mathbf{r}$]

11. *Solved Sum.*

Evaluate $\int \bar{f}.d\bar{r}$ along the straight line joining (0, 0) and (1, 1).

(*i*) if $\bar{f} = x\bar{j} - y\bar{i}$

(*i*) if $\bar{f} = x^2\bar{i} - y^2\bar{j}$

These sums pertain to two-dimension. Hence

$$\bar{r} = x\bar{i} + y\bar{j}.$$

The equation of the line joining (0, 0) and (1, 1) is $y = x$

Taking $x = t$, we get the parametric equations of the line as

$$x = t, y = t.$$

Clearly $t = 0$ gives the point (0, 0) and $t = 1$ gives the point (1, 1). Denote the line segment joining these two points by C. Now, on C, t varies from 0 to 1 and

$$dx = dt, dy = dt.$$

(*i*)
$$\begin{aligned}\bar{f}\cdot d\bar{r} &= (x\bar{j} - y\bar{i})\cdot d(x\bar{i} + y\bar{j})\\ &= (-y\bar{i} + x\bar{j})\cdot(dx\bar{i} + dy\bar{j}) = -y\,dx + x\,dy\\ &= -t\,dt + t\,dt = 0\,dt.\end{aligned}$$

$$\therefore \qquad \int_C \bar{f}\cdot d\bar{r} = \int_0^1 0\,dt = 0.$$

(ii) $$\bar{f}\cdot d\bar{r} = (x^2\bar{i}+y^2\bar{j})\cdot(dx\bar{i}+dy\bar{j}) = x^2dx+y^2dy$$
$$= t^2dt+t^2dt = 2t^2dt.$$

$\therefore$ $$\int_C \bar{f}\cdot d\bar{r} = \int_0^1 2t^2\,dt = \frac{2}{3}.$$

12. *Solved sum.* Compute $\int_C \bar{F}\cdot d\bar{r}$ along $y^2 = 4x$ from (0, 0) to (4, 4) if

(i) $\bar{F} = x\bar{j} - y\bar{i}$ (ii) $\bar{F} = 3x^2y\bar{i} + (x^3-2y^2)\bar{j}$

Let C be the arc of the parabola $y^2 = 4x$ with end points (0, 0) and (4, 4). The parametric equations of C are

$$x = t^2, y = 2t. \qquad dx = 2t\,dt, dy = 2dt.$$

Now $t = 0$ gives (0, 0) and $t = 2$ gives (4, 4). So, for C, t varies from 0 to 2.

(i) Now, on C, we have

$$\bar{F}\cdot d\bar{r} = (x\bar{j}-y\bar{i})\cdot d(x\bar{i}+y\bar{j})$$
$$= (-y\bar{i}+x\bar{j})\cdot(dx\bar{i}+dy\bar{j}) = -y\,dx + x\,dy$$
$$= -(2t)(2t\,dt)+(t^2)(2dt)$$
$$= -4t^2\,dt + 2t^2\,dt = -2t^2\,dt.$$

$\therefore$ $$\int_C \bar{F}\cdot d\bar{r} = \int_0^2 (-2t^2)\,dt = -2\left[\frac{t^3}{3}\right]_0^2 = -\frac{16}{3}.$$

(ii) On C, we have

$$\bar{F}\cdot d\bar{r} = [3x^2y\bar{i}+(x^3-2y^2)\bar{j}]\cdot d(x\bar{i}+y\bar{j})$$
$$= [3x^2y\bar{i}+(x^3-2y^2)\bar{j}]\cdot(dx\bar{i}+dy\bar{j})$$
$$= 3x^2y\,dx + (x^3-2y^2)\,dy$$
$$= 3(t^2)^2(2t)(2t\,dt) + [(t^2)^3 - 2(2t)^2](2\,dt)$$
$$= 12t^6\,dt + [t^6 - 8t^2](2\,dt) = (14t^6 - 16t^2)\,dt.$$

$\therefore$ $$\int_C \bar{F}\cdot d\bar{r} = \int_0^2 (14t^6-16t^2)\,dt = \left[14\cdot\frac{t^7}{7} - 16\cdot\frac{t^3}{3}\right]_0^2$$
$$= 2\times 2^7 - \frac{16}{3}\times 2^3 = 256 - \frac{128}{3} = \frac{640}{3}.$$

13. *Solved sum.* Evaluate $\int_C 3x(x+2y)dx + (3x^2-y^3)dy$, where C is the line joining (0,0) and (1,2).

C is the line joining (0, 0) to (1, 2). Its slope is 2 and its equation is $y = 2x$. Setting $x = t$, we get the parametric equations of C as $x = t, y = 2t$, where t varies from 0 to 1. Now $dx = dt, dy = 2\,dt$.

$\therefore$ $$\int_C 3x(x+2y)dx + (3x^2-y^3)dy$$
$$= \int_0^1 3t(t+4t)\,dt + (3t^2-8t^3)2\,dt$$

$$= \int_0^1 (21t^2 - 16t^3)\,dt = \left[7t^3 - 4t^4\right]_0^1 = 3.$$

14. *Solved sum.* If $\bar{F} = 3xy\bar{i} - y^3\bar{j}$, compute $\int_C \bar{F}\cdot d\bar{r}$ along $y = 2x^2$ from (0, 0) to (1, 2).

Let C be the arc in question. Taking $x = t$, we get the parametric equations of C as

$$x = t, y = 2t^2.$$

Now $t = 0$ gives the end point (0, 0) and $t = 1$ gives the end point (1, 2) and, for C, t varies from 0 to 1 and

$dx = dt, dy = 4t\,dt.$

$$\begin{aligned} \therefore \quad \bar{F}\cdot d\bar{r} &= (3xy\bar{i} - y^3\bar{j})\cdot d(x\bar{i} + y\bar{j}) \\ &= (3xy\bar{i} - y^3\bar{j})\cdot(dx\bar{i} + dy\bar{j}) \\ &= 3xy\,dx - y^3\,dy = 3(t)(2t^2)\,dt - (2t^2)^3(4t\,dt) \\ &= 6t^3\,dt - 8t^6(4t\,dt) = (6t^3 - 32t^7)dt. \end{aligned}$$

$$\therefore \quad \int_C \bar{F}\cdot d\bar{r} = \int_0^1 (6t^3 - 32t^7)\,dt = \left[6\cdot\frac{t^4}{4} - 32\cdot\frac{t^8}{8}\right]_0^1$$

$$= \frac{6}{4} - \frac{32}{8} = \frac{3}{2} - 4 = -\frac{5}{2}.$$

15. *Solved sum.* Evaluate the line integral

$$\int_C \bar{F}\cdot d\bar{r},$$

where $\bar{F} = yz\bar{i} + zx\bar{j} - xy\bar{k}$ and C is the three arcs having end points $O(0, 0, 0)$ and P(2, 4, 8) is defined below.

(*i*) C is the arc whose parametric equations are

$$x = t, y = t^2, z = t^3.$$

(*ii*) C is the rectilinear path formed by the line segments

OA, AB, BP

where A, B, P are (2, 0, 0), (2, 4, 0), (2, 4, 8).

(*iii*) C is the straight line segment OP.

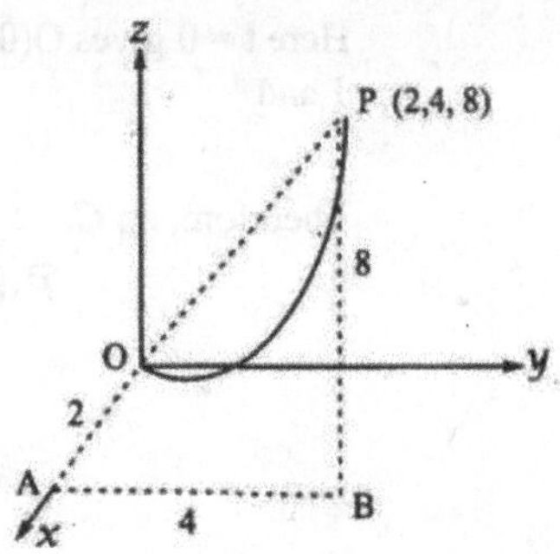

Case (*i*). This sum pertains to three-dimension. So

$$\bar{r} = x\bar{i} + y\bar{j} + z\bar{k}.$$

Now the parametric equations of C are

$$x = t, y = t^2, z = t^3.$$

Here $t = 0$ gives $O(0, 0, 0)$ and $t = 2$ gives $P(2, 4, 8)$. So, for C, t varies from 0 to 2 and

$$dx = dt, \quad dy = 2t\,dt, \quad dz = 3t^2\,dt.$$

$$\begin{aligned} \bar{F}\cdot d\bar{r} &= (yz\bar{i} + zx\bar{j} - xy\bar{k})\cdot(dx\bar{i} + dy\bar{j} + dz\bar{k}) \\ &= yz\,dx + zx\,dy - xy\,dz. \\ &= (t^2.t^3)dt + (t^3.t)2t\,dt - (t.t^2)\,3t^2\,dt \\ &= [t^5 + 2t^5 - 3t^5]\,dt = (0)\,dt. \end{aligned}$$

$$\therefore \quad \int_C \bar{F}\cdot d\bar{r} = \int_0^2 (0)\,dt = 0.$$

Case (ii). The parametric equations of OA, AB, BP and the range over which the parameter varies are

OA: $x = t, y = 0, z = 0$; t: From 0 to 2

AB: $x = 2, y = t, z = 0$; t: From 0 to 4

BP: $x = 2, y = 4, z = t$; t: From 0 to 8.

$$\text{Now} \quad \bar{F}\cdot d\bar{r} = (yz\bar{i} + zx\bar{j} - xy\bar{k}).(dx\bar{i} + dy\bar{j} + dz\bar{k})$$
$$= yz\,dx + zx\,dy - xy\,dz.$$

So the value of $\bar{F}\cdot d\bar{r}$ on OA, AB, BP are respectively

$$0, 0, -8\,dt.$$

$$\therefore \quad \int_{OA} \bar{F}\cdot d\bar{r} = 0, \int_{AB} \bar{F}\cdot d\bar{r} = 0. \qquad (1)$$

$$\int_{BP} \bar{F}\cdot d\bar{r} = \int_0^8 (-8)\,dt = -8[t]_0^8 = -64. \qquad (2)$$

Adding all the three integrals in (1) and (2).

$$\int_C \bar{F}\cdot d\bar{r} = 0 + 0 + (-64) = -64.$$

Case (iii). Here C is the straight line segment OP. Its equations are

$$\frac{x-0}{2-0} = \frac{y-0}{4-0} = \frac{z-0}{8-0} = t.$$

Hence the parametric equations of C are

$$x = 2t, y = 4t, z = 8t.$$

Here t = 0 gives O(0, 0, 0) and t = 1 gives $P(2, 4, 8)$. So the parameter t varies from 0 to 1 and

$$dx = 2\,dt, dy = 4\,dt, dz = 8\,dt.$$

Therefore, on C,

$$\bar{F}\cdot d\bar{r} = (yz\bar{i} + zx\bar{j} - xy\bar{k})\cdot(dx\bar{i} + dy\bar{j} + dz\bar{k})$$
$$= yz\,dx + zx\,dy - xy\,dz$$
$$= (4t\,.\,8t)2\,dt + (8t\,.\,2t)4\,dt - (2t\,.\,4t)8\,dt$$
$$= 64t^2\,dt.$$

$$\therefore \quad \int_C \bar{F}\cdot d\bar{r} = 64\int_0^1 t^2\,dt = 64\left[\frac{t^3}{3}\right]_0^1 = \frac{64}{3}.$$

16. *Solved sum.* If $\bar{F} = xz\bar{i} + yz\bar{j} + z^2\bar{k}$, evaluate $\int_C \bar{F}\cdot d\bar{r}$ from the point (0, 0, 0) to (1, 1, 1) where C is given by $x = t, y = t^2, z = t^3$.

Now the parametric equations of C are

$$x = t, y = t^2, z = t^3.$$

Here $t = 0$ gives (0, 0, 0) and $t = 1$ gives (1, 1, 1). So, for C, t varies from 0 to 1 and $dx = dt, dy = 2t\,dt, dz = 3t^2\,dt$

$$\overline{F}\cdot d\overline{r} = (xz\overline{i}+yz\overline{j}+z^2\overline{k})\cdot(dx\overline{i}+dy\overline{j}+dz\overline{k})$$
$$= xz\,dx + yz\,dy + z^2dz$$
$$= (t.t^3)\,dt + (t^2.t^3)\,2t dt + (t^3)^2\,3t^2\,dt$$
$$= (t^4+2t^6+3t^8)dt$$

$$\therefore \quad \int_C \overline{F}\cdot d\overline{r} = \int_0^1 (t^4+2t^6+3t^8)\,dt.$$
$$= \left[\frac{t^5}{5}+2\frac{t^7}{7}+3\frac{t^9}{9}\right]_0^1 = \frac{86}{105}.$$

17. *Solved sum.* Evaluate the integral
$$\mathbf{I} = \int_C x\,dx + y\,dy + z\,dz,$$

where C is the circle $x^2+y^2+z^2=a^2, z=0.$

The parametric equations of the circle are
$$x = a\cos t, y = a\sin t, z = 0,$$
where t varies from 0 to 2π. Also
$$dx = -a\sin t\,dt, dy = a\cos t\,dt, dz = 0$$
$$\therefore \quad \mathbf{I} = \int_C x\,dx + y\,dy + z\,dz$$
$$= \int_0^{2\pi} (a\cos t)(-a\sin t\,dt)+(a\sin t)(a\cos t\,dt)+0$$
$$= a^2\int_0^{2\pi} (-\sin t\cos t + \sin t\cos t)\,dt$$
$$= a^2\int_0^{2\pi} 0\,dt = 0.$$

18. *Solved sum.* Evaluate $\int_C x\,dx + y\,dy$, where C is the ellipse $x^2+4y^2=4$.

The equation of the ellipse is
$$\frac{x^2}{2^2}+\frac{y^2}{1^2}=1.$$

So the parametric equations are $x = 2\cos t, y = \sin t$, where t varies from 0 to 2π. Therefore
$$\mathbf{I} = \int_C x\,dx + y\,dy$$
$$= \int_0^{2\pi} (2\cos t)(-2\sin t\,dt)+(\sin t)(\cos t\,dt)$$
$$= -\int_0^{2\pi} 3\sin t\cos t\,dt = -\frac{3}{2}\int_0^{2\pi} \sin 2t\,dt$$
$$= \frac{3}{4}[\cos 2t]_0^{2\pi} = 0.$$

19. If $\bar{F}=(5xy-6x^2)\bar{i}+(2y-4x)\bar{j}$ and C is the arc of the cubical parabola $y=x^3$ from (1, 1) to (2, 8), then show that

$$\int_C \bar{F}\cdot d\bar{r}=35.$$

[**Hint:** Parametric equations of $C: x=t, y=t^3$, where t varies from 1 to 2; $dx=dt, dy=3t^2\,dt$.
But $\bar{F}\cdot d\bar{r}=(5xy-6x^2)\,dx+(2y-4x)\,dy$.
On C, $\bar{F}\cdot d\bar{r}=(5t^4-6t^2)\,dt+(6t^5-12t^3)dt$. Integrate this from 1 to 2]

20. If $\bar{F}=(2x+y)\bar{i}+(3y-x)\bar{j}$ and C is the arc of the cubical parabola $y=x^3$ from (1, 1) to (2,8), then show that

$$\int_C \bar{F}\cdot d\bar{r}=90.$$

[**Hint:** $\bar{F}\cdot d\bar{r}=\int_1^2 (2t+t^3)\,dt+(3t^3-t)(3t^2\,dt)=90$]

21. Evaluate $\int_C \bar{F}\cdot d\bar{r}$ when $\bar{F}=x^2\bar{i}+xy\bar{j}$ from (0, 0) to (1, 1) along $y^2=x$.

[**Hint:** C is given by $y=t, x=t^2$, where t varies from 0 to 1. So $dx=2t\,dt, dy=dt$. Thus

$$\bar{F}\cdot d\bar{r}=x^2dx+xy\,dy=t^4(2tdt)+t^3(dt)=(2t^5+t^3)\,dt$$

$$\therefore \int_C \bar{F}\cdot d\bar{r}=\int_0^1 (2t^5+t^3)\,dt=\left[\frac{t^6}{3}+\frac{t^4}{4}\right]_0^1=\frac{7}{12}]$$

22. If $\bar{F}=2xy\bar{i}+y\bar{j}+xz\bar{k}$ and C is the arc of the parabola $y^2=2x$, $z=0$ from (0, 0, 0) to (1/2, 1, 0), then show that

$$\int_C \bar{F}\cdot d\bar{r}=\frac{7}{10}.$$

[**Hint:** Now $\bar{F}\cdot d\bar{r}=2xy\,dx+y\,dy+xz\,dz$. Choosing $y=t$, the parametric equations of C are $x=\frac{t^2}{2}$, $y=t, z=0$, $dx=t\,dt$, $dy=dt$, $dz=0$.

Now t varies from 0 to 1. On C, $\bar{F}\cdot d\bar{r}=(t^4+t)\,dt$]

23. If $\bar{F}=(2x+y)\bar{i}+(3y-x)\bar{j}$ and C, is formed by the straight lines OA and AB, where O, A, B are (0, 0), (2, 0), (3, 2), then show that $\int_C \bar{F}\cdot d\bar{r}=11$.

[**Hint:** $\bar{F}\cdot d\bar{r}=(2x+y)dx+(3y-x)dy$. Now
OA: $x=t, y=0, dx=dt, dy=0$; t varies from 0 to 2; $\bar{F}\cdot d\bar{r}=2t\,dt$; I = 4.
AB: Equation of AB is $y=2x-4$. Choosing $x=t$, parametric equations are $x=t, y=2t-4$, where t varies from 2 to 3; $dx=dt, dy=2\,dt$; $\bar{F}\cdot d\bar{r}=(14t-28)\,dt$; I = 7]

24. If $\bar{F}=(3x^2+6y)\bar{i}-14yz\bar{j}+20xz^2\bar{k}$, evaluate $\int_C \bar{F}\cdot d\bar{r}$, where C is the following three different curves whose end points are $O(0, 0, 0)$ and $P(1, 1, 1)$:

(*i*) Curve whose parametric equations are
$x = t, y = t^2, z = t^3$.

(*ii*) Straight lines *OA*, *AB*, *BP*, where *A* is (1, 0, 0), *B* is (1, 1, 0).

(*iii*) Straight line *OP*.
[**Hint:** $\bar{F}\cdot d\bar{r} = (3x^2 + 6y)dx - 14yz\, dy + 20\, xz^2\, dz.$]

(*i*) $x = t, y = t^2, z = t^3; dx = dt, dy = 2t\, dt,$
$dz = 3t^2\, dt.$, t varies from 0 to 1;
$\bar{F}\cdot d\bar{r} = 9t^2\, dt - 28t^6\, dt + 60t^9\, dt; \mathbf{I} = 5.$

(*ii*) $OA: x = t, y = 0, z = 0$ t: from 0 to 1.
AB: x = 1, y = t, z = 0 t: from 0 to 1.
BP: x = 1, y = 1, z = t t: from 0 to 1.
On *OA*: $\bar{F}\cdot d\bar{r} = 3t^2 dt,\quad \mathbf{I} = 1$
On *AB*: $\bar{F}\cdot d\bar{r} = 0,\quad \mathbf{I} = 0$
On *BP*: $\bar{F}\cdot d\bar{r} = 20t^2 dt,\quad \mathbf{I} = 20/3.$ Sum = 23/3.

(*iii*) $x = y = z = t; dx = dy = dz = dt;$ t varies from 0 to 1.
$\bar{F}\cdot d\bar{r} = (-11t^2 + 6t + 20t^3)\, dt; \mathbf{I} = 13/3$]

25. If $\bar{F} = y\bar{i} - z\bar{j} + x\bar{k}$ and C is the arc of the curve $\bar{r} = \cos t\,\bar{i} + \sin t\,\bar{j} + t\bar{k}$, where t takes the values from 0 to 2π, show that

$$\int_C \bar{F}\cdot d\bar{r} = -\pi.$$

[**Hint:** $\mathbf{I} = \int_0^{2\pi} (-\sin^2 t - t\cos t + \cos t)\, dt$

$$= -\int_0^{2\pi} \frac{1-\cos 2t}{2} dt - [t\sin t + \cos t]_0^{2\pi} + [\sin t]_0^{2\pi} = -\pi\,]$$

26. *Solved sum.* Show that the work done in moving a particle in a force field

$$\bar{F} = 3xy\bar{i} + (x+y)\bar{j} - z\bar{k}$$

along the curve $x = t + 1, y = t - 1, z = t^2$ from (2, 0, 1) to (4, 2, 9) is – 12.

The parametric equations of *C* are

$$x = t + 1, y = t - 1, z = t^2.$$

Evidently $t = 1$ gives (2, 0, 1) and $t = 3$ gives (4, 2, 9).

Now

$$dx = dt, dy = dt, dz = 2t\, dt.$$

$$\text{Workdone} = \int_C \bar{F}\cdot d\bar{r} = \int_C 3xy\, dx + (x+y)dy - z\, dz.$$

$$= \int_1^3 3(t^2 - 1)\, dt + (2t)\, dt - t^2(2t\, dt)$$

$$= \int_1^3 (-2t^3 + 3t^2 + 2t - 3)\, dt$$

$$= \left[-\frac{t^4}{2}+t^3+t^2-3t\right]_1^3 = -12.$$

27. Find the work done in moving a particle in a force field $\bar{F} = 3xy\bar{i} - 5z\bar{j} + 10x\bar{k}$ along the curve $x = t^2 + 1, y = 2t^2, z = t^3$ from (2, 2, 1) to (5, 8, 8).

[**Hint:** The points are given by $t = 1, t = 2$ and $dx = 2\,dt, dy = 4t\,dt, dz = 3t^2\,dt$.

$$\text{Work done} = \int_C \bar{F}\cdot d\bar{r} = \int_C 3xy\,dx - 5z\,dy + 10x\,dz$$

$$= \int_1^2 (12t^5 + 10t^4 + 12t^3 + 30t^2)\,dt = 303\,]$$

28. Find the work done in moving a particle in a force field

$$\bar{F} = 3x^2\bar{i} + (2xz - y)\bar{j} + z\bar{k}$$

along the straight line from (0, 0, 0) to (2, 1, 3).

[**Hint:** The cartesian and parametric equations of the line are

$$\frac{x}{2} = \frac{y}{1} = \frac{z}{3};\; x = 2t, y = t, z = 3t,$$

where t varies from 0 to 1; $dx = 2\,dt, dy = dt, dz = 3\,dt$.

$$\text{Workdone} = \int_0^1 (36t^2 + 8t)\,dt = 16\,]$$

29. Find the work done by the force

$$\bar{F} = 3(x + yz)\bar{i} + 7xz^2\bar{j} - 5yz^2\bar{k}$$

when it moves a particle from (0, 0, 0) to (1, 1, 1) along the curve $x = t^3, y = t^2, z = t$.

[**Hint:** The points are given by $t = 0, t = 1$ and

$$dx = 3t^2\,dt, dy = 2t\,dt, dz = dt.$$

$$\text{Workdone} = \int_C \bar{F}\cdot d\bar{r} = \int_C 3(x + yz)dx + 7xz^2\,dy - 5yz^2\,dz$$

$$= \int_0^1 (18t^5 + 14t^6 - 5t^4)\,dt = 4\,]$$

30. *Solved sum.* Show that the integral of

$$\mathbf{F} = (3x^2 + 6xy)\mathbf{i} + (3x^2 - y^3)\mathbf{j}$$

is independent of the path of integration. Find $\int \mathbf{F}.d\mathbf{r}$ along any curve joining (0, 0) and (1, 2). **(1982)**

$$\nabla\times\mathbf{F} = \begin{vmatrix} \mathbf{i} & \mathbf{j} & \mathbf{k} \\ \frac{\partial}{\partial x} & \frac{\partial}{\partial y} & \frac{\partial}{\partial z} \\ 3x^2 + 6xy & 3x^2 - y^3 & 0 \end{vmatrix} = 0.$$

Therefore the integral of **F** is independent of the path of integration. We shall find its integral from (0, 0) to (1, 2) along the straight line $y = 2x$ joining them. Then x varies from 0 to 1and the integral is

$$\int_{(0,0)}^{(1,2)} \{3x^2 + 6x(2x)\}dx + \{3x^2 - (2x)^3\}\, dy$$

$$= \int_0^1 15x^2 dx + \int_0^1 (3x^2 - 8x^3)(2dx) = 3.$$

31. *Solved sum.* Show that the vector field **f**, where $\mathbf{f} = (y + y^2 + z^2)\,\mathbf{i} + (x + z + 2xy)\mathbf{j} + (y + 2zx)\,\mathbf{k}$, is conservative and find its scalar potential. **(1974)**

It can easily be checked that $\nabla \times \mathbf{f} = \mathbf{0}$. So **f** is conservative and a scalar potential ϕ exists such that $\nabla\phi = \mathbf{f}$. Hence.

$$\phi_x = y + y^2 + z^2,\ \phi_y = x + z + 2xy,\ \phi_z = y + 2xz.$$

Integrating these with respect to x, y, z respectively,

$$\phi = xy + xy^2 + xz^2 + f_1(y, z),$$
$$\phi = xy + yz + xy^2 + f_2(z, x)$$
$$\phi = yz + xz^2 + f_3(x, y)$$
$$\therefore \quad \phi = xy + yz + xy^2 + xz^2 + C.$$

Second method. Alternatively $\mathbf{f}.\, d\mathbf{r} = (\nabla\phi)\,.\, d\mathbf{r} = d\phi$.

$$d\phi = \mathbf{f}\,.\,d\mathbf{r} = (y + y^2 + z^2)\,dx + (x + z + 2xy)\,dy + (y + 2\,zx)\,dz$$
$$= (y^2dx + 2xy\,dy) + (ydx + xdy) + (zdy + ydz) + (2zxdz + z^2dx)$$
$$= d(xy^2) + d(xy) + d(yz) + d(xz^2).$$
$$\therefore \quad \phi = xy^2 + xy + yz + xz^2 + C.$$

32. (*i*) show that $\mathbf{F} = (2xy + z^3)\mathbf{i} + x^2\mathbf{j} + 3xz^2\mathbf{k}$ is conservative and evaluate $\int \mathbf{F}\cdot d\mathbf{r}$ along any curve joining (1, –2, 1) and (3, 1, 4). **(1977)**

(*ii*) Show that $\int \mathbf{r}\cdot d\mathbf{r}$ along any closed curve is zero. **(1985)**

[**Hint:** (*i*) $\nabla\times\mathbf{F} = 0; \mathbf{F} = \nabla\phi; \int \mathbf{F}\cdot d\mathbf{r} = \int(\nabla\phi)\cdot d\mathbf{r} = \int d\phi = 202$ since $\phi = x^2y + xz^3 + C$. Or use the equation of the straight line $\mathbf{r} = (1 - t)\,(\mathbf{i} - 2\mathbf{j} + \mathbf{k}) + t(3\mathbf{i} + \mathbf{j} + 4\mathbf{k})$.

(*ii*) Use book work 3.2]

3.5. SURFACE INTEGRAL

(i) PLANE SURFACE

33. *Solved sum.* Evaluate the integral $\iint_S \mathbf{A}\cdot\mathbf{n}\, dS$ if $\mathbf{A} = 4y\,\mathbf{i} + 18z\,\mathbf{j} - x\mathbf{k}$ and *S* is the surface of the portion of the plane $3x + 2y + 6z = 6$ contained in the first octant.

Let *ABC* be the given surface *S* (Fig. 3.8.2). Then the projection *R* of *S* on the *xOy* plane is *OAB*. Now

$$\mathbf{n} = \frac{\nabla(3x+2y+6z-6)}{|\nabla(3x+2y+6z-6)|} = \frac{3\mathbf{i}+2\mathbf{j}+6\mathbf{k}}{|3\mathbf{i}+2\mathbf{j}+6\mathbf{k}|}$$

$$= \frac{3}{7}\mathbf{i} + \frac{2}{7}\mathbf{j} + \frac{6}{7}\mathbf{k}.$$

$$\iint_S A\cdot n\, dS = \iint_R \mathbf{A}\cdot\mathbf{n}\frac{dx\,dy}{|\mathbf{n}.\mathbf{k}|}\left(\text{or} \iint_R A\cdot\nabla\phi\frac{dx\,dy}{|\nabla\phi\cdot\mathbf{k}|}\right)$$

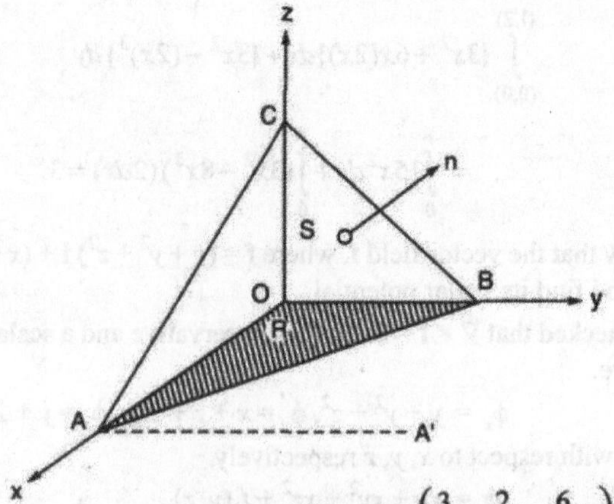

$$= \iint_R \frac{(4y\mathbf{i}+18z\mathbf{j}-x\mathbf{k})\cdot\left(\frac{3}{7}\mathbf{i}+\frac{2}{7}\mathbf{j}+\frac{6}{7}\mathbf{k}\right)}{\left|\left(\frac{3}{7}\mathbf{i}+\frac{2}{7}\mathbf{j}+\frac{6}{7}\mathbf{k}\right)\cdot\mathbf{k}\right|}dx\,dy$$

$$= \iint_R (2y+6z-x)\,dx\,dy = \iint_R (6-4x)\,dx\,dy$$

since $6z = -3x - 2y + 6$ from the equation of the plane. Let AA' be parallel to the y-axis. Then R lies between the y-axis and AA', where A is (2.0.0). Thus $0 \le x \le 2$. With this restriction on the x coordinate of a point of R, the y coordinate varies from $y = 0$ to $y = 3(2-x)/2$ since R is bounded by OA and AB. Thus

$$\iint_S \mathbf{A}\cdot\mathbf{n}\,dS = \int_0^2 \int_{y=0}^{y=3(2-x)/2} (6-4x)\,dy\,dx$$

$$= \int_0^2 (6-4x)\,y \Big|_{y=0}^{y=3(2-x)/2} dx$$

$$= 3\int_0^2 (6-7x+2x^2)\,dx = 10.$$

***Remark.* Alternatively, when the projection OAC of S on the xOz plane is considered,**

$$\iint_S \mathbf{A}\cdot\mathbf{n}dS = \iint_{OAC} \mathbf{A}\cdot\mathbf{n}\frac{dx\,dz}{|\mathbf{n}\cdot\mathbf{j}|} = 3\iint_{OAC}(2y+6z-x)\,dx\,dz$$

$$= 6\iint_{OAC}(3-2x)dx\,dz = 6\int_0^2\int_0^{1-(x/2)}(3-2x)\,dz\,dx$$

$$= 6\int_0^2 (3-2x)(1-\frac{1}{2}x)\,dx = 10.$$

34. **Evaluate $\iint_S \mathbf{A}\cdot\mathbf{n}\,dS$ if $A = (x+y^2)\mathbf{i} - 2x\mathbf{j} + 2yz\mathbf{k}$ and S is the surface of the plane $2x + y + 2z = 6$ in the first octant.** **[Ans. 81]**

[Hint: $\mathbf{n} = \frac{2}{3}\mathbf{i}+\frac{1}{3}\mathbf{j}+\frac{2}{3}\mathbf{k},\ dS = \frac{dx\,dy}{|\mathbf{n}\cdot\mathbf{k}|} = \frac{3dx\,dy}{2}$,

$\mathbf{A}\cdot\mathbf{n} = 2y(6-2x)/3$ since $z = (6-2x-y)/2$. The integral is

$$2\int_0^3\int_{y=0}^{6-2x}(3-x)\,y\,dy\,dx = \int_0^3(108-108x+36x^2-4x^3)\,dx = 81.]$$

35. Evaluate $\iint \mathbf{A}\cdot\mathbf{n}\,dS$ if $\mathbf{A} = 18z\,\mathbf{i} - 12\,\mathbf{j} + 3y\,\mathbf{k}$ and S is the surface $2x + 3y + 6z = 12$ in the first octant. **(1980)**

[**Hint:** $n = \frac{2}{7}\mathbf{i} + \frac{3}{7}\mathbf{j} + \frac{6}{7}\mathbf{k}, dS = \frac{dx\,dy}{|\mathbf{n}\cdot\mathbf{k}|} = \frac{7\,dx\,dy}{6}$, $\mathbf{A}\cdot\mathbf{n} = 12(3-x)/7$ since $z = (12-2x-3y)/6$.

The integral is $2\int_0^6\int_{y=0}^{(12-2x)/3}(3-x)\,dy\,dx = 2\int_0^6(3-x)\frac{12-2x}{3}dx = 24.]$

36. Find the area of the surface S of the portion of the plane $3x + 2y + 6z = 6$ contained in the first octant. [Ans. 7/2]

[**Hint:** $Area = \iint_S dS = \int_0^2\int_{y=0}^{y=3(2-x)/2} 1\frac{dy\,dx}{|\mathbf{n}\cdot\mathbf{k}|} = \frac{7}{2}.]$

37. S is the surface of the portion of the plane $2x + 2y + z = 4$, in the first octant and $\mathbf{A} = 2x\mathbf{j} - xz\mathbf{k}$. Show that

$$\iint_S \mathbf{A}\cdot\mathbf{n}\,dS = 4, \int_S \mathbf{r}\cdot\mathbf{n}\,dS = 8, \iint_S \mathbf{A}\,dS = 4(2\mathbf{j}-\mathbf{k}),$$

$$\iint_S \mathbf{A}\times\mathbf{n}\,dS = \frac{16}{3}i - \frac{8}{3}\mathbf{j} - \frac{16}{3}\mathbf{k}.$$

38. *Solved sum.* Evaluate $\iint_S \mathbf{F}\cdot\mathbf{n}\,dS$ if $\mathbf{F} = (x+y)\mathbf{i} + x\mathbf{j} + z\mathbf{k}$ and S is the surface of the cube bounded by the planes $x = 0, x = 1, y = 0, y = 1, z = 0, z = 1$.

Suppose the faces whose equations are $x = 0, x = 1, y = 0, y = 1, z = 0, z = 1$ are respectively named $S_1, S_2, S_3, S_4, S_5, S_6$ and $\mathbf{n}$ denotes the unit vectors normal to them, drawn outwardly (Fig. 3.8.3). Now, on $S_1, x = 0$, $\mathbf{F} = y\mathbf{i} + z\mathbf{k}$, $\mathbf{n} = -i$.

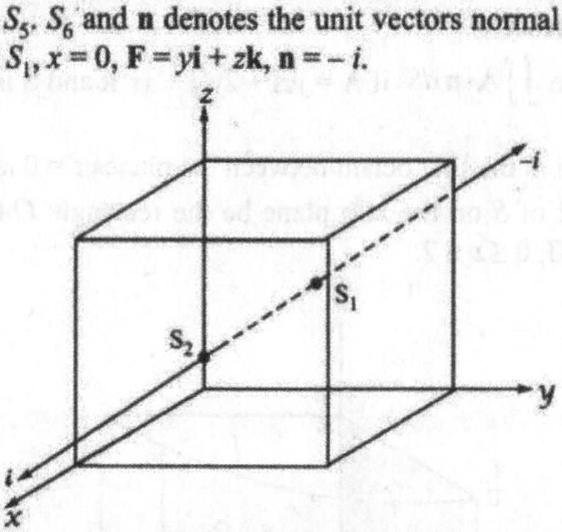

Fig. 3.8.3

$$\therefore \quad \iint_{S_1}\mathbf{F}\cdot\mathbf{n}\,dS = -\int_0^1\int_0^1 y\,dy\,dz = -\frac{1}{2}.$$

On S_2, $x = 1, \mathbf{F} = (1+y)\,\mathbf{i} + \mathbf{j} + z\mathbf{k}, \mathbf{n} = \mathbf{i}.$

$$\therefore \quad \iint_{S_2} \mathbf{F}\cdot\mathbf{n}\,dS = \int_0^1\!\!\int_0^1 (1+y)\,dy\,dz = \frac{3}{2}.$$

On S_3, $y = 0, \mathbf{F} = x\mathbf{i} + x\mathbf{j} + z\mathbf{k}, \mathbf{n} = -\mathbf{j}.$

$$\therefore \quad \iint_{S_3} \mathbf{F}\cdot\mathbf{n}\,dS = -\int_0^1\!\!\int_0^1 x\,dx\,dz = -\frac{1}{2}.$$

On S_4, $y = 1, \mathbf{F} = (x+1)\,\mathbf{i} + x\mathbf{j} + z\mathbf{k}, \mathbf{n} = \mathbf{j}.$

$$\therefore \quad \iint_{S_4} \mathbf{F}\cdot\mathbf{n}\,dS = \int_0^1\!\!\int_0^1 x\,dx\,dz = \frac{1}{2}.$$

On S_5, $z = 0, \mathbf{F} = (x+y)\mathbf{i} + x\mathbf{j}, \mathbf{n} = -\mathbf{k}.$

$$\therefore \quad \iint_{S_5} \mathbf{F}\cdot\mathbf{n}\,dS = \int_0^1\!\!\int_0^1 0\,dx\,dy = 0.$$

On S_6, $z = 1, \mathbf{F} = (x+y)\mathbf{i} + x\mathbf{j} + \mathbf{k}, \mathbf{n} = \mathbf{k}.$

$$\therefore \quad \iint_{S_6} \mathbf{F}\cdot\mathbf{n}\,dS = \int_0^1\!\!\int_0^1 1\,dx\,dy = 1.$$

Adding, the integral is $-\frac{1}{2} + \frac{3}{2} + \left(-\frac{1}{2}\right) + \frac{1}{2} + 0 + 1 = 2.$

Remark. The above surface integral can be evaluated as a single volume integral by using divergence theorem (vide next chapter).

39. Show that $\iint_S \mathbf{r}\cdot\mathbf{n}\,dS = 3$, where S is the surface of the cube bounded by the planes $x = 0, x = 1, y = 0, y = 1, z = 0, z = 1.$

Remark. Here also the method by divergence theorem will be easier.

(ii) CYLINDRICAL SURFACE

40. *Solved sum.* Evaluate $\iint_S \mathbf{A}\cdot\mathbf{n}\,dS$ if $\mathbf{A} = yz\mathbf{i} + 2y^2\mathbf{j} + xz^2\mathbf{k}$ and S is the surface of the cylinder $x^2 + y^2 = 9$ contained in the first octant between the planes $z = 0$ and $z = 2$.

Let the projection R of S on the xOz plane be the rectangle $OABC$ (Fig. 3.8.4). Now R is specified by $0 \le x < 3, 0 \le z \le 2$.

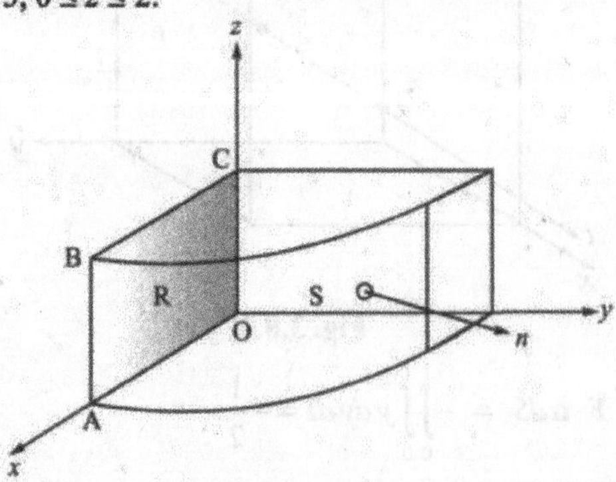

The unit normal on S is

$$\mathbf{n} = \frac{\nabla(x^2+y^2)}{|\nabla(x^2+y^2)|} = \frac{2x\mathbf{i}+2y\mathbf{j}}{|2x\mathbf{i}+2y\mathbf{j}|} = \frac{x\mathbf{i}+y\mathbf{j}}{\sqrt{x^2+y^2}} = \frac{x\mathbf{i}+y\mathbf{j}}{3}.$$

$$\iint_S \mathbf{A}\cdot\mathbf{n}\,dS = \iint_R \frac{\mathbf{A}.\mathbf{n}}{\mathbf{n}.\mathbf{j}}\,dx\,dz = \iint_R \frac{\frac{1}{3}(xyz+2y^3)}{\frac{1}{3}y}\,dx\,dz$$

$$= \iint_R (xz+2y^2)\,dx\,dz = \iint_R \{xz+2(9-x^2)\}\,dx\,dz$$

$$= \int_0^3 dx\int_0^2 (xz+18-2x^2)\,dz = \int_0^3 \left.\frac{xz^2}{2}+18z-2x^2z\right|_0^2 dx$$

$$= \int_0^3 (2x+36-4x^2)\,dx = 81.$$

41. Evaluate $\iint_S \mathbf{A}\cdot\mathbf{n}\,dS$ if $\mathbf{A} = z\mathbf{i}+x\mathbf{j}-y^2z\mathbf{k}$ and S is the surface of the cylinder $x^2+y^2=1$ in the first octant between the planes $z=0$ and $z=2$. **(1980)**

[Hint: $\mathbf{n} = x\mathbf{i}+y\mathbf{j}$, $\mathbf{A}.\mathbf{n} = xz+xy$, $\mathbf{n}.\mathbf{j} = y$; $y^2 = 1-x^2$;

$$\int_{x=0}^{1}\int_{z=0}^{2}\left(\frac{xz}{\sqrt{1-x^2}}+x\right)dz\,dx = \int_0^1\left(\frac{2x}{\sqrt{1-x^2}}+2x\right)dx$$

$$= \left.1-2\sqrt{1-x^2}+x^2\right]_0^1 = 3.]$$

42. *Solved Sum:* If the parametric equation of a surface S is $\mathbf{r} = \mathbf{r}(u, v)$ and $\mathbf{F}$ is a point function defined on S, Show that

$$\iint_S \mathbf{F}\cdot\mathbf{n}\,dS = \iint_{R_{uv}} \mathbf{F}\cdot\left(\frac{\partial\mathbf{r}}{\partial u}\times\frac{\partial\mathbf{r}}{\partial v}\right)du\,dv,$$

where R_{uv} is the region in the *uv* plane which corresponds to *S*.

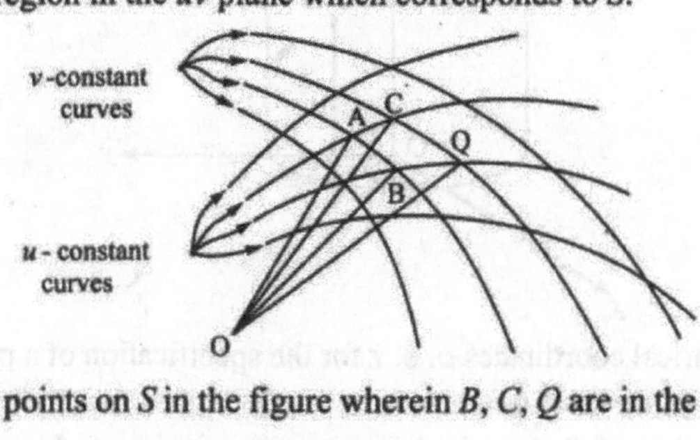

Let A, B, C, Q be points on S in the figure wherein B, C, Q are in the neighbourhood of A such that

$$\overline{OA} = \mathbf{r}(u, v),$$
$$\overline{OB} = \mathbf{r}(u+du, v),$$
$$\overline{OC} = \mathbf{r}(u, v+dv),$$
$$\overline{OQ} = \mathbf{r}(u+du, v+dv).$$

Then the area, say dS, of the figure ABQC lying on S, corresponds to the area R_{uv} in the uv plane. Now we get, to the first approximations, the following results :

$$d\mathbf{r} = \frac{\partial \mathbf{r}}{\partial u}du + \frac{\partial \mathbf{r}}{\partial v}dv,$$

$$\overline{AB} = \frac{\partial \mathbf{r}}{\partial u}du + \frac{\partial \mathbf{r}}{\partial v}(0) = \frac{\partial \mathbf{r}}{\partial u}du,$$

$$\overline{AC} = \frac{\partial \mathbf{r}}{\partial u}(0) + \frac{\partial \mathbf{r}}{\partial v}dv = \frac{\partial \mathbf{r}}{\partial v}dv,$$

$$dS = \left|\overline{AB}\times\overline{AC}\right| = \left|\frac{\partial \mathbf{r}}{\partial u}\times\frac{\partial \mathbf{r}}{\partial v}\right|du\,dv.$$

So, from (1.9.2) $$\mathbf{n}\,dS = \left(\frac{\partial \mathbf{r}}{\partial u}\times\frac{\partial \mathbf{r}}{\partial v}\right)du\,dv$$

if the direction of $\frac{\partial \mathbf{r}}{\partial u}\times\frac{\partial \mathbf{r}}{\partial v}$ is chosen as the direction of the normal on the positive side of S.

Thus the result follows as stated.

43. *Solved sum.* Evaluate $\iint_S \mathbf{A}\cdot\mathbf{n}\,dS$ over the structure S of the region bounded by $x^2 + y^2 = 4$, $z = 0$, $z = 3$ if $\mathbf{A} = 4x\mathbf{i} - 2y^2\mathbf{j} + z^2\mathbf{k}$. **(1980)**

Let S be divided into three parts S_1, S_2, S_3, where S_1 denotes the curved surface, S_2 denotes the plane surface lying in the plane $z = 0$, and S_3 denotes the plane surface lying in the plane $z = 3$ (Fig. 3.8.6).

Method 1 to calculate on S_1

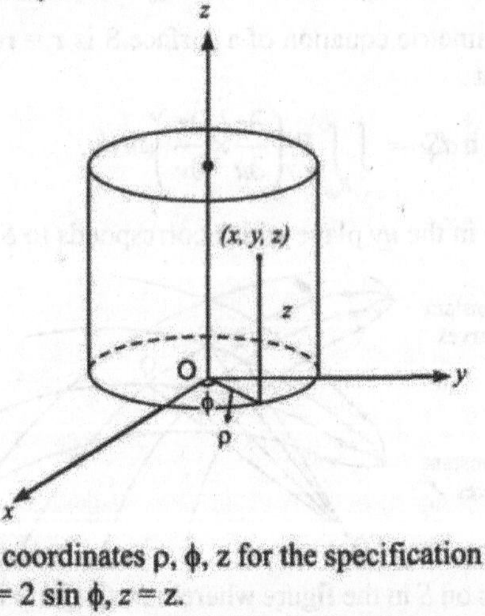

If we use the cylindrical coordinates ρ, ϕ, z for the specification of a point (x, y, z) on S_1, then $\rho = 2$ and $x = 2\cos\phi$, $y = 2\sin\phi$, $z = z$.

So the vector equation of S_1 in cylindrical coordinates is $\mathbf{r} = 2\cos\phi\,\mathbf{i} + 2\sin\phi\,\mathbf{j} + z\mathbf{k}$, $0 \le \phi \le 2\pi$, $0 \le z \le 3$ and on S_1, $\mathbf{A} = 4(2\cos\phi)\,\mathbf{i} - 2(2\sin\phi)^2\,\mathbf{j} + z^2\mathbf{k}$. Now

$$\frac{\partial \mathbf{r}}{\partial \phi}\times\frac{\partial \mathbf{r}}{\partial z} = \begin{vmatrix} \mathbf{i} & \mathbf{j} & \mathbf{k} \\ -2\sin\phi & 2\cos\phi & 0 \\ 0 & 0 & 1 \end{vmatrix} = 2(\cos\phi\,\mathbf{i} + \sin\phi\,\mathbf{j})$$

and, using the previous solved sum,

$$\iint_{S_1} \mathbf{A}\cdot\mathbf{n}\,dS = \int_{z=0}^{z=3}\int_{\phi=0}^{\phi=2\pi} \mathbf{A}\cdot\left(\frac{\partial \mathbf{r}}{\partial \phi}\times\frac{\partial \mathbf{r}}{\partial z}\right)d\phi\,dz$$

$$= \int_{z=0}^{z=3}\int_{\phi=0}^{\phi=2\pi} 16(\cos^2\phi - \sin^3\phi)\,d\phi\,dz = 48\pi$$

$$\text{since } \int_{\phi=0}^{\phi=2\pi} \cos^2\phi\,d\phi = \pi,\quad \int_{\phi=0}^{\phi=2\pi} \sin^3\phi\,d\phi = 0.$$

Method 2 to calculate on S_1

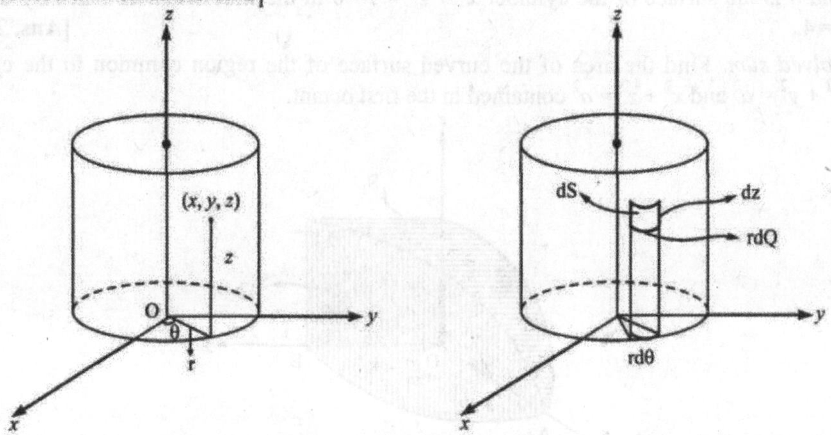

Fig. 3.8.6

If we use the cylindrical coordinates r, ϕ, z for the specification of a point (x, y, z) on S_1, then $r = 2$ and $x = 2\cos\theta, y = 2\sin\theta, z = z$.

The elementary lengths due to inifinitesimal changes $dr, d\theta, dz$ in r, θ, z are $dr, rd\theta, dz$. Now $dr = 0$ since r is a constant on S and $dS = (rd\theta)(dz) = 2\,d\theta\,dz$.

But $$\mathbf{n} = \frac{\nabla\phi}{|\nabla\phi|} = \frac{\nabla(x^2+y^2-4)}{|\nabla(x^2+y^2-4)|} = \frac{x\mathbf{i}+y\mathbf{j}}{2}.$$

On S_1, $$\mathbf{A}\cdot\mathbf{n} = 2x^2 - y^3 = 8\cos^2\theta - 8\sin^3\theta.$$

$\therefore$ $$\iint_{S_1} \mathbf{A}\cdot\mathbf{n}\,dS = \int_{\theta=0}^{2\pi}\int_{z=0}^{3} 8(\cos^2\theta - \sin^3\theta)(2dz\,d\theta)$$

$$= 48\int_{\theta=0}^{2\pi}(\cos^2\theta - \sin^3\theta)\,d\theta$$

$$= 48(\pi - 0) = 48\pi.$$

But $$\iint_{S_2} \mathbf{A}\cdot\mathbf{n}\,dS = \iint_{S_2}(4x\mathbf{i} - 2y^2\mathbf{j} + z^2\mathbf{k})\cdot(-k)\,dxdy = 0,$$

$$\iint_{S_3} \mathbf{A}\cdot\mathbf{n}\,dS = \iint_{S_3}(4x\mathbf{i} - 2y^2\mathbf{j} + 3^2\mathbf{k}).(\mathbf{k})\,dxdy$$

$$= 9\iint_{S_3} dx\,dy = 9(\pi 2^2) = 36\pi.$$

So the surface integral on S equals $48\pi + 0 + 36\pi = 84\pi$.

44. Evaluate $\iint_S \mathbf{A}\cdot\mathbf{n}\,dS$ if $\mathbf{A} = y\mathbf{i} - x\mathbf{j} + z\mathbf{k}$ and S is the surface of the region bounded by the cylinder $x^2 + z^2 - 9 = 0$ and planes $x = 0, y = 0, z = 0$, and $y = 4$. **[Ans. 9π]**

45. Evaluate $\iint_S \mathbf{A}\cdot\mathbf{n}\,dS$ if $\mathbf{A} = y\mathbf{i} - x\mathbf{j} + z\mathbf{k}$.

and S is the surface of the cylinder $x^2 + z^2 - 9 = 0$ in the first octant bounded by the plane $y=4$. **[Ans. $24 + 9\pi$]**

46. *Solved sum.* Find the area of the curved surface of the region common to the cylinders $x^2 + y^2 = a^2$ and $x^2 + z^2 = a^2$ contained in the first octant.

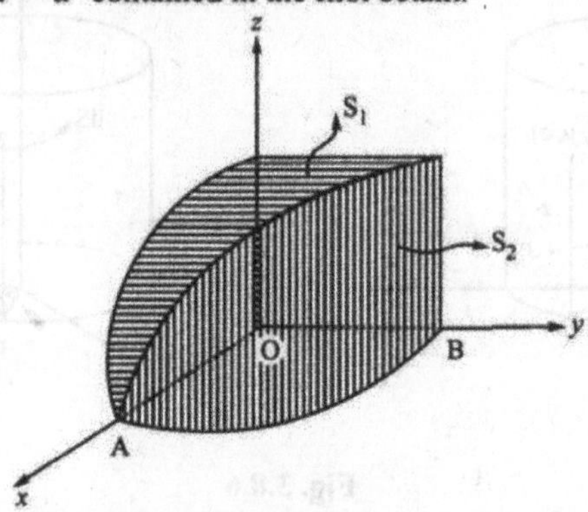

Fig. 3.8.7

Let the curved surface belonging to the cylinder $x^2 + z^2 = a^2$ be S_1 and that belonging to the cylinder $x^2 + y^2 = a^2$ be S_2 (Fig. 3.8.7). Now, in the figure, the projection of S_1 on the xy plane is OAB. For the surface S_1

$$n = \frac{\nabla(x^2 + z^2 - a^2)}{|\nabla(x^2 + z^2 - a^2)|} = \frac{x\mathbf{i} + z\mathbf{k}}{\sqrt{x^2 + z^2}} = \frac{x\mathbf{i} + z\mathbf{k}}{a}.$$

$$\text{Area of } S_1 = \iint_{S_1} dS = \iint_{OAB} \frac{dx\,dy}{\mathbf{n}\cdot\mathbf{k}} = \iint_{OAB} \frac{a}{z}\,dx\,dy$$

$$= a\int_0^a \int_0^{\sqrt{a^2-x^2}} \frac{1}{\sqrt{a^2 - x^2}}\,dy\,dx$$

$$= a\int_0^a \left[\frac{1}{\sqrt{a^2 - x^2}}\,y\right]_0^{\sqrt{a^2-x^2}} dx$$

$$= a\int_0^a 1\,dx = a^2.$$

By symmetry, area of S_1 = area of S_2. So the required area $= 2a^2$.

***(iii)* SPHERICAL SURFACE**

47. Evaluate $\iint_S \mathbf{r}\cdot\mathbf{n}\,dS$ if S is the surface of the sphere $x^2+y^2+z^2-a^2=0$. **[Ans.** $4\pi a^2$**]**

***(iv)* HEMISPHERICAL SURFACE**

48. *Solved sum.* Evaluate $\iint_S \mathbf{A}\cdot\mathbf{n}\,dS$, if $\mathbf{A}=x\mathbf{i}+y\mathbf{j}-2z\mathbf{k}$ and S is the surface of the sphere $x^2+y^2+z^2=a^2$ above the xOy plane. **(1977, 84)**

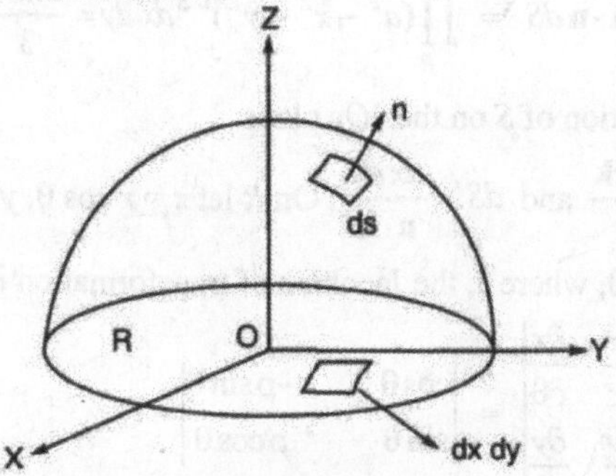

Fig. 3.8.8

If **n** is the unit vector normal to the elementary surface area of the sphere $x^2+y^2+z^2=a^2$ (Fig 3.8.8), then

$$\mathbf{n}=\frac{\nabla(x^2+y^2+z^2-a^2)}{|\nabla(x^2+y^2+z^2-a^2)|}=\frac{2(x\mathbf{i}+y\mathbf{j}+z\mathbf{k})}{2\sqrt{x^2+y^2+z^2}}=\frac{x\mathbf{i}+y\mathbf{j}+z\mathbf{k}}{a}$$

$$dS=\frac{dx\,dy}{|\mathbf{n}\cdot\mathbf{k}|}=\frac{dx\,dy}{(z/a)}=\frac{a\,dx\,dy}{\sqrt{a^2-x^2-y^2}}.$$

On S,
$$\mathbf{A}\cdot\mathbf{n}=\frac{x^2+y^2-2z^2}{a}=\frac{x^2+y^2-2(a^2-x^2-y^2)}{a}$$

$$=\frac{3x^2+3y^2-2a^2}{a}$$

$$\iint_S \mathbf{A}\cdot\mathbf{n}\,dS=\iint_R \frac{3(x^2+y^2)(x^2+y^2)-2a^2}{\sqrt{a^2-(x^2+y^2)}}\,dx\,dy,$$

where R is the circular area which is the projection of S on xOy plane.

$$=\int_{r=0}^{a}\int_{\theta=0}^{2\pi}\frac{3r^2-2a^2}{\sqrt{a^2-r^2}}\,r\,dr\,d\theta,$$

using polar coordinates so that $x=r\cos\theta$, $y=r\sin\theta$

$$=\int_0^a \frac{3r^2-2a^2}{\sqrt{a^2-r^2}}\,[\theta]_0^{2\pi}\,r\,dr$$

$$=-2\pi\int_a^0[3(a^2-t^2)-2a^2]\,dt,\ \text{if } a^2-r^2=t^2$$

$$= 2\pi\int_0^a (a^2-3t^2)\,dt = 2\pi[a^3-a^3] = 0.$$

Remark. This sum is solved in the next chapter by using divergence theorem.

49. S is the surface of the hemisphere $x^2+y^2+z^2=a^2, z\ge 0$ and $\mathbf{A}=z\mathbf{k}$. Show that

$$\iint \mathbf{A}\cdot\mathbf{n}\,dS = \iint_R (a^2-x^2-y^2)^{1/2}\,dx\,dy = \frac{2\pi a^3}{3},$$

where R is the projection of S on the xOy plane.

[Hint: $\mathbf{n}=\dfrac{x\mathbf{i}+y\mathbf{j}+z\mathbf{k}}{a}$ and $dS\ \dfrac{dx\,dy}{\mathbf{n}\cdot\mathbf{k}}$. On R let $x=r\cos\theta, y=r\sin\theta$.

Since $dx\,dy = J\,dr\,d\theta$, where J, the Jacobian of transformation is

$$\begin{vmatrix}\dfrac{\partial x}{\partial r} & \dfrac{\partial x}{\partial\theta}\\[2mm] \dfrac{\partial y}{\partial r} & \dfrac{\partial y}{\partial\theta}\end{vmatrix} = \begin{vmatrix}\cos\theta & -\rho\sin\theta\\ \sin\theta & \rho\cos\theta\end{vmatrix} = r$$

Then the integral is $\displaystyle\int_{r=0}^{a}\int_{\theta=0}^{2\pi}(a^2-r^2)^{1/2}\,r\,d\theta\,dr = 2\pi\int_0^a (a^2-r^2)^{1/2}\,r\,dr.$]

50. Find the area of the surface S of the hemisphere $x^2+y^2+z^2=a^2, z\ge 0$.

[Hint:
$$\iint_S dS = \iint_{Rxy} 1\,\frac{dx\,dy}{\mathbf{n}\cdot\mathbf{k}}$$
$$= a\iint_{Rxy}\frac{1}{\sqrt{a^2-x^2-y^2}}\,dx\,dy = 2\pi a^2\ ]$$

(v) PARABOLIC CYLINDER

51. *Solved sum.* Evaluate $\iint_S \mathbf{A}\cdot\mathbf{n}\,dS$ if $\mathbf{A}=y\mathbf{i}-x\mathbf{j}+z\mathbf{k}$ and if S is the surface of the parabolic cylinder $y^2-4x=0$ in the first octant bounded by the planes $x=4$ and $z=3$.

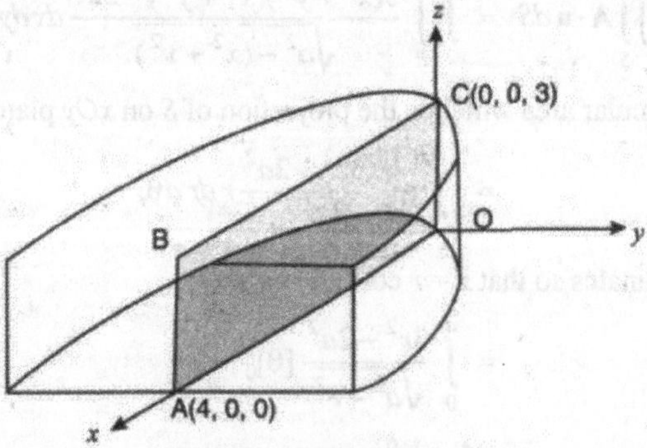

Fig. 3.8.9

Now,
$$\mathbf{n} = \frac{\nabla(y^2-4x)}{|\nabla(y^2-4x)|} = \frac{2(-2\mathbf{i}+y\mathbf{j})}{2\sqrt{4+y^2}}.$$

$$dS = \frac{dx\,dz}{|\mathbf{n}\cdot\mathbf{j}|} = \frac{\sqrt{4+y^2}}{y}dx\,dz$$

$$\mathbf{A.n}\,dS = (y\mathbf{i} - x\mathbf{j} + z\mathbf{k}) \cdot \frac{-2\mathbf{i}+y\mathbf{j}}{\sqrt{4+y^2}} \cdot \frac{\sqrt{4+y^2}}{y}dx\,dz$$

$$= -(2+x)\,dx\,dz$$

Now the projection of *S* on *xOz* plane is the rectangle *OABC* (Fig. 3.8.9) in which x varies from 0 to 4 and z varies from 0 to 3.

$$\therefore \qquad \iint_S \mathbf{A}\cdot\mathbf{n}\,dS = -\int_{x=0}^{4}\int_{z=0}^{3}(2+x)\,dz\,dx$$

$$= -3\int_0^4 (2+x)\,dx = -3\left[2x+\frac{x^2}{2}\right]_0^4 = -48.$$

52. *Solved sum.* **Evaluate $\iint_S \mathbf{A}\cdot\mathbf{n}\,dS$ if $\mathbf{A} = y\mathbf{i} - x\mathbf{j} + z\mathbf{k}$ and *S* is the surface of the parabolic cylinder $z^2 - 4x = 0$ in the first octant bounded by the planes $x = 3$ and $y = 4$.**

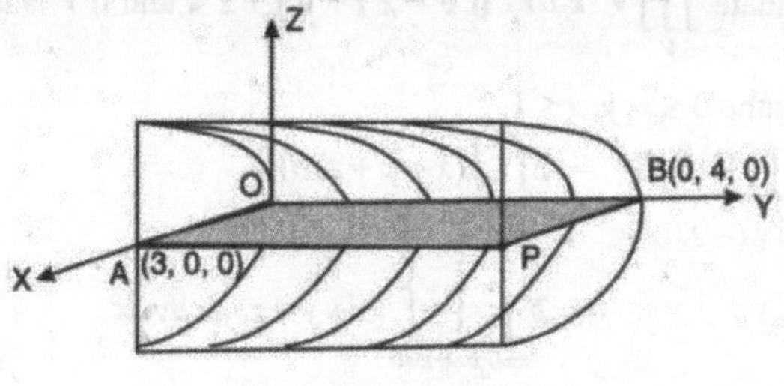

Fig. 3.8.10

Now
$$\mathbf{n} = \frac{-2\mathbf{i}+z\mathbf{k}}{\sqrt{4+z^2}}.$$

$$dS = \frac{dx\,dy}{|\mathbf{n}\cdot\mathbf{k}|} = \frac{\sqrt{4+z^2}}{2}x\,dy.$$

$$\mathbf{A.n}\,dS = \frac{-2y+z^2}{z}dx\,dy = \frac{-2y+4x}{2\sqrt{x}} \qquad (\because \text{ on } S,\ z^2 = 4x)$$

Further the projection of *S* on the *xOy* plane is the rectangle *OAPB* in which *x* varies from 0 to 3 and *y* varies from 0 to 4 (Fig. 3.8.10).

$$\therefore \qquad \iint_S \mathbf{A}\cdot\mathbf{n}\,dS = \int_{x=0}^{3}\int_{y=0}^{4}\frac{-2y+4x}{2\sqrt{x}}dy\,dx$$

$$= \frac{1}{2}\int_0^3\int_0^4(-2x^{-1/2}y + 4x^{1/2})\,dy\,dx$$

$$= \frac{1}{2}\int_0^3\left\{-2x^{-1/2}\left(\frac{4^2}{2}\right)+4x^{1/2}(4)\right\}dx$$

$$= 8\int_0^3 (-x^{-1/2} + x^{1/2})\,dx$$

$$= 8\left[-2x^{1/2} + \frac{2}{3}x^{3/2}\right]_0^3 = 0.$$

53. A surface S has a projection R on the xOy plane. Show that the area of S is

$$\iint_R \left\{1 + \left(\frac{\partial z}{\partial x}\right)^2 + \left(\frac{\partial z}{\partial y}\right)^2\right\}^{1/2} dx\,dy$$

or

$$\iint_R \left[\left\{\left(\frac{\partial F}{\partial x}\right)^2 + \left(\frac{\partial F}{\partial y}\right)^2 + \left(\frac{\partial F}{\partial z}\right)^2\right\}^{1/2} \bigg/ \left|\frac{\partial F}{\partial z}\right|\right] dx\,dy$$

according as the equation of S is

$$z = f(x, y) \text{ or } F(x, y, z) = 0.$$

3.6 Volume integral

54. *Solved sum.* Evaluate $\iiint_V \nabla\cdot\mathbf{F}\,dV$ if $\mathbf{F} = x^2\mathbf{i} + y^2\mathbf{j} + z^2\mathbf{k}$ and if V is the volume of the region enclosed by the cube $0 \le x, y, z \le 1$. **(1981)**

$$\iiint_V \nabla\cdot\mathbf{F}\,dV = 2\iiint_V (x+y+z)\,dV$$

$$= 2\int_{x=0}^{1}\int_{y=0}^{1}\int_{z=0}^{1} (x+y+z)\,dz\,dy\,dx$$

$$= 2\int_{x=0}^{1}\int_{y=0}^{1} \left(x+y+\frac{1}{2}\right)dy\,dx$$

$$= 2\int_{x=0}^{1} \left(x+\frac{1}{2}+\frac{1}{2}\right)dx = 3.$$

Second method. If V is the volume of the cube and $(\bar{x}, \bar{y}, \bar{z})$ is the mass centre of the cube, then

$$V = 1 \text{ and } \bar{x} = \bar{y} = \bar{z} = \frac{1}{2}.$$

Therefore

$$\iiint_V \nabla\cdot\mathbf{F}\,dV = 2\left[\iiint_V x\,dV + \iiint_V y\,dV + \iiint_V z\,dV\right]$$

$$= 2(V\bar{x} + V\bar{y} + V\bar{z}) = 3.$$

55. *Solved sum.* Evaluate $\iiint_V 45x^2y\,dV$, where V is the closed region bounded by the planes $x = 0, y = 0, z = 0, 4x + 2y + z = 8$.

Considering the x coordinates of the points in the region, we see that, the region is bounded by the planes

$$x = 0, x = 2.$$

Considering the y coordinates, we see that the region is further bounded by the planes

$$y = 0, 4x + 2y = 8$$

i.e., $$y = 0, y = 2(2 - x).$$

Considering the z coordinates, we see that the region is further bounded by the planes

$$z = 0, 4x + 2y + z = 8$$

i.e., $$z = 0, z = 4(2 - x) - 2y.$$

$\therefore$ $$\mathbf{I} = \iiint_V \phi\, dV = \iiint_V \phi\, dz\, dy\, dx = \iiint_V 45x^2 y\, dz\, dy\, dx$$

$$= 45 \int_{x=0}^{x=2} x^2 \int_{y=0}^{y=2(2-x)} y \int_{z=0}^{z=4(2-x)-2y} dz\, dy\, dx$$

$$= 45 \int_{x=0}^{x=2} x^2 \int_{y=0}^{y=2(2-x)} y[z]_0^{4(2-x)-2y}\, dy\, dx$$

$$= 45 \int_{x=0}^{x=2} x^2 \int_{y=0}^{y=2(2-x)} y[4(2-x) - 2y]\, dy\, dx$$

$$= 45 \int_{x=0}^{x=2} x^2 \int_{y=0}^{y=2(2-x)} \left[4(2-x)y - 2y^2\right] dy\, dx$$

$$= 45 \int_{x=0}^{x=2} x^2 \left[2(2-x)y^2 - \frac{2y^3}{3}\right]_0^{2(2-x)} dx$$

$$= 45 \int_{x=0}^{x=2} x^2 \left[8(2-x)^3 - \frac{16}{3}(2-x)^3\right] dx$$

$$= 45 \int_0^2 x^2 \cdot \frac{8(2-x)^3}{3}\, dx \qquad \text{... (1)}$$

$$= 120 \int_0^2 x^2 (8 - 12x + 6x^2 - x^3)\, dx$$

$$= 120 \left[\frac{8x^3}{3} - 3x^4 + \frac{6x^5}{5} - \frac{x^6}{6}\right]_0^2$$

$$= 120 \left[\frac{64}{3} - 48 + \frac{192}{5} - \frac{32}{3}\right]$$

$$= 120 \times 16 \left[\frac{4}{3} - 3 + \frac{12}{5} - \frac{2}{3}\right]$$

$$= 120 \times 16 \times \frac{1}{15} = 128.$$

Note. Alternatively, from (1), we have

$$\mathbf{I} = 120 \int_0^2 (2-x)^3 x^2\, dx.$$

Using Bernoulli's formula, we get

$$\mathbf{I} = 120\left[\frac{(2-x)^4}{4(-1)}(x^2)-\frac{(2-x)^5}{4\times5}(2x)+\frac{(2-x)^6}{4\cdot5\cdot6(-1)}(2)\right]_0^2$$

$$= 120\left[0-0+0-0+0+\frac{2^7}{4\cdot5\cdot6}\right]=2^7=128.$$

56. *Solved sum.* Evaluate $\iiint_V \mathbf{F}\,dV$, where $\mathbf{F} = 2xz\mathbf{i} - x\mathbf{j} + y^2\mathbf{k}$ and V is the volume of the region enclosed by the cylinder $x^2 + y^2 = a^2$ between the planes $z = 0, z = c$. (1974)

$$\iiint_V \mathbf{F}dV = 2\mathbf{i}\iiint_V xz\,dV + -\mathbf{j}\iiint_V x\,dV + \mathbf{k}\iiint_V y^2 dV$$

We shall denote the integrals on the right by I_1, I_2, I_3. Then, using cylindrical coordinates,

$$I_1 = 2\int_{r=0}^{a}\int_{\theta=0}^{2\pi}\int_{z=0}^{c} r\cos\theta\, z(r\,dz\,d\theta\,dr)$$

$$= 2\int_{r=0}^{a}\int_{\theta=0}^{2\pi} r^2\cos\theta\left[\frac{1}{2}z^2\right]_0^c d\theta\,dr$$

$$= 2\int_{r=0}^{a} r^2\left(\frac{1}{2}C^2\right)[\sin\theta]_0^{2\pi}\,dr = 0$$

$$I_2 = \int_{r=0}^{a}\int_{\theta=0}^{2\pi}\int_{z=0}^{c} r\cos\theta(r\,dz\,d\theta\,dr)$$

$$= \left[\frac{1}{3}r^3\right]_0^a[\sin\theta]_0^{2\pi}[z]_0^c = 0 \qquad (\text{Also } I_2 = V\bar{x} = V(0) = 0)$$

$$I_3 = \int_{r=0}^{a}\int_{\theta=0}^{2\pi}\int_{z=0}^{c} r^2\sin^2\theta(r\,dz\,d\theta\,dr)$$

$$= \left[\frac{1}{4}r^4\right]_0^a\left[\frac{1}{2}\theta-\frac{\sin 2\theta}{4}\right]_0^{2\pi}[z]_0^c = \frac{a^4c\pi}{4}.$$

Therefore the integral is $0\mathbf{i} + 0\mathbf{j} + \frac{1}{4}a^4c\pi\mathbf{k} = \frac{1}{4}a^4c\pi\mathbf{k}$.

57. Evaluate $\iiint_V \nabla\cdot\mathbf{A}\,dV$ *if* $\mathbf{A} = 2x^2y\mathbf{i} - y^2\mathbf{j} + 4xz^2\mathbf{k}$ and V is the region in the first octant bounded by the cylinder $y^2 + z^2 = 9$ and the plane $x = 2$. (1983)

[**Hint:** x varies from 0 to 2. Expressing y and z in r and θ, $y = r\cos\theta$ and $z = r\sin\theta$, where $0 \le r \le 3$ and $0 \le \theta \le \frac{1}{2}\pi$. Since $dV = dx(r\,dr\,d\theta)$. the integral is

$$\int_0^2\int_0^3\int_0^{\frac{1}{2}\pi}(4xr\cos\theta - 2r\cos\theta + 8xr\sin\theta)\,r d\theta\,dr\,dx$$

$$= \int_0^2\int_0^3 (4xr^2 - 2r^2 + 8xr^2)\,dr\,dx$$

$$= \int_0^2 (36x - 18 + 72x)\,dx = 180.]$$

58. ***Solved sum.*** **Evaluate the volume integral over the region common to the cylinders $x^2+y^2=a^2$ and $x^2+z^2=a^2$ contained in the first octant if $\mathbf{A}=xy\mathbf{i}-3y^2z\mathbf{j}$:**

$$\iiint_V \nabla\times\mathbf{A}\,dV.$$

Now $\nabla\times\mathbf{A}=3y^2\mathbf{i}-x\mathbf{k}$. So from Fig. 3.8.7, the integral is

$$\mathbf{i}\int_{x=0}^{a}\int_{y=0}^{\sqrt{a^2-x^2}}\int_{z=0}^{\sqrt{a^2-x^2}} 3y^2\,dz\,dy\,dx - k\int_{x=0}^{a}\int_{y=0}^{\sqrt{a^2-x^2}}\int_{z=0}^{\sqrt{a^2-x^2}} x\,dz\,dy\,dx$$

$$= \mathbf{i}\int_0^a\int_{y=0}^{\sqrt{a^2-x^2}} 3y^2\sqrt{a^2-x^2}\,dy\,dx - \mathbf{k}\int_0^a\int_{y=0}^{\sqrt{a^2-x^2}} x\sqrt{a^2-x^2}\,dy\,dx$$

$$= \mathbf{i}\int_0^a (a^2-x^2)^2\,dx - k\int_0^a x(a^2-x^2)\,dx$$

$$= \frac{8a^5}{15}\mathbf{i} - \frac{a^4}{4}\mathbf{k}.$$

CHAPTER

4

Integral Theorems

4.1. Integral theorems

The study of line integrals, surface integrals and volume integrals provides the ground for the study of the following integral theorems:

(*i*) Green's theorem in the plane, (*ii*) Gauss' divergence theorem,

(*iii*) Stoke's theorem.

In the previous chapter we studied integration of scalar and vector point functions. This provides the ground for the study of two central theorems on integration, namely,

(*i*) Gauss' divergence theorem, (*ii*) Stokes's theorem

The divergence theorem enables us to convert a surface integral on a closed surface into a volume integral and the Stokes' theorem enable us to convert a line integral into a surface integral. Such a conversion often simplifies the evaluation of the original integral.

4.2. Gauss' divergence theorem

Book work 4.1. (Gauss' divergence theorem). If V is the volume of a closed surface S and **A**, a vector point function with continous derivatives in V, then

$$\iint_S \mathbf{A}\cdot\hat{\mathbf{n}}\,dS = \iiint_V \nabla\cdot\mathbf{A}\,dV \qquad (4.2.1)$$

Where $\hat{\mathbf{n}}$ is the unit normal to the surface S outwardly.

Gauss' theorem in words. We know that, if $\hat{e}$ is a unit vector, then $\bar{A}\cdot\hat{e}$ is the component of $\bar{A}$ in the direction of $\hat{e}$. Now $\bar{A}\cdot\hat{n}$ is the component of $\bar{A}$ in the direction normal to S. Thus the Gauss's theorem, in words, is

The surface integral of the normal component of $\bar{A}$ taken over the closed surface S is equal to the volume integral of divergence of $\bar{A}$ taken over the volume enclosed by S.

To prove this we first assume that S is a closed surface such that no line parallel to a coordinate axis intersects it in more than two points (Fig. 4.2.1). Let the cylinder whose generators are parallel to the z-axis and which envelops the surface S, touch S along the curve C′ and intersect the xy plane along the curve C. Now the curve C′ divides the surface S into two parts, say S_1 and S_2, whose equations are $z = f_1(x, y)$, $z = f_2(x, y)$ as shown in the figure. If $\mathbf{A} = A_1\mathbf{i} + A_2\mathbf{j} + A_3\mathbf{k}$, and R is the area enclosed by C in the xy plane, then considering the volume integral of $\frac{\partial A_3}{\partial z}$ over the region enclosed by S, we get

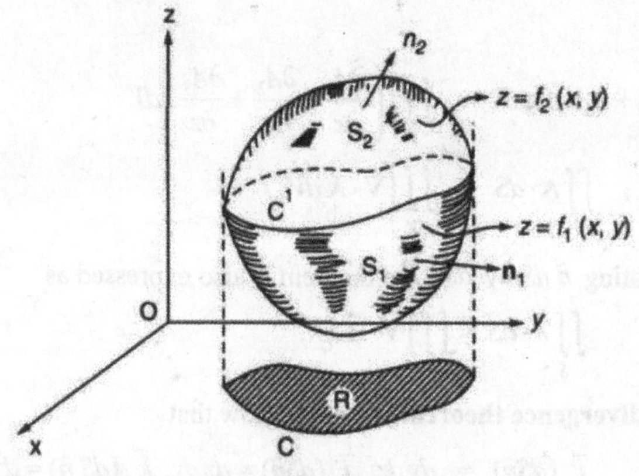

Fig. 4.2.1

$$\iiint_V \frac{\partial A_3}{\partial z} dV = \iint_R \left[\int_{z=f_1(x,y)}^{z=f_2(x,y)} \frac{\partial A_3}{\partial z} dz \right] dx\, dy$$

$$= \iint_R \left[A_3(x,y,z) \Big|_{z=f_1(x,y)}^{z=f_2(x,y)} \right] dx\, dy$$

$$= \iint_R [A_3(x,y,f_2(x,y)) - A_3(x,y,f_1(x,y))] dx\, dy \quad \text{... (4.2.2)}$$

Also, if the unit normals to the surface S at points respectively on S_1 and S_2 are denoted by $\mathbf{n}_1$ and $\mathbf{n}_2$, then

$$\iint_S A_3(x,y,z)\mathbf{k} \cdot d\mathbf{S} = \iint_S A_3(x,y,z)\mathbf{k} \cdot \mathbf{n}\, dS$$

$$= \iint_{S_1} A_3(x,y,z)\mathbf{k} \cdot \mathbf{n}_1 dS + \iint_{S_2} A_3(x,y,z)\mathbf{k} \cdot \mathbf{n}_2\, dS$$

$$= \iint_R A_3(x,y,f_1(x,y))(-1)\, dx\, dy + \iint_R A_3(x,y,f_2(x,y))\, dx\, dy$$

since $\quad \mathbf{k}.\mathbf{n}_2\, dS = \cos\theta_2\, dS = dx\, dy$

$\quad\quad\quad \mathbf{k}.\mathbf{n}_1\, dS = \cos\theta_1\, dS = -\, dx\, dy,$

the angle θ_1 between $\mathbf{k}$ and $\mathbf{n}_1$ being obtuse.

$$\therefore \quad \iint_S A_3(x,y,z)\mathbf{k} \cdot d\mathbf{S} = \iint_R [A_3(x,y,f_2(x,y)) - A_3(x,y,f_1(x,y))] dx\, dy \quad \text{(4.2.3.)}$$

Thus, from (4.2.2) and (4.2.3).

$$\iint_S (A_3\mathbf{k}) \cdot \mathbf{n}\, dS = \iiint_V \frac{\partial A_3}{\partial z} dV.$$

Similarly $\quad \iint_S (A_1\mathbf{i}) \cdot \mathbf{n} dS = \iiint_V \frac{\partial A_1}{\partial x} dV,$

$$\iint_S (A_2\mathbf{j}) \cdot \mathbf{n}\, dS = \iiint_V \frac{\partial A_2}{\partial y} dV.$$

Adding,

$$\iint_S (A_1\mathbf{i}+A_2\mathbf{j}+A_3\mathbf{k})\cdot\mathbf{n}\,dS = \iiint_V \left(\frac{\partial A_1}{\partial x}+\frac{\partial A_2}{\partial y}+\frac{\partial A_3}{\partial z}\right)dV$$

or

$$\iint_S \mathbf{A}\cdot d\mathbf{S} = \iiint_V \nabla\cdot\mathbf{A}\,dV.$$

Remark 1. Denoting $\hat{n}\,dS$ by $d\bar{S}$, the theorem is also expressed as

$$\iint_S \bar{A}\cdot d\bar{S} = \iiint_V \nabla\cdot\bar{A}\,dV.$$

Scalar form of divergence theorem. Also we know that

$$\bar{i}\cdot(dS\hat{n}) = dy\,dz,\ \bar{j}\cdot(dS\hat{n}) = dz\,dx,\ \bar{k}\cdot(dS\,\hat{n}) = dx\,dy.$$

So the theorem is expressed in scalar form as

$$\iint_S A_1 dy\,dz + A_2 dz\,dx + A_3\,dx\,dy$$

$$= \iiint_V \left(\frac{\partial A_1}{\partial x}+\frac{\partial A_2}{\partial y}+\frac{\partial A_3}{\partial z}\right)dx\,dy\,dz.$$

Remark. The result can be extended to surfaces which are intersected by a straight line parallel to a coordinate axis in more than two points, by subdividing the regions by plane surfaces into regions whose surfaces will be intersected by straight lines parallel to the coordinate axis in two points only (vide Green's theorem in plane, section 4.4).

4.3. Integral theorems derived from the divergence theorem

Now we prove the following two results as the immediate consequences of the divergence theorem :

(*i*) $\iint_S \phi\,dS = \iiint_V \nabla\phi\,dV,$ ... (4.3.1)

(*ii*) $\iint_S \mathbf{A}\times d\mathbf{S} = -\iiint_V \nabla\times\mathbf{A}\,dV,$ ... (4.3.2)

where ϕ is a scalar point function defined in the region V enclosed by the closed surface S.

(*i*) Let **a** be an arbitrary constant vector. Now, choosing $\phi\mathbf{a}$ for **A** in (4.2.1), we get

$$\iint_S (\phi\mathbf{a})\cdot d\mathbf{S} = \iiint_V \nabla\cdot(\phi\mathbf{a})\,dV$$

$$= \iiint_V \{(\nabla\phi)\cdot\mathbf{a}+\phi(\nabla\cdot\mathbf{a})\}dV \text{ by } (2.7.1)$$

$$= \iiint_V (\nabla\phi)\cdot\mathbf{a}\,dV + 0$$

since **a** is constant vector.

$$\therefore \quad \left(\iint_S \phi\,d\mathbf{S}\right)\cdot\mathbf{a} - \left(\iiint_V (\nabla\phi)\,dV\right)\cdot\mathbf{a} = 0$$

or

$$\left(\iint_S \phi\,d\mathbf{S} - \iiint_V \nabla\phi\,dV\right)\cdot\mathbf{a} = 0.$$

Since **a** is arbitrary,

$$\iint_S \phi d\mathbf{S} - \iiint_V \nabla\phi dV = 0$$

or

$$\iint_S \phi d\mathbf{S} = \iiint_V \nabla\phi dV.$$

(*ii*) Next, replacing **A** with **a** × **A** in (4.2.1), we get

$$\iint_S (\mathbf{a}\times\mathbf{A})\cdot d\mathbf{S} = \iiint_V \nabla\cdot(\mathbf{a}\times\mathbf{A})\, dV.$$

Interchanging the dot and cross on the left side and using (2.7.4) on the right side,

$$\iint_S \mathbf{a}\cdot(\mathbf{A}\times d\mathbf{S}) = \iiint_V [\mathbf{A}\cdot(\nabla\times\mathbf{a}) - \mathbf{a}\cdot(\nabla\times\mathbf{A})]dV.$$

$$\therefore \quad \iint_S \mathbf{a}\cdot(\mathbf{A}\times d\mathbf{S}) = 0 - \iiint_V \mathbf{a}\cdot(\nabla\times\mathbf{A})dV$$

or

$$\mathbf{a}\cdot\left[\iint_S \mathbf{A}\times d\mathbf{S} + \iiint_V (\nabla\times\mathbf{A})dV\right] = 0.$$

But **a** is arbitrary. Therefore

$$\iint_S \mathbf{A}\times d\mathbf{S} + \iiint_V \nabla\times\mathbf{A}\, dV = 0$$

or

$$\iint_S A\times d\mathbf{S} = -\iiint_V (\nabla\times\mathbf{A})\, dV.$$

4.4 Green's theorem in plane.

Book work 4.2 **(Green's theorem in plane).** If C is a simple closed curve in the xOy plane bounding an area R and $M(x, y)$ and $N(x, y)$ are continuous functions of x and y having continuous derivatives in R, then

$$\oint Mdx + Ndy = \iint_R \left(\frac{\partial N}{\partial x} - \frac{\partial M}{\partial y}\right) dx\, dy, \qquad \text{... (4.4.1)}$$

where C is traced in the sense shown in Figs. 4.4.1 and 4.4.2.

In this result, LHS is a line integral along C and RHS is a double integral on the region R enclosed by C. The sense of orientation of C should be anticlockwise (counter clockwise).

It is to be noted that, for a man walking along C in the sense in which C is described, the area R lies to his left. To prove the theorem suppose that the equations of the arcs APB and AQB are respectively $y = f_1(x)$ and $y = f_2(x)$ (Fig. 4.4.1). Then

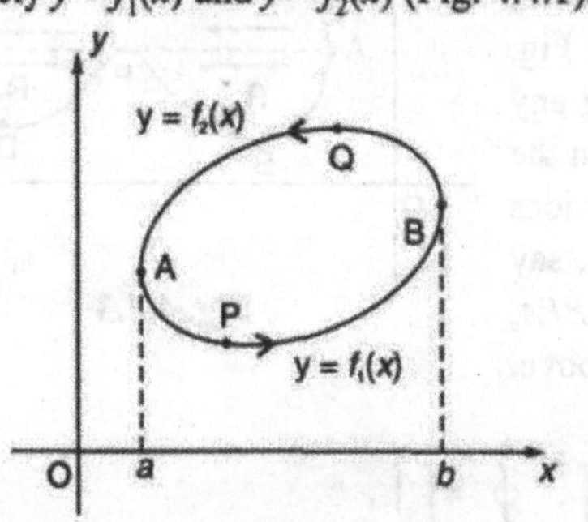

Fig. 4.4.1

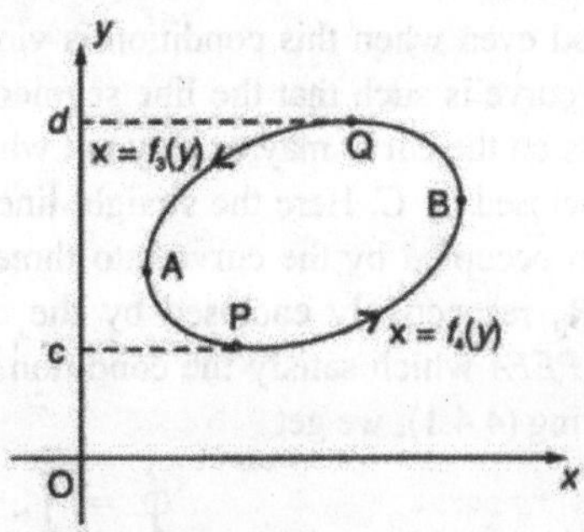

Fig. 4.4.2

$$\iint_R \left(-\frac{\partial M}{\partial y}\right) dx\, dy = \int_{x=a}^{x=b}\left[\int_{y=f_1(x)}^{y=f_2(x)}\left(-\frac{\partial M}{\partial y}\right) dy\right] dx$$

$$= \int_{x=a}^{x=b}\left[-M(x,y)\Big|_{y\ =\ f_1(x)}^{y\ =\ f_2(x)}\right] dx$$

$$= \int_{x=a}^{x=b} [M(x, f_1(x)) - M(x, f_2(x))] dx$$

$$= \int_{APB} M(x,y)\, dx - \int_{AQB} M(x,y)\, dx$$

$$= \int_{APB} M\, dx + \int_{BQA} M dx = \oint M dx. \qquad \text{... (4.4.2)}$$

Similarly, if the equations of PAQ and PBQ are $x = f_3(y)$ and $x = f_4(y)$ (Fig. 4.4.2), then

$$\iint_R \frac{\partial N}{\partial x} dx\, dy = \int_{y=c}^{y=d}\left[\int_{x=f_3(y)}^{x=f_4(y)} \frac{\partial N}{\partial x} dx\right] dy$$

$$= \int_{y=c}^{y=d}\left[N(x,y)\Big|_{x\ =\ f_3(y)}^{x\ =\ f_4(y)}\right] dy$$

$$= \int_{y=c}^{y=d} [N(f_4(y), y) - N(f_3(y), y)] dy$$

$$= \int_{PBQ} N(x,y)\, dy - \int_{PAQ} N(x,y)\, dy$$

$$= \int_{PBQ} N\, dy + \int_{QAP} N\, dy = \oint_C N\, dy. \qquad \text{... (4.4.3)}$$

Adding (4.4.2) and (4.4.3), we get the result as stated.

Remark 1. In the above work, in taking the equations of C to be $y = f_1(x)$, $y = f_2(x)$ or $x = f_3(y)$, $x = f_4(y)$, it is assumed that C encloses a region R which is such that the line segment joining any two points on C, lies wholly in R. However, it can be proved that the result (4.4.1) holds good even when this condition is violated. In Fig. 4.4.3 the curve is such that the line segment joining any two points on the curve may or may not wholly lie in the region enclosed by C. Here the straight line APE divides the region occupied by the curve into three regions, say R_1, R_2, R_3 respectively enclosed by the curves $ABPA$, $PDEP$, $APEFA$ which satisfy the condition stated above. So, applying (4.4.1), we get

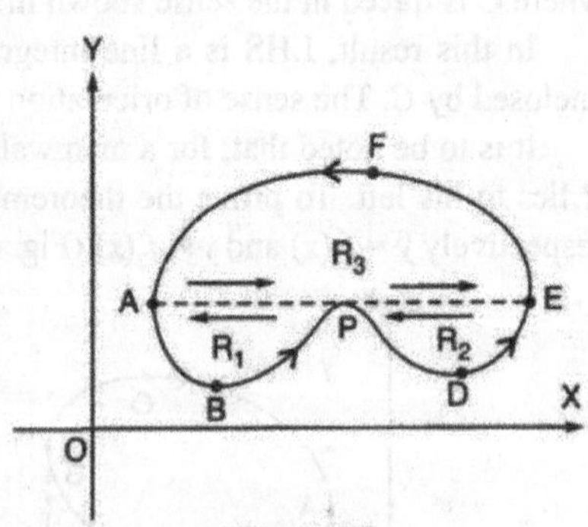

Fig. 4.4.3

$$\oint_{ABPA} = \iint_{R_1}, \quad \oint_{PDEP} = \iint_{R_2}, \quad \oint_{AEFA} = \iint_{R_3}.$$

Adding,

$$\oint_{ABPA} + \oint_{PDEP} + \oint_{AEFA} = \iint_{R_1} + \iint_{R_2} + \iint_{R_3} = \iint_{R}.$$

But

$$\oint_{ABPA} + \oint_{PDEP} + \oint_{AEFA} = \left[\int_{ABP} + \int_{PA}\right] + \left[\int_{PDE} + \int_{EP}\right]\left[\int_{AP} + \int_{PE} + \int_{EFA}\right]$$

$$= \int_{ABP} + \int_{PDE} + \int_{EFA} = \oint_C$$

since $\int_{PA} + \int_{AP} = 0$, etc.

4.5 Stokes' theorem

Book work 4.3 **(Strokes' theorem). If** ***S*** **is a surface bounded by a simple closed curve** ***C*****, then**

$$\oint_C \mathbf{A}\cdot d\mathbf{r} = \iint_S (\nabla\times\mathbf{A})\cdot d\mathbf{S} = \iint_S (\nabla\times\mathbf{A})\cdot\hat{\mathbf{n}}\, dS \qquad (4.5.1)$$

where A has continuous derivatives on S (Fig. 4.5.1) and $\hat{\mathbf{n}}$ is the unit vector normal to S.

Here the orientation of C on S is in the sense such that, for an insect which crawls on S along C in the sense of the orientation, S is to its left. Consider

$$\oint_C \mathbf{A}\cdot d\mathbf{r} = \oint_C (A_1\mathbf{i} + A_2\mathbf{j} + A_3\mathbf{k})\cdot(dx\mathbf{i} + dy\mathbf{j} + dz\mathbf{k})$$

$$= \oint_C A_1 dx + A_2 dy + A_3 dz$$

Note. If LHS $\int_C A_1 dx + A_2 dy + A_3\, dz$, then $\overline{A} = A_1\overline{i} + A_2\overline{j} + A_3\overline{k}$

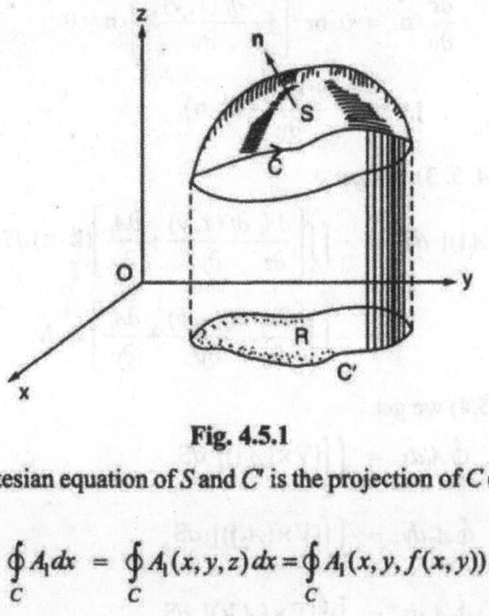

Fig. 4.5.1

If $z = f(x, y)$ is the cartesian equation of S and C' is the projection of C on the xy plane, enclosing a region R, then

$$\oint_C A_1 dx = \oint_C A_1(x, y, z)\, dx = \oint_C A_1(x, y, f(x, y))\, dx$$

$$= \oint_C M(x,y)\,dx, \text{ where } M = A_1(x, y, f(x,y))$$

$$= -\iint_R \frac{\partial M(x,y)}{\partial y} dx\,dy \text{ by (4.4.1)}$$

$$= -\iint_R \left[\frac{\partial A_1}{\partial y} + \frac{\partial A_1}{\partial z}\frac{\partial f(x,y)}{\partial y}\right] dx\,dy \qquad \text{... (4.5.2)}$$

since $M(x, y) = A_1(x, y, z)$ and $z = f(x, y)$, and thus

$$\frac{\partial M(x,y)}{\partial y} = \frac{\partial A_1}{\partial y} + \frac{\partial A_1}{\partial z}\frac{\partial f(x,y)}{\partial y}.$$

But $$\iint_S [\nabla \times (A_1\mathbf{i})] \cdot d\mathbf{S} = \iint_S \begin{vmatrix} \mathbf{i} & \mathbf{j} & \mathbf{k} \\ \frac{\partial}{\partial x} & \frac{\partial}{\partial y} & \frac{\partial}{\partial z} \\ A_1 & 0 & 0 \end{vmatrix} . d\mathbf{S}$$

$$= \iint_S \left(\frac{\partial A_1}{\partial z}\mathbf{j} - \frac{\partial A_1}{\partial y}\mathbf{k}\right) \cdot \mathbf{n}\,dS$$

$$= \iint_S \left[\frac{\partial A_1}{\partial z}(\mathbf{j}\cdot\mathbf{n}) - \frac{\partial A_1}{\partial y}(\mathbf{k}\cdot\mathbf{n})\right] dS. \qquad \text{... (4.5.3)}$$

Now the position vector of any point on S is

$$\mathbf{r} = x\mathbf{i} + y\mathbf{j} + z\mathbf{k} = x\mathbf{i} + y\mathbf{j} + f(x, y)\mathbf{k}$$

and $\frac{\partial \mathbf{r}}{\partial y}$, being tangential to S, is perpendicular to $\mathbf{n}$. Hence

$$\frac{\partial \mathbf{r}}{\partial y} \cdot \mathbf{n} = 0 \text{ or } \left[\mathbf{j} + \frac{\partial f(x,y)}{\partial y}\mathbf{k}\right] \cdot \mathbf{n} = 0.$$

Thus $$\mathbf{j.n} = -\frac{\partial f(x,y)}{\partial y}(\mathbf{k}\cdot\mathbf{n}).$$

Substituting this in (4. 5. 3), we get

$$\iint_S [\nabla \times (A_1\mathbf{i})] \cdot d\mathbf{S} = -\iint_S \left[\frac{\partial A_1}{\partial z}\frac{\partial f(x,y)}{\partial y} + \frac{\partial A_1}{\partial y}\right](\mathbf{k}\cdot\mathbf{n})\,dS$$

$$= -\iint_R \left[\frac{\partial A_1}{\partial z}\frac{\partial f(x,y)}{\partial y} + \frac{\partial A_1}{\partial y}\right] dx\,dy. \qquad \text{(4.5.4)}$$

From (4.5.2) and (4.5.4) we get

$$\oint_C A_1 dx = \iint_S [\nabla \times (A_1\mathbf{i})] \cdot d\mathbf{S}.$$

Similarly $$\oint_C A_2 dy = \iint_S [\nabla \times (A_2\mathbf{j})] \cdot d\mathbf{S},$$

$$\oint_C A_3 dz = \int_S\oint [\nabla \times (A_3\mathbf{k})] \cdot d\mathbf{S}.$$

By addition we get

$$\oint_C A_1 dx + A_2 dy + A_3 dz = \iint_S [\nabla \times (A_1\mathbf{i} + A_2\mathbf{j} + A_3\mathbf{k})] \cdot d\mathbf{S}$$

or

$$\oint_C \mathbf{A} \cdot d\mathbf{r} = \iint_S (\nabla \times \mathbf{A}) \cdot d\mathbf{S}.$$

Remark. It is noteworthy that in the above proof the surface S is taken in such a manner that **k . n** is positive. However, if S is chosen in the other way, then the orientation of the boundary is opposite to that of C.

4.6 Integral theorems derived from Stokes' theorem

Now we prove the following two results as the immediate consequences of Stokes' theorem:

(*i*)
$$\oint_C \phi\, d\mathbf{r} = -\iint_S (\nabla\phi) \times d\mathbf{S} \qquad \text{... (4.6.1)}$$

(*ii*)
$$\oint_C \mathbf{A} \times d\mathbf{r} = -\iint_S (\mathbf{n} \times \nabla) \times \mathbf{A}\, dS$$

$$= -\iint_S (d\mathbf{S} \times \nabla) \times \mathbf{A}. \qquad \text{... (4.6.2)}$$

Here C and S are a closed curve and a surface as in Stokes' theorem.

(*i*) Let **a** be any constant vector. Now choosing $\phi\mathbf{a}$ for **A** in (4.5.1), we get

$$\oint_C (\phi\mathbf{a}) \cdot d\mathbf{r} = \iint_S [\nabla \times (\phi\mathbf{a})] \cdot d\mathbf{S}$$

or

$$\oint_C \phi\, d\mathbf{r} \cdot \mathbf{a} = \iint_S [(\nabla\phi) \times \mathbf{a} + \phi(\nabla \times \mathbf{a})] \cdot d\mathbf{S} \text{ by (2.7.2)}$$

$$= \iint_S [(\nabla\phi) \times \mathbf{a}] \cdot d\mathbf{S} + 0 = \left(-\iint_S (\nabla\phi) \times d\mathbf{S} \right) \cdot \mathbf{a}.$$

But **a** is arbitrary. So

$$\oint_C \phi\, d\mathbf{r} = -\iint_S (\nabla\phi) \times d\mathbf{S}.$$

(*ii*) Next, replacing **A** with $\mathbf{a} \times \mathbf{A}$ in (4.5.1), we get

$$\oint_C (\mathbf{a} \times \mathbf{A}) \cdot d\mathbf{r} = \iint_S [\nabla \times (\mathbf{a} \times \mathbf{A})] \cdot d\mathbf{S}$$

$$= \iint_S [(\mathbf{A} \cdot \nabla)\mathbf{a} - (\mathbf{a} \cdot \nabla)\mathbf{A} + (\nabla \cdot \mathbf{A})\mathbf{a} - (\nabla \cdot \mathbf{a})\mathbf{A}] \cdot d\mathbf{S} \text{ by (2.7.3)}$$

$$= \iint_S [0 - (\mathbf{a} \cdot \nabla)\mathbf{A} + (\nabla \cdot \mathbf{A})\mathbf{a} + 0] \cdot d\mathbf{S}$$

$$= \iint_S [(\nabla \cdot \mathbf{A})\mathbf{a} - (\mathbf{a} \cdot \nabla)\mathbf{A}] \cdot d\mathbf{S}$$

$$= \iint_S [(\nabla \cdot \mathbf{A})\mathbf{a} - (\mathbf{a} \cdot \nabla)\mathbf{A}].\, \mathbf{n}\, dS$$

$$= \iint_S [(\mathbf{a} \cdot \mathbf{n})(\nabla \cdot \mathbf{A}) - (\mathbf{a} \cdot \nabla)(\mathbf{A} \cdot \mathbf{n})]\, dS$$

$$= \iint_S [\mathbf{a}\cdot\{\mathbf{n}(\nabla\cdot\mathbf{A})\} - \mathbf{a}\cdot\{\nabla(\mathbf{A}\cdot\mathbf{n})\}]dS$$

$$= \mathbf{a}.\iint_S [\mathbf{n}(\nabla\cdot\mathbf{A}) - \nabla(\mathbf{A}\cdot\mathbf{n})]dS$$

$$= \mathbf{a}\cdot\left[-\iint_S (\mathbf{n}\times\nabla)\times\mathbf{A}\right]d\mathbf{S}$$

$$= \mathbf{a}\cdot\left[-\iint_S (d\mathbf{S}\times\nabla)\times\mathbf{A}\right] \quad \text{... (4.6.3)}$$

since it can be verified that

(*i*) $(\mathbf{a}\,.\,\mathbf{n})\,(\nabla.\mathbf{A}) = \mathbf{a}\,.\,\{\mathbf{n}(\nabla\,.\,\mathbf{A})\}$

(*ii*) $(\mathbf{a}\,.\,\nabla)\,(\mathbf{A}\,.\,\mathbf{n}) = \mathbf{a}\,.\,\{\nabla\,(\mathbf{A}\,.\,\mathbf{n})\}$.

(*iii*) $\mathbf{n}(\nabla\,.\,\mathbf{A}) - \nabla(\mathbf{A}\,.\,\mathbf{n}) = -(\mathbf{n}\times\nabla)\times\mathbf{A}$.

$$\oint_C (\mathbf{a}\times\mathbf{A})\cdot d\mathbf{r} = \oint_C \mathbf{a}\cdot(\mathbf{A}\times d\mathbf{r}) = \mathbf{a}\cdot\oint_C \mathbf{A}\times d\mathbf{r}. \quad (4.6.4)$$

Since **a** is arbitrary, we get from (4.6.3) and (4.6.4) the result (4.6.2).

4.7 Operational meanings of ∇, ∇., ∇ × in terms of surface integrals

In Chapter II we used rectangular triad of unit vectors **i, j, k** and partial derivatives with respect to *x, y, z* to define the gradient, divergence and curl at a point. Now we proceed to show that these definitions can also be provided in terms of surface integrals as

$$(\nabla\phi)_P = \lim_{\Delta V\to 0}\frac{1}{\Delta V}\iint_{\Delta S}\phi\, dS, \quad \text{... (4.7.1)}$$

$$(\nabla\,.\,\mathbf{A})_P = \lim_{\Delta V\to 0}\frac{1}{\Delta V}\iint_{\Delta S}\mathbf{A}\cdot d\mathbf{S}, \quad \text{... (4.7.2)}$$

$$(\nabla\,.\,\mathbf{A})_P = -\lim_{\Delta V\to 0}\frac{1}{\Delta V}\iint_{\Delta S}\mathbf{A}\times d\mathbf{S}, \quad \text{... (4.7.3)}$$

where ϕ and **A** are defined in an infinitesimal region of volume ΔV around the point **P**, enclosed by ΔS and the limit is taken by shrinking the region to the point *P*.

First, we shall establish (4.7.2). For that, we use the divergence theorem on ΔV and obtain that

$$\iint_{\Delta S}\mathbf{A}\cdot d\mathbf{S} = \iiint_{\Delta V}\nabla\cdot\mathbf{A}\, dV. \quad \text{... (4.7.4)}$$

But, by mean value theorem of integrals, we see that

$$\iiint_{\Delta V}\nabla\cdot\mathbf{A}\, dV = \iiint_{\Delta V}\overline{\nabla\cdot\mathbf{A}}\, dV$$

$$= \overline{\nabla\cdot\mathbf{A}}\iiint_{\Delta V} dV = (\overline{\nabla\cdot A})\Delta V$$

where $\overline{\nabla\cdot\mathbf{A}}$ is some value $\nabla\,.\,\mathbf{A}$ lying between the minimum of $\nabla\,.\,\mathbf{A}$ and the maximum of $\nabla\,.\,\mathbf{A}$ in the volume ΔV. So (4.7.4) becomes

$$\overline{\nabla\cdot\mathbf{A}} = \frac{1}{\Delta V}\iint_{\Delta S}\mathbf{A}\cdot d\mathbf{S}.$$

Now, taking limit as $\Delta V \to 0$, in such a way that the region will always contain P, we get that, at P,

$$\nabla \cdot \mathbf{A} = \lim_{\Delta V \to 0} \frac{1}{\Delta V} \iint_{\Delta S} \mathbf{A} \cdot d\mathbf{S}. \tag{4.7.2}$$

Further, replacing **A** with $\phi\mathbf{a}$, where **a** is an arbitrary constant vector, we get

$$\nabla \cdot (\phi\mathbf{a}) = \lim_{\Delta V \to 0} \frac{1}{\Delta V} \iint_{\Delta S} (\phi\mathbf{a}) \cdot d\mathbf{S}$$

or
$$(\nabla\phi) \cdot \mathbf{a} + \phi(\nabla \cdot \mathbf{a}) = \left[\lim_{\Delta V \to 0} \frac{1}{\Delta V} \iint_{\Delta S} \phi d\mathbf{S}\right] \mathbf{a}$$

i.e.,
$$(\nabla\phi) \cdot \mathbf{a} + 0 = \left[\lim_{\Delta V \to 0} \frac{1}{\Delta V} \iint_{\Delta S} \phi \mathbf{S}\right] \cdot \mathbf{a}.$$

$$\nabla\phi = \lim_{\Delta V \to 0} \frac{1}{\Delta V} \iint_{\Delta S} \phi d\mathbf{S}. \tag{4.7.1}$$

Next, replacing **A** in (4.7.2) with $\mathbf{a} \times \mathbf{A}$, we get

$$\nabla \cdot (\mathbf{a} \times \mathbf{A}) = \lim_{\Delta V \to 0} \frac{1}{\Delta V} \iint_{\Delta S} (\mathbf{a} \times \mathbf{A}) \cdot d\mathbf{S}.$$

or
$$\mathbf{A} \cdot (\nabla \times \mathbf{a}) - \mathbf{a} \cdot (\nabla \times \mathbf{A}) = \lim_{\Delta V \to 0} \frac{1}{\Delta V} \iint_{\Delta S} (\mathbf{a} \times \mathbf{A}) \cdot d\mathbf{S}$$

i.e.,
$$0 - \mathbf{a} \cdot (\nabla \times \mathbf{A}) = \lim_{\Delta V \to 0} \frac{1}{\Delta V} \iint_{\Delta S} (\mathbf{a} \times \mathbf{A}) \cdot d\mathbf{S}$$

i.e.,
$$\mathbf{a} \cdot (-\nabla \times \mathbf{A}) = \lim_{\Delta V \to 0} \frac{1}{\Delta V} \iint_{\Delta S} \mathbf{a} \cdot (\mathbf{A} \times d\mathbf{S})$$

i.e.,
$$\mathbf{a} \cdot (-\nabla \times \mathbf{A}) = \mathbf{a} \cdot \left[\lim_{\Delta V \to 0} \frac{1}{\Delta V} \iint_{\Delta S} (\mathbf{A} \times d\mathbf{S})\right]$$

$\therefore$
$$\nabla \times \mathbf{A} = -\left[\lim_{\Delta V \to 0} \frac{1}{\Delta V} \iint_{\Delta S} (\mathbf{A} \times d\mathbf{S})\right]. \tag{4.7.3}$$

From the results (4.7.1), (4.7.2) and (4.7.3) it is evident that the gradient, divergence and curl do not depend on any particular coordinate system.

Lastly, it is important to note that the three results (4.7.1), (4.7.2) and (4.7.3) can be taken, to start with, as the definitions of gardient, divergence and curl and that all the integral theorems can be obtained therefrom.

4.8 Examples

4.2 Gauss divergence theorem

1. *Solved sum.* Show that, if S is the surface of the rectangular parallelopiped formed by the planes $x = 0, x = a; y = 0, y = b; z = 0, z = c$, then

$$\iint_S \bar{r}\cdot\hat{n}\,dS = 3\,abc.$$

Divergence theorem: $\iint_S \bar{A}\cdot\hat{n}\,dS = \iiint_V \nabla\cdot\bar{A}\,dV.$

$$\iint_S \bar{r}\cdot\hat{n}\,dS = \iiint_V \nabla\cdot(x\bar{i}+y\bar{j}+z\bar{k})\,dV = \iiint_V (1+1+1)\,dV$$

$= 3 \times$ (Volume of the parallelopiped)

$= 3abc$

because the length, breadth, height of the rectangular parallelopiped are a, b, c.

2. *Solved sum*. Show that $\iint_S \mathbf{r}\cdot\mathbf{n}\,dS = 4\pi a^3$ if S is the surface of the sphere $x^2+y^2+z^2=a^2$. **(1981, 83, 85)**

$$\iint_S \mathbf{r}\cdot\mathbf{n}\,dS = \iiint_V \nabla\cdot\mathbf{r}\,dV \text{ by divergence theorem}$$

$$= \iiint_V \left(\mathbf{i}\frac{\partial}{\partial x}+\mathbf{j}\frac{\partial}{\partial y}+\mathbf{k}\frac{\partial}{\partial z}\right)\cdot(x\mathbf{i}+y\mathbf{j}+z\mathbf{k})\,dV$$

$$= \iiint_V (1+1+1)\,dV = 3V = 4\pi a^3 \qquad \left(\because V=\frac{4}{3}\pi a^3\right).$$

3. *Solved sum*. Show that the volume V of the region enclosed by the surface S is $\frac{1}{3}\iint_S \mathbf{r}\cdot d\mathbf{S}$. **(1981, 83)**

By Gauss Divergence theorem

$$\frac{1}{3}\iint_S \mathbf{r}\cdot d\mathbf{S} = \frac{1}{3}\iiint_V \nabla\cdot\mathbf{r}\,dV = \frac{1}{3}\iiint_V (1+1+1)\,dV = V.$$

4. Evaluate $\iint_S \mathbf{r}\cdot d\mathbf{S}$, where

(*i*) S is the sphere $x^2+y^2+z^2=9$. **(1977)**

(*ii*) S is the cube bounded by $x=-1, x=1, y=-1, y=1, z=-1, z=1$.

[Hint: $\iint_S \mathbf{r}\cdot d\mathbf{S} = \iiint_V \nabla\cdot\mathbf{r}\,dV = \iiint_V 3\,dV = 3V.$

(*i*) $3\cdot\frac{4\pi}{3}\cdot 3^3 = 108\pi.$ (*ii*) $3V = 3.2.2.2 = 24.$**]**

5. *Solved sum.* Show, for a closed surface S enclosing a region of volume V, that $\iint_S (ax\mathbf{i}+by\mathbf{j}+cz\mathbf{k}).\,d\mathbf{S} = (a+b+c)V.$

$$\iint_S (ax\mathbf{i}+by\mathbf{j}+cz\mathbf{k})\cdot d\mathbf{S} = \iiint_V \nabla\cdot(ax\mathbf{i}+by\mathbf{j}+cz\mathbf{k})\,dV = \iiint_V \left(\frac{\partial}{\partial x}ax+\frac{\partial}{\partial y}by+\frac{\partial}{\partial z}cz\right)dV$$

$$= \iiint_V (a+b+c)\,dV = (a+b+c)\iiint_V dV = (a+b+c)V.$$

6. *Solved sum.* If $\mathbf{F} = x^3\mathbf{i} + y^3\mathbf{j} = z^3\mathbf{k}$ and S is the surface of the sphere $x^2 + y^2 + z^2 = a^2$, show that

$$\iint_S \mathbf{F}\cdot d\mathbf{S} = \frac{12}{5}\pi a^5. \qquad \textbf{(1980, 81, 83)}$$

$$\iint_S \mathbf{F}\cdot d\mathbf{S} = \iiint_V \nabla\cdot\mathbf{F}\,dV \qquad \text{(Divergence theorem)}$$

$$= \iiint_V (3x^2 + 3y^2 + 3z^2)\,dx\,dy\,dz$$

$$= \int_{r=0}^{a}\int_{\theta=0}^{\pi}\int_{\phi=0}^{2\pi} (3r^2)(r^2 \sin\theta\, dr\, d\theta\, d\phi),$$

changing to spherical coordinates

$$= 3\int_{r=0}^{a}\int_{\theta=0}^{\pi} r^4 \sin\theta [\phi]_0^{2\pi}\, d\theta\, dr$$

$$= 6\pi\int_{r=0}^{a} r^4[-\cos\theta]_0^{\pi}\, dr = 12\pi\left[\frac{r^5}{5}\right]_0^a = \frac{12}{5}\pi a^5.$$

7. *Solved sum.* Show that $\iint_S \mathbf{F}\cdot d\mathbf{S} = 0$ if $\mathbf{F} = x\mathbf{i} + y\mathbf{j} - 2z\mathbf{k}$ and S is the surface of the upper hemisphere $x^2 + y^2 + z^2 = a^2$. **(1977, 84)**

Method 1. Let S' be the diametral plane which lies in the plane $z = 0$ and V, the volume bounded by S and S'. Then, by divergence theorem,

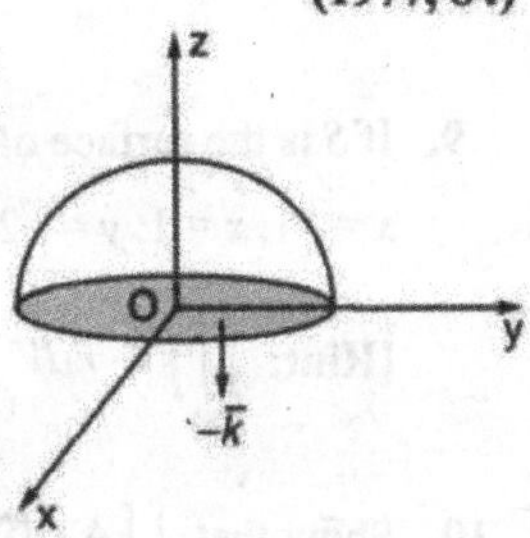

$$\iint_S \bar{A}\cdot\hat{n}\,dS + \iint_{S'} \bar{A}\cdot\hat{n}\,dS = \iiint_V \nabla\cdot\bar{A}\,dV. \qquad \text{... (1)}$$

For S', the outward unit normal is $-\bar{k}$ and so $\hat{n} = -\hat{k}$. Also, on S', $z = 0$. Therefore

$$\iint_{S'} \bar{A}\cdot\hat{n}dS = \iint_{S'} (x\bar{i} + y\bar{j} - 2z\bar{k})\cdot(-\bar{k})\,dS$$

$$= 2\iint z\,dS = 2\iint 0\,dS = 0. \qquad \text{... (2)}$$

$$\iiint_V \nabla\cdot\bar{A}\,dV = \iiint_V \left(\frac{\partial x}{\partial x} + \frac{\partial y}{\partial y} - 2\frac{\partial z}{\partial z}\right)dV$$

$$= \iiint_V (2-2)\,dV = 0. \qquad \text{... (3)}$$

From (1), (2), (3) we get the result as stated.

Method 2. If S' is the surface of the bounding diametral plane, by divergence theorem,

$$\iint_S \mathbf{F}\cdot d\mathbf{S} + \iint_{S'} \mathbf{F}\cdot d\mathbf{S} = \iiint_V \nabla\cdot\mathbf{F}\,dV.$$

But $\nabla\,.\,\mathbf{F} = 1 + 1 - 2 = 0$. So the integral on the right is zero.

But $\iint_{S'} \mathbf{F}\cdot d\mathbf{S} = \iint_{S'} \mathbf{F}\cdot(-\mathbf{k}\,dx\,dy) = \int\int_{S'} 2z\,dx\,dy = 0$

because z on S' is zero. Hence $\iint_S \mathbf{F}\cdot d\mathbf{S}$ also is zero.

8. Solved sum. Evaluate $\iint_S \mathbf{A}\cdot d\mathbf{S}$ if $\mathbf{A} = 2xy\mathbf{i} + xz^2\mathbf{j} + xz\mathbf{k}$ and S is the surface of the parallelepiped formed by the planes $x = 0, x = 2, y = 0, y = 1, z = 0, z = 3$. **(1976, 83)**

$$\iint_S \mathbf{A}\cdot d\mathbf{S} = \iiint_V \nabla\cdot\mathbf{A}\,dV \text{ (Divergence theorem)}$$

$$= \iiint_S (2y + z^2 + x)\,dV$$

$$= 2V\bar{y} + \iiint_V z^2\,dV + V\bar{x},$$

where $\bar{x}, \bar{y}$ are the x,y coordinates of the mass centre of the parallelepiped.

$$= 2(6)\left(\frac{1}{2}\right) + \int_0^2\int_0^1\int_0^3 z^2\,dz\,dy\,dx + 6(1)$$

since mass centre $(\bar{x}, \bar{y}, \bar{z})$ is (1, 1/2, 3/2); Volume of the parallelepiped V = 2.1.3 = 6.

$$= 12 + 9\int_0^2\int_0^1 dy\,dx = 12 + 9\int_0^2 dx = 30,$$

9. If S is the surface of the rectangular parallelepiped formed by the planes $x = -1, x = 1; y = -2, y = 2; z = -3, z = 3$, then show that $\iint_S \bar{r}\cdot\hat{n}\,dS = 144$.

[Hint: $\iiint_V \nabla\cdot\bar{r}\,dV = \iiint_V 3\,dV = 3V; V = 2\times4\times6$]

10. Show that $\iint_S \mathbf{A}\cdot d\mathbf{S} = \frac{3}{2}$ if $\mathbf{A} = 4xz\mathbf{i} - y^2\mathbf{j} + yz\mathbf{k}$ and S is the surface of the cube bounded by $x = 0, x = 1, y = 0, y = 1$, $z = 0, z = 1$. **(1986)**

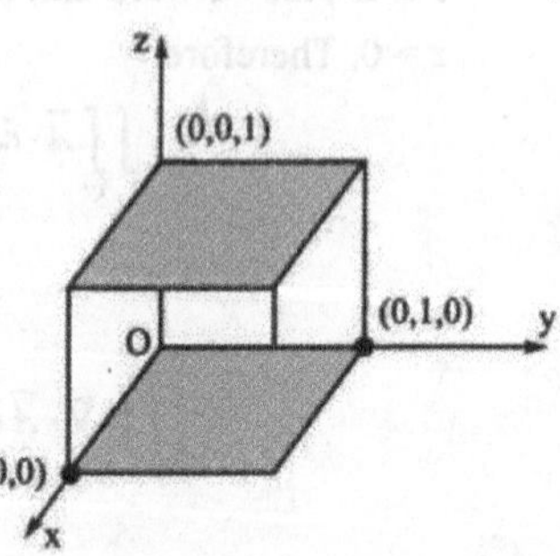

The edges of the cube are of length 1. So its volume is 1 and its mass centre $(\bar{x}, \bar{y}, \bar{z})$ is (1/2, 1/2, 1/2). In this example we are going to use the formulas

$$\iiint_V x\,dV = \bar{x}V, \text{ etc.}$$

Divergence theorem: $\iint_S \bar{A}\cdot\hat{n}\,dS = \iiint_V \nabla\cdot\bar{A}\,dV$.

$$\iint_S \bar{A}\cdot\hat{n}\,dS = \iiint_V \left(\frac{\partial}{\partial x}4xz - \frac{\partial y^2}{\partial y} + \frac{\partial yz}{\partial z}\right)dV$$

$$= \iiint_V (4z - 2y + y)\,dV = \iiint_V (4z - y)\,dV$$

$$= 4\bar{z}\,V = \bar{y}\,V = 4(1/2)(1) - (1/2)(1) = 3/2.$$

11. *Solved sum.* Using divergence theorem, evaluate $\int_S \bar{A}\cdot\hat{n}\,dS$ where $\bar{A} = 4xz\bar{i} - y^2\bar{j} + yz\bar{k}$ and

S is the surface of the cube bounded by the planes.

$$x = 0, x = 2, y = 0, y = 2, z = 0, z = 2.$$

See the previous sum. Now

$$\bar{x} = 1, \bar{y} = 1, \bar{z} = 1.$$

$$\iint_S \bar{A}\cdot\hat{n}\,dS = 4\bar{z}\,V - \bar{y}\,V = 3V = 3\times 2^3 = 24$$

12. Evaluate $\iint_S \mathbf{A}\cdot d\mathbf{S}$, where $\mathbf{A} = xz\mathbf{i} - yz\mathbf{j} + 2z^2\mathbf{k}$ and

S is the surface of the region bounded by the surfaces

(*i*) $x = 0, x = 1, y = 0, y = 2, z = 0, z = 3;$

(*ii*) $x = 0, y = 0, z = 0, 4x + 2y + z = 8;$

(*iii*) $x = 0, y = 0, z = 0, z = 2, x^2 + y^2 - 9 = 0;$

(*iv*) $x = 0, y = 0, z = 0, x^2 + y^2 = a^2, x^2 + z^2 = a^2;$

(*v*) $x = 0, y = 0, y = 3, z = 25, z = x^2.$

[**Hint:** The equivalent volume integrals in terms of cartesian coordinates are

(*i*) $$\int_{x=0}^{x=1}\int_{y=0}^{y=2}\int_{z=0}^{z=3} (4z)\,dz\,dy\,dx = 36;$$

$$\bar{z} = 3/2, V = 6; \iiint_V 4z\,dV = 4\bar{z}\,V = 36.$$

(*ii*) $$\int_{x=0}^{x=2}\int_{y=0}^{y=4-2x}\int_{z=0}^{z=8-4x-2y} (4z)\,dz\,dy\,dx = \frac{256}{3}.$$

(*iii*) $$\int_{x=0}^{x=3}\int_{y=0}^{y=\sqrt{9-x^2}}\int_{z=0}^{z=2} (4z)\,dz\,dy\,dx = 18\pi.$$

Alternatively using $x = r\cos\theta, y = r\sin\theta, z = z, dV = (rd\theta)\,dr\,dz$, for the cylinder in question, $0 \le r \le 3, 0 \le \theta \le \pi/2, 0 \le z \le 2.$

$$\iiint_V \nabla\cdot\bar{A}\,dV = \int_{r=0}^{r=3}\int_{\theta=0}^{\theta=\pi/2}\int_{z=0}^{z=2} 4z(r\,d\theta)(dr)\,dz$$

$$= \int_{r=0}^{3}\int_{\theta=0}^{\pi/2}\left[2z^2\right]_0^2 r\,d\theta\,dr$$

$$= 8\int_{r=0}^{3} r\left[\theta\right]_0^{\pi/2} dr = 4\pi\left[\frac{r^2}{2}\right]_0^3 = 18\pi.$$

Or $$\iiint_V \nabla\cdot\bar{A}\,dV = \int_{x=0}^{x=3}\int_{y=0}^{y=\sqrt{9-x^2}}\int_{z=0}^{z=2} 4z\,dz\,dy\,dx = 18\pi.$$

(iv) $\int\limits_{x=0}^{x=a}\int\limits_{y=0}^{y=\sqrt{a^2-x^2}}\int\limits_{z=0}^{z=\sqrt{a^2-x^2}}(4z)\,dz\,dy\,dx=\frac{3\pi a^4}{8}$.

(v) $\int\limits_{x=0}^{x=5}\int\limits_{y=0}^{y=3}\int\limits_{z=x^2}^{z=25}(4z)\,dz\,dy\,dx=15{,}000.$]

13. *Solved sum.* Evaluate $\iint\limits_S x^2\,dy\,dz+y^2\,dz\,dx+2z(xy-x-y)\,dx\,dy$, where S is the surface of the cube $0\le x\le 1, 0\le y\le 1, 0\le z\le 1$.

Divergence theorem: $\iint\limits_S \bar{A}\cdot\hat{n}\,dS=\iiint\limits_V \nabla\cdot\bar{A}dV$.

Rewriting the given integral in the vector form by using the results.

$$dy\,dz=\bar{i}\cdot dS\hat{n}, dz\,dx=\bar{j}\cdot dS\,\hat{n}, dx\,dy=\bar{k}\cdot dS\hat{n},$$

we have

$$\iint\limits_S x^2dy\,dz+y^2dz\,dx+2z(xy-x-y)dx\,dy$$

$$=\iint\limits_S\left[x^2\bar{i}+y^2\bar{j}+2z(xy-x-y)\bar{k}\right]\cdot\hat{n}\,dS$$

$$=\iiint\limits_V\nabla\cdot\left[x^2\bar{i}+y^2\bar{j}+2z(xy-x-y)\bar{k}\right]dV \text{ by Divergence theorem}$$

$$=\iiint\limits_V\left[\frac{\partial x^2}{\partial x}+\frac{\partial y^2}{\partial y}+\frac{\partial 2z(xy-x-y)}{\partial z}\right]dV$$

$$=\iiint\limits_V[2x+2y+2(xy-x-y)]dV$$

$$=\iiint\limits_V 2xy\,dV=2\int_0^1\left[\int_0^1\left(\int_0^1 xy\,dz\right)dy\right]dx$$

$$=2\int_0^1\left[\int_0^1\left(xy[z]_0^1\right)dy\right]dx=2\int_0^1\left[\int_0^1 xy\,dy\right]dx$$

$$=2\int_0^1 x\left[\frac{y^2}{2}\right]_0^1dx=2\int_0^1 x\left[\frac{1}{2}\right]dx$$

$$=\int_0^1 x\,dx=\left[\frac{x^2}{2}\right]_0^1=\frac{1}{2}.$$

14. *Solved sum.* Evaluate $\iint\limits_S x^3dy\,dz+x^2y\,dz\,dx+x^2z\,dx\,dy$, where S is the surface of the cube bounded by the planes.

$$x=0, y=0, z=0, x=a, y=a, z=a.$$

Rewriting the equation in the vector form, using $dy\,dz=\bar{i}\cdot dS\hat{n}, dz\,dx=\bar{j}\cdot dS\hat{n}, dx\,dy=\bar{k}\cdot dS\,\hat{n}$, we get the integral as

$$\iint_S (x^3\bar{i} + x^2y\bar{j} + x^2z\bar{k})\cdot\hat{n}dS = \iiint_V \nabla\cdot(x^3\bar{i} + x^2y\bar{j} + x^2z\bar{k})\ dV \text{ by Divergence theorem}$$

$$= \iiint_V \left(\frac{\partial x^3}{\partial x} + \frac{\partial x^2y}{\partial y} + \frac{\partial x^2z}{\partial z}\right)dV = \iiint_V (3x^2 + x^2 + x^2)dV$$

$$= \int_0^a\left[\int_0^a\left(\int_0^a 5x^2dz\right)dy\right]dx = \int_0^a\left[\int_0^a (5x^2z)_0^a\, dy\right]dx = \int_0^a\left[\int_0^a 5x^2a\,dy\right]dx$$

$$= \int_0^a \left[5x^2ay\right]_0^a dx = \int_0^a 5x^2a^2dx = 5a^2\left[\frac{x^3}{3}\right]_0^a = \frac{5}{3}a^5.$$

15. *Solved sum.* Apply Gauss' divergence theorem to evaluate

$$\iint_S (x^3 - yz)\,dy\,dz - 2x^2y\,dz\,dx + z\,dx\,dy$$

over the surface of the cube bounded by the coordinate planes and the planes $x = y = z = a$.

Rewriting the equation in the vector form, using $dy\,dz = \bar{i}\cdot dS\,\hat{n}, dz\,dx = \bar{j}\cdot dS\hat{n}, dx\,dy = \bar{k}\cdot dS\,\hat{n}$, we get the integral as

$$\iint_S [(x^3 - yz)\bar{i} - 2x^2y\bar{j} + z\bar{k}]\cdot dS\,\hat{n}$$

$$= \iiint_V \nabla\cdot\left[(x^3 - yz)\bar{i} - 2x^2y\bar{j} + z\bar{k}\right]dV$$

$$= \iiint_V (3x^2 - 2x^2 + 1)\,dV = \iiint_V (x^2 + 1)\,dV$$

$$= \int_0^a\int_0^a\int_0^a (x^2 + 1)\,dx\,dy\,dz = \int_0^a\int_0^a\left[\frac{x^3}{3} + x\right]_0^a dy\,dz$$

$$= \left(\frac{a^3}{3} + a\right)\int_0^a\int_0^a 1\,dy\,dz = \left(\frac{a^3}{3} + a\right)\times a\times a = \frac{a^5}{3} + a^3.$$

16. *Solved sum.* If $\bar{A} = 6z\bar{i} + (2x + y)\bar{j} - x\bar{k}$ and S is the surface of the cylindrical solid bounded by $y = 0, y = 8; x = 0, z = 0; x^2 + z^2 = 9$, then show that

$$\iint_S \bar{A}\cdot\hat{n}\,dS = 18\,\pi.$$

For the solid y varies from 0 to 8. So the height of the cylinder is 8. It is evident that the base radius is 3.

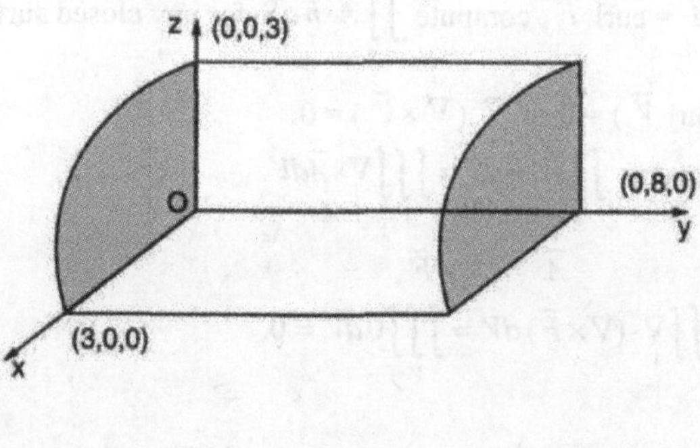

One-fourth of the cylinder which lies in the first octant is under consideration. Its volume is $\frac{1}{4}(\pi r^2 h) = \frac{1}{4}(\pi \cdot 3^2 \cdot 8) = 18\pi.$

Divergence theorem: $\iint_S \bar{A}\cdot\hat{n}\,dS = \iiint_V \nabla\cdot\bar{A}\,dV.$

$$\iint_S \bar{A}\cdot\hat{n}\,dS = \iiint_V \left(\frac{\partial(6z)}{\partial x}+\frac{\partial(2x+y)}{\partial y}-\frac{\partial x}{\partial z}\right)dV$$

$$= \iiint_V (0+1-0)\,dV = V = 18\pi.$$

17. *Solved sum.* Evaluate $\iint_S x\,dy\,dz + y\,dz\,dx + z\,dx\,dy$ over the surface of the sphere

$$x^2+y^2+z^2 = a^2.$$

Rewriting the integral in vector form, using

$$dy\,dz = \bar{i}\cdot dS\,\hat{n},\ dz\,dx = \bar{j}\cdot dS\hat{n},\ dx\,dy = \bar{k}\cdot dS\hat{n},$$

we have

$$\iint_S x\,dy\,dz + y\,dz\,dx + z\,dx\,dy = \iint_S (x\bar{i}+y\bar{j}+z\bar{k})\cdot\hat{n}\,dS$$

$$= \iiint_V \nabla\cdot(x\bar{i}+y\bar{j}+z\bar{k})\,dV \text{ by divergence theorem}$$

$$= \iiint_V \left(\frac{\partial x}{\partial x}+\frac{\partial y}{\partial y}+\frac{\partial z}{\partial z}\right)dV$$

$$= \iiint_V (1+1+1)\,dV = 3\iiint_V dV$$

$$= 3(\text{Volume of the sphere}) = 3\times\frac{4}{3}\pi a^3 = 4\pi a^3.$$

18. Apply divergence theorem to evaluate $\iint_S (x+z)\,dy\,dz + (y+z)\,dz dx + (x+y)\,dx\,dy,$ where S is the surface of the sphere $x^2+y^2+z^2 = a^2$.

[**Hint:** $\iint_S (...) = \iiint_V \nabla\cdot[(x+z)\bar{i}+(y+z)\bar{j}+(x+y)\bar{k}]\,dV$

$$= \iiint_V (1+1+0)\,dV = \frac{8}{3}\pi a^3]$$

19. *Solved sum.* If $\bar{A} = \text{curl } \bar{F}$, compute $\iint_S \bar{A}\cdot\hat{n}\,dS$ for any closed surface S.

A result. Div (curl $\bar{V}$) = 0 or $\nabla.(\nabla\times\bar{V}) = 0.$

Divergence theorem: $\iint_S \bar{A}\cdot\hat{n}\,dS = \iiint_V \nabla\cdot\bar{A}\,dV.$

Since $\bar{A} = \nabla\times\bar{F},$

$$\iint_S \bar{A}\cdot\hat{n}\,dS = \iiint_V \nabla\cdot(\nabla\times\bar{F})\,dV = \iiint_V 0\,dV = 0.$$

Verification of divergence theorem.

20. *Solved sum.* Verify the divergence theorem for

$$\mathbf{A} = (x+y)\mathbf{i} + x\mathbf{j} + z\mathbf{k}$$

taken over the region V of the cube bounded by the planes $x=0, x=1, y=0, y=1, z=0, z=1$. Denote the faces respectively by $S_1, S_2, S_3, S_4, S_5, S_6$ and the outward unit normal by $\hat{n}$ and the surface integral by I.

$$S_1: x = 0,\ \hat{n} = -\hat{i}, \mathrm{I} = -\int_0^1\int_0^1 y\,dy\,dz = -\frac{1}{2}.$$

$$S_2: x = 1,\ \hat{n} = \overline{i},\ \mathrm{I} = \int_0^1\int_0^1 (1+y)\,dy\,dz = \frac{3}{2}.$$

$$S_3: y = 0,\ \hat{n} = -\overline{j},\ \mathrm{I} = -\int_0^1\int_0^1 x\,dx\,dz = -\frac{1}{2}.$$

$$S_4: y = 1,\ \hat{n} = \overline{j},\ \mathrm{I} = \int_0^1\int_0^1 x\,dx\,dz = \frac{1}{2}.$$

$$S_5: z = 0,\ \hat{n} = -\overline{k}, \mathrm{I} = \int_0^1\int_0^1 0\,dx\,dy = 0.$$

$$S_6: z = 1,\ \hat{n} = \overline{k},\ \mathrm{I} = \int_0^1\int_0^1 1\,dx\,dy = 1.$$

$$\therefore \quad \iint_S \overline{A}\cdot\hat{n}\,dS = -\frac{1}{2}+\frac{3}{2}-\frac{1}{2}+\frac{1}{2}+0+1 = 2.$$

If S is the surface of the cube, $\iint_S \mathbf{A}\cdot d\mathbf{S} = 2$

$$\iiint_V (\nabla\cdot\mathbf{A})\,dV = \iiint_V \left(\mathbf{i}\frac{\partial}{\partial x}+\mathbf{j}\frac{\partial}{\partial y}+\mathbf{k}\frac{\partial}{\partial z}\right)\cdot\{(x+y)\mathbf{i}+x\mathbf{j}+z\mathbf{k}\}\,dV$$

$$= \iiint_V \left[\frac{\partial}{\partial x}(x+y)+\frac{\partial(x)}{\partial y}+\frac{\partial(z)}{\partial z}\right]dV$$

$$= \iiint_V 2\,dV = 2\iiint_V dV = 2V = 2(1.1.1) = 2 \qquad (\because V = 1)$$

Thus divergence theorem stands verified.

21. *Solved sum.* Verify divergence theorem for $\overline{f} = x^2\overline{i} + y^2\overline{j} + z^2\overline{k}$ taken over the cube bounded by the planes.

$$x = 0, x = 1; y = 0, y = 1; z = 0, z = 1.$$

Now we have to verify that

$$\iint_S \overline{f}\cdot\hat{n}\,dS = \iiint_V \nabla\cdot\overline{f}\,dV.$$

Denote the faces of the cube lying in the planes

$$x = 0, x = 1, y = 0, y = 1, z = 0, z = 1$$

by $S_1, S_2, S_3, S_4, S_5, S_6$. Let $\hat{n}$ denote the unit vectors normal to them drawn outwardly.

On S_1: $\hat{n} = -\bar{i}$ and $x = 0$.

$$\therefore \quad \iint_{S_1} \bar{f}\cdot\hat{n}\,dS = \iint_{S_1} (x^2\bar{i}+y^2\bar{j}+z^2\bar{k})\cdot(-i)\,dS$$

$$= \iint_{S_1} (-x^2)\,dS = \iint_{S_1} (-0^2)\,dS = 0.$$

On S_2: $\hat{n} = \bar{i}$ and $x = 1$.

$$\iint_{S_2} \bar{f}\cdot\hat{n}\,dS = \iint_{S_2} (x^2\bar{i}+y^2\bar{j}+z^2\bar{k})\cdot\bar{i}\,dS$$

$$= \iint_{S_2} x^2\,dS = \iint_{S_2} 1^2\,dS = S_2 = 1\times 1 = 1.$$

Similarly, for other surfaces, we have

$$\iint_{S_3} \bar{f}\cdot\hat{n}\,dS = 0, \iint_{S_4} \bar{f}\cdot\hat{n}\,dS = 1, \iint_{S_5} \bar{f}\cdot\hat{n}\,dS = 0, \iint_{S_6} \bar{f}\cdot\hat{n}\,dS = 1.$$

$$\therefore \quad \iint_{S} \bar{f}\cdot\hat{n}\,dS = (0+1)+(0+1)+(0+1) = 3. \qquad (1)$$

On the other hand, we have

$$\iiint_V \nabla\cdot\bar{f}\,dV = \iiint_V \left(\frac{\partial x^2}{dx}+\frac{\partial y^2}{\partial y}+\frac{\partial z^2}{\partial z}\right)dV$$

$$= 2\iiint_V (x+y+z)\,dV = 2[\bar{x}V+\bar{y}V+\bar{z}V],$$

where $(\bar{x},\bar{y},\bar{z})$ is the mass centre of V. But

$$\bar{x} = 1/2, \bar{y} = 1/2, \bar{z} = 1/2.$$

Also, we have V = length × breadth × height = 1 × 1 × 1 = 1.

$$\therefore \quad \iiint_V \nabla\cdot\bar{f}\,dV = 2\left[\frac{1}{2}\times 1+\frac{1}{2}\times 1+\frac{1}{2}\times 1\right] = 3. \qquad (2)$$

(1) and (2) are equal. This verifies divergence theorem.

22. Verify Gauss' divergence theorem for

$\bar{F} = x^2\bar{i}+2y^2\bar{j}+3z^2\bar{k}$ taken over the cube bounded by the planes.

$$x = 0, x = 1, y = 0, y = 1, z = 0, z = 1.$$

[**Hint:** Denote the faces respectively by $S_1, S_2, S_3, S_4, S_5, S_6$ and the outward unit normal by $\hat{n}$ and the surface integral by **I**.]

S_1: $\quad x = 0,\ \hat{n} = -\bar{i},\ \mathbf{I} = -\int_0^1\int_0^1 x^2\,dS = -\int_0^1\int_0^1 0^2\,dy\,dz = 0.$

S_2: $\quad x = 1,\ \hat{n} = -\bar{i},\ \mathbf{I} = \int_0^1\int_0^1 x^2\,dS = \int_0^1\int_0^1 1^2\,dy\,dz = 1.$

S_3: $\quad y = 0,\ \hat{n} = -\bar{j},\ \mathbf{I} = -2\int_0^1\int_0^1 y^2\,dS = -2\int_0^1\int_0^1 0^2\,dx\,dz = 0.$

S_4: $\quad y = 1, \hat{n} = \bar{j}, \ \mathbf{I} = 2\int_0^1\int_0^1 y^2 dS = 2\int_0^1\int_0^1 1^2 dx\,dz = 2.$

S_5: $\quad z = 0, \hat{n} = -\bar{k}, \ \mathbf{I} = -3\int_0^1\int_0^1 z^2 dS = -3\int_0^1\int_0^1 0^2\, dx\,dy = 0.$

S_6: $\quad z = 1, \hat{n} = \bar{k}, \ \mathbf{I} = 3\int_0^1\int_0^1 z^2 dS = 3\int_0^1\int_0^1 1^2 dx\,dy = 3.$

$$\therefore \quad \iint_S \bar{F}\cdot\hat{n}\,dS = 0+1+0+2+0+3 = 6.$$

$$\iiint_V \nabla\cdot\bar{F}dV = \iiint_V (2x+4y+6z)\,dV = 2\bar{x}V + 4\bar{y}V + 6\bar{z}V$$

$$= 1+2+3 = 6, \text{ since } \bar{x} = \bar{y} = \bar{z} = 1/2\,]$$

23. *Solved sum.* Verify divergence theorem for $\vec{F} = 4xz\bar{i} - y^2\bar{j} + yz\bar{k}$ taken over the cube bounded by

$$x = 0, x = 1; y = 0, y = 1; z = 0, z = 1.$$

Now we have to verify that

$$\iint_S \bar{F}\cdot\hat{n}\,dS = \iiint_V \nabla\cdot\bar{F}\,dV.$$

Denote the faces lying in the planes $x = 0, x = 1, y = 0, y = 1, z = 0, z = 1$ by $S_1, S_2, S_3, S_4, S_5, S_6$. Let $\hat{n}$ denote the unit vectors normal to them drawn outwardly. Then,

On S_1:

$$\hat{n} = -\bar{i}, x = 0.$$

$$\therefore \quad \iint_{S_1} \bar{F}\cdot\hat{n}\,dS = \iint_{S_1} (4xz\bar{i}) - y^2\bar{j} + yz\bar{k})\cdot(-i)\,dS$$

$$= \iint_{S_1} (-4xz)\,dS = \iint_{S_1} [-4(0)z]\,dS = 0.$$

On S_2: $\quad \hat{n} = \hat{i}, x = 1.$

$$\therefore \quad \iint_{S_2} \bar{F}\cdot\hat{n}\,dS = \iint_{S_2} (4xz\bar{i} - y^2\bar{j} + yz\bar{k})\cdot\bar{i}\,dS$$

$$= \iint_{S_2} (4xz)\,dS = 4\iint_{S_2} (1)\,z\,dS$$

$$= 4\int_0^1\left[\int_0^1 z\,dz\right]dy = 4\int_0^1\left[\frac{z^2}{2}\right]_0^1 dy$$

$$= 4\int_0^1 \frac{1}{2}dy = 2[y]_0^1 = 2.$$

On S_3: $\quad \hat{n} = -\bar{j}, y = 0.$

$$\therefore \quad \iint_{S_3} \bar{F}\cdot\hat{n}\,dS = \iint_{S_3} (4xz\bar{i} - y^2\bar{j} + yz\bar{k})\cdot(-\bar{j})\,dS$$

$$= \iint_{S_3} y^2 dS = \iint_{S_3} 0^2\, dS = 0.$$

On S_4: $\hat{n} = \bar{j}, \quad y = 1.$

$$\therefore \quad \iint_{S_4} \bar{F}\cdot\hat{n}\,dS = \iint_{S_4} (4xz\bar{i} - y^2\bar{j} + yz\bar{k})\cdot\bar{j}dS$$

$$= \iint_{S_4} (-y^2)\,dS = -\iint_{S_4} 1dS = -S_4 = -1\times 1 = -1.$$

On S_5: $\hat{n} = -\bar{k}\ , z = 0.$

$$\therefore \quad \iint_{S_5} \bar{F}\cdot\hat{n}\,dS = \iint_{S_5} (4xz\bar{i} - y^2\bar{j} + yz\bar{k})\cdot(-\bar{k})\,dS$$

$$= \iint_{S_5} (-yz)\,dS = \iint_{S_5} 0dS = 0.$$

On S_6: $\hat{n} = \bar{k}, \quad z = 1.$

$$\therefore \quad \iint_{S_6} \bar{F}\cdot\hat{n}\,dS = \iint_{S_6} (4xz\bar{i} - y^2\bar{j} + yz\bar{k})\cdot\bar{k}\,dS$$

$$= \iint_{S_6} yz\,dS = \iint_{S_6} y(1)\,dS = \int_0^1\left[\int_0^1 y\,dy\right]dx = \int_0^1\left[\frac{y^2}{2}\right]_0^1 dx$$

$$= \int_0^1 \frac{1}{2}dx = \left[\frac{x}{2}\right]_0^1 = \frac{1}{2}.$$

$$\therefore \quad \iint_S \bar{F}\cdot\hat{n}\,dS = (0+2) + (0-1) + \left(0 + \frac{1}{2}\right) = \frac{3}{2}. \qquad \text{... (1)}$$

Next, we have

$$\iiint_V \nabla\cdot\bar{F}dV = \iiint\left[\frac{\partial(4xz)}{\partial x} - \frac{\partial(y^2)}{\partial y} + \frac{\partial(yz)}{\partial z}\right]dV$$

$$= \iiint_V (4z - 2y + y)\,dV = \iiint_V (4z - y)\,dV$$

$$= 4\bar{z}\,V - \bar{y}\,V.$$

where ($\bar{x}, \bar{y}, \bar{z}$) is the mass centre of V. But

$$\bar{x} = 1/2, \quad \bar{y} = 1/2, \quad \bar{z} = 1/2.$$

Also, we have

$$V = \text{length} \times \text{breadth} \times \text{height} = 1\times 1\times 1 = 1.$$

$$\therefore \quad \iiint_V \nabla\cdot\bar{F}dV = 4\cdot\frac{1}{2}\cdot 1 - \frac{1}{2}\cdot 1 = \frac{3}{2}. \qquad \text{... (2)}$$

(1) and (2) are equal. This verifies divergence theorem.

24. Verify Gauss' divergence theorem for

$$\bar{F} = (2x - z)\bar{i} + x^2y\bar{j} - xz^2\bar{k}$$

over the cube bounded by the planes

$$x = 0, x = 1, y = 0, y = 1, z = 0, z = 1.$$

[**Hint:** Denote the faces respectively by $S_1, S_2, S_3, S_4, S_5, S_6$ and the outward unit normal by $\hat{n}$ and the surface integral by **I**.]

S_1: $\quad x = 0,\ \hat{n} = -\bar{i},\ \mathrm{I} = -\int_0^1\int_0^1 (2x - z)\,dS = \int_0^1\int_0^1 z\,dy\,dz = \frac{1}{2}.$

S_2: $\quad x = 1,\ \hat{n} = \bar{i},\ \mathrm{I} = \int_0^1\int_0^1 (2x - z)\,dS = \int_0^1\int_0^1 (2 - z)\,dy\,dz = \frac{3}{2}.$

S_3: $\quad y = 0,\ \hat{n} = -\bar{j},\ \mathrm{I} = -\int_0^1\int_0^1 x^2 y\,dS = -\int_0^1\int_0^1 0\,dx\,dz = 0.$

S_4: $\quad y = 1,\ \hat{n} = \bar{j},\ \mathrm{I} = \int_0^1\int_0^1 x^2 y\,dS = \int_0^1\int_0^1 x^2\,dx\,dz = \frac{1}{3}.$

S_5: $\quad z = 0,\ \hat{n} = -\bar{k},\ \mathrm{I} = \int_0^1\int_0^1 xz^2\,dS = \int_0^1\int_0^1 0^2\,dx\,dy = 0.$

S_6: $\quad z = 1,\ \hat{n} = \bar{k},\ \mathrm{I} = -\int_0^1\int_0^1 xz^2\,dS = -\int_0^1\int_0^1 x\,dx\,dy = -\frac{1}{2}.$

$$\therefore \quad \iint_S \bar{F}\cdot\hat{n}\,dS = \frac{1}{2} + \frac{3}{2} + 0 + \frac{1}{3} + 0 - \frac{1}{2} = \frac{11}{6}.$$

$$\iiint_V \nabla\cdot\bar{F}\,dV = \int_0^1\int_0^1\int_0^1 (2 + x^2 - 2xz)\,dz\,dy\,dx$$

$$= \int_0^1\int_0^1 \left[2z + x^2 z - xz^2\right]_0^1 dy\,dx = \frac{11}{6}]$$

25. *Solved sum.* Verify Gauss' divergence theorem for $\bar{f} = (x^3 - yz)\bar{i} - 2x^2y\bar{j} + 2\bar{k}$ over the cube bounded by $x = 0, y = 0, z = 0, x = a, y = a, z = a$.

Now we have to verify that

$$\iint_S \bar{f}\cdot\hat{n}\,dS = \iiint_V \nabla\cdot\bar{f}\,dV$$

Denote the faces lying in the planes

$$x = 0, x = a; y = 0, y = a; z = 0, z = a$$

by $S_1, S_2, S_3, S_4, S_5, S_6$. Let $\hat{n}$ denote the unit vectors normal to them drawn outwardly. Then,

On S_1,

$$\hat{n} = -\bar{i}, \quad x = 0.$$

$$\therefore \quad \iint_{S_1} \bar{f}\cdot\hat{n}\,dS = \iint_{S_1} \left[(x^3 - yz)\bar{i} - 2x^2y\bar{j} + 2\bar{k}\right]\cdot(-i)\,dS$$

$$= \iint_{S_1} (x^3 - yz)(-1)\,dS = \iint_{S_1} yz\,dS \text{ since } x = 0$$

$$= \int_0^a\int_0^a yz\,dy\,dz = \int_0^a z\left[\frac{y^2}{2}\right]_0^a dz = \frac{a^2}{2}\int_0^a z\,dz$$

$$= \frac{a^2}{2} \cdot \frac{a^2}{2} = \frac{a^4}{4}.$$

On S_2: $\hat{n} = \bar{i},\ x = a.$

$$\therefore \iint_{S_2} \bar{f} \cdot \hat{n}\, dS = \iint_{S_1} \left((x^3 - yz)\bar{i} - 2x^2 y\bar{j} + 2\bar{k}\right) \cdot \bar{i}\, dS$$

$$= \iint_{S_1} (x^3 - yz)\, dS = \int_0^a\!\!\int_0^a (a^3 - yz)\, dy\, dz$$

$$= a^3 \int_0^a\!\!\int_0^a dy\, dz - \int_0^a\!\!\int_0^a yz\, dy\, dz$$

$$= a^3(a^2) - \left(\frac{a^2}{2}\right)\left(\frac{a^2}{2}\right)$$

$$= a^5 - a^4/4.$$

On S_3: $\hat{n} = -\bar{j},\ y = 0.$

$$\therefore \iint_{S_3} \bar{f} \cdot \hat{n}\, dS = \iint_{S_3} \left[(x^3 - yz)\bar{i} - 2x^2 y\bar{j} + 2\bar{k}\right](-\bar{j})\, dS$$

$$= \iint_{S_3} 2x^2 y\, dS = \iint_{S_3} 2x^2(0)\, dS = 0 \text{ since } y = 0.$$

On S_4: $\hat{n} = \bar{j},\ y = a.$

$$\therefore \iint_{S_4} \bar{f} \cdot \hat{n}\, dS = \iint_{S_4} \left[(x^3 - yz)\bar{i} - 2x^2 y\bar{j} + 2\bar{k}\right] \cdot \bar{j}\, dS$$

$$= \iint_{S_4} (-2x^2 y)\, dS = \iint_{S_4} (-2x^2 a)\, dS$$

$$= -2a \int_0^a\!\!\int_0^a x^2\, dx\, dz = -2a \cdot \frac{a^3}{3} \cdot a = -\frac{2a^5}{3}.$$

On S_5: $\hat{n} = -\bar{k},\ z = 0.$

$$\therefore \iint_{S_5} \bar{f} \cdot \hat{n}\, dS = \iint_{S_5} \left((x^3 - yz)\bar{i} - 2x^2 y\bar{j} + 2\bar{k}\right) \cdot (-\bar{k})\, dS$$

$$= \iint_{S_5} (-2)\, dS = -2\int_0^a\!\!\int_0^a dx\, dy = -2a^2.$$

On S_6: $\hat{n} = \bar{k},\ z = a.$

$$\therefore \iint_{S_6} \bar{f} \cdot \hat{n}\, dS = \iint_{S_6} \left[(x^3 - yz)\bar{i} - 2x^2 y\bar{j} + 2\bar{k}\right] \cdot [\bar{k}]\, dS$$

$$= \iint_{S_6} 2\, dS = 2\int_0^a\!\!\int_0^a dx\, dy = 2a^2.$$

$$\therefore \quad \iint_S \bar{f}\cdot\hat{n}\,dS = \left(\frac{a^4}{4}+a^5-\frac{a^4}{4}\right)+\left(0-\frac{2a^5}{3}\right)+(-2a^2+2a^2)$$

$$= a^5-\frac{2}{3}a^5=\frac{1}{3}a^5 \quad \text{... (1)}$$

Next, we have

$$\therefore \quad \iiint_V \nabla\cdot\bar{f}dV = \iiint_V\left[\frac{\partial}{\partial x}(x^3-yz)+\frac{\partial}{\partial y}(-2x^2y)+\frac{\partial}{\partial z}(2)\right]dV$$

$$= \iint_V(3x^2-2x^2)dV = \iiint_V x^2 dV$$

$$= \int_0^a\int_0^a\int_0^a x^2 dx\,dy\,dz = \frac{a^3}{3}\cdot a\cdot a = \frac{a^5}{3}. \quad (2)$$

(1) and (2) are equal. This verifies divergence theorem.

26. Verify Guass' divergence theorem for $\bar{F}=x^3\bar{i}+y^3\bar{j}+z\bar{k}$ taken over the cube bounded by $x=0, x=a, y=0, y=a, z=0, z=a$.

[**Hint:** Denote the faces respectively by $S_1, S_2, S_3, S_4, S_5, S_6$ and the outward unit normal by $\hat{n}$ and the surface integral by I.]

S_1: $\quad x=0,\ \hat{n}=-\bar{i},\ \text{I}=-\int_0^a\int_0^a x^3 dS=-\int_0^a\int_0^a 0^3\,dy\,dz=0.$

S_2: $\quad x=a,\ \hat{n}=\bar{i},\ \text{I}=\int_0^a\int_0^a x^3 dS=\int_0^a\int_0^a a^3 dy\,dz=a^5.$

S_3: $\quad y=0,\ \hat{n}=-\bar{j},\ \text{I}=-\int_0^a\int_0^a y^3\,dS=\int_0^a\int_0^a 0^3\,dx\,dz=0.$

S_4: $\quad y=a,\ \hat{n}=\bar{j},\ \text{I}=\int_0^a\int_0^a y^3 dS=\int_0^a\int_0^a a^3 dx\,dz=a^5.$

S_5: $\quad z=0,\ \hat{n}=-\bar{k},\ \text{I}=-\int_0^a\int_0^a z\,dS=-\int_0^a\int_0^a 0\,dx\,dy=0.$

S_6: $\quad z=a,\ \hat{n}=\bar{k},\ \text{I}=\int_0^a\int_0^a z\,dS=\int_0^a\int_0^a a\,dx\,dy=a^3.$

$$\therefore \quad \iint_S \bar{F}\cdot\hat{n}\,dS = 0+a^5+0+a^5+0+a^3=2a^5+a^3.$$

$$\iiint_V \nabla\cdot\bar{F}\,dV = \iiint_V(3x^2+3y^2+1)dV$$

$$= \int_0^a\int_0^a\int_0^a(3x^2+3y^2+1)dx\,dy\,dz = 2a^5+a^3]$$

27. Verify Gauss' divergence theorem for

$\bar{A} = (x^2 - yz)\bar{i} + (y^2 - zx)\bar{j} + (z^2 - xy)\bar{k}$ and the surface $0 \le x \le a; 0 \le y \le b; 0 \le z \le c$.

[**Hint:** Denote the faces by $S_1, S_2, S_3, S_4, S_5, S_6$ and the outward unit normal by $\hat{n}$ and the respective surface integrals by $I_1, I_2, I_3, I_4, I_5, I_6$. Then we have

$$S_1: \quad x = 0,\ \hat{n} = -\hat{i},\ I_1 = -\int_0^c\int_0^b (0^2 - yz)\,dy\,dz$$

$$S_2: \quad x = a,\ \hat{n} = \hat{i},\ I_2 = \int_b^c\int_0^b (a^2 - yz)\,dy\,dz.$$

$$\therefore \quad I_1 + I_2 = \int_0^c\int_0^b a^2\,dy\,dz = a^2bc.$$

Thus, by symmetry, $I_3 + I_4 = ab^2c$, $I_5 + I_6 = abc^2$ and the surface integral on the parallelepiped is $abc(a + b + c)$.

$$\iiint_V \nabla \cdot \vec{A}\,dV = \iiint_V (2x + 2y + 2z)\,dV = 2(\bar{x}\,V + \bar{y}\,V + \bar{z}\,V)$$

$$= 2\left[\frac{a}{2}V + \frac{b}{2}V + \frac{c}{2}V\right]$$

$$= (a + b + c)\,V = abc(a + b + c)$$

28. *Solved sum.* Verify divergence theorem for the vector

$$\mathbf{A} = (x^2 - yz)\mathbf{i} + (y^2 + zx)\,\mathbf{j} + (z^2 - xy)\,\mathbf{k}$$

taken over the parallelepiped defined by

$$0 \le x \le a, 0 \le y \le b, 0 \le z \le c. \qquad \textbf{(1979)}$$

Suppose the faces whose equations are $x = 0, x = a, y = 0, y = b, z = 0, z = c$ are respectively named $S_1, S_2, S_3, S_4, S_5, S_6$ and **n** denotes the unit vectors normal to them drawn outwardly (Fig. 3.8.3). Then, on S_1, $x = 0$, $\mathbf{n} = -\mathbf{i}$ and **A.n** is

$$[\{0 - yz\}\mathbf{i} + \{y^2 - z\,(0)\}\mathbf{j} + \{z^2 - (0)y\}\,\mathbf{k}]\,.\,(-\mathbf{i}) = yz,$$

$$\therefore \quad \iint_{S_1} \mathbf{A}\cdot\mathbf{n}\,dS = \int_0^b\int_0^c yz\,dz\,dy$$

But, on S_2, $x = a$, $\mathbf{n} = \mathbf{i}$ and $\mathbf{A}\,.\,\mathbf{n}$ is $[(a^2 - yz)\mathbf{i} + (...)\mathbf{j} + (...)\mathbf{k}]\,.\,\mathbf{i} = a^2 - yz.$

$$\therefore \quad \iint_{S_2} \mathbf{A}\cdot\mathbf{n}\,dS = \int_0^b\int_0^c (a^2 - yz)\,dz\,dy.$$

$$\therefore \iint_{S_1} \mathbf{A}\cdot\mathbf{n}\,dS + \iint_{S_2} \mathbf{A}\cdot\mathbf{n}\,dS = \int_0^b\int_0^c (yz + a^2 - yz)\,dz\,dy$$

$$= \int_0^b\int_0^c a^2\,dz\,dy = a^2\int_0^b c\,dy = a^2bc.$$

Similarly the sum of the integrals along S_3 and $S_4 = ab^2c$ and that of the integrals along S_5 and S_6 is abc^2. Adding the six integrals we get that

$$\iint_S \mathbf{A}\cdot\mathbf{n}\,dS = a^2bc + ab^2c + abc^2 = abc(a + b + c). \qquad ...(1)$$

But $\iiint_V \nabla\cdot\mathbf{A}\,dV = \iiint_V (2x+2y+2z)\,dV$

$$= 2[(V\,\bar{x}+V\,\bar{y}+V\,\bar{z})],$$

where ($\bar{x},\bar{y},\bar{z}$) is the mass centre of the parallelepiped and V its volume

$$= 2(abc)\left[\frac{1}{2}a+\frac{1}{2}b+\frac{1}{2}c\right]$$

$$= abc\,(a+b+c) \qquad ...(2)$$

By (1) and (2) divergence theorem stands verified.

29. *Solved sum.* Verify the divergence theorem for $\mathbf{A} = 4x\mathbf{i} - 2y^2\mathbf{j} + z^2\mathbf{k}$ taken over the cylindrical region bounded by the surfaces $x^2+y^2=4$, $z=0$, $z=3$. **(1980, 82, 83)**

If S is the surface of the cylindrical region.

$\iint_S \mathbf{A}\cdot d\mathbf{S} = 84\pi$ (vide sum 23, Ch. 3).

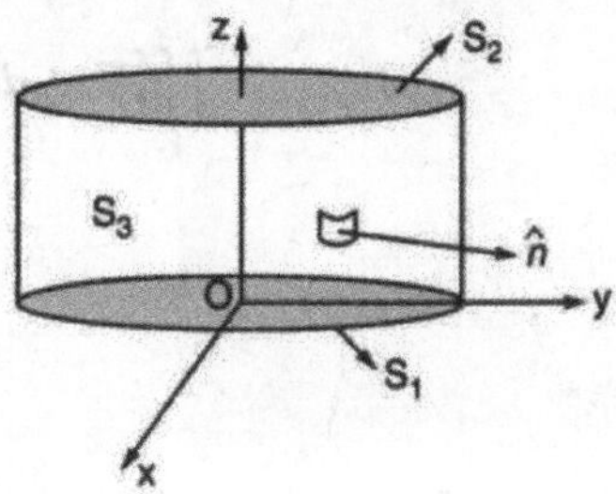

Let us have the following assumptions:

S_1 : Bottom circular face in the plane $z = 0$.

S_2 : Top circular face in the plane $z = 3$.

S_3 : Curved surface.

On S_1: $z = 0,\ \hat{n} = -\bar{k}$;

$$\bar{A}\cdot\bar{n} = -\bar{A}\cdot\bar{k} = -z^2 = -0^2 = 0.$$

$$\therefore \quad \iint_{S_1} \bar{A}\cdot\hat{n}\,dS = \iint_{S_1} 0\,dS = 0. \qquad ...(1)$$

On S_2: $z = 3,\ \hat{n} = \bar{k};\ \bar{A}\cdot\hat{n} = \bar{A}\cdot\bar{k} = z^2 = 9.$

$$\iint_{S_2} \bar{A}\cdot\hat{n}\,dS = 9\iint_{S_2} dS = 9[\pi(radius)^2] = 9(\pi 2^2) = 36\pi. \qquad ...(2)$$

On S_3:

$$\hat{n} = \frac{\nabla(x^2+y^2-4)}{|\nabla(x^2+y^2-4)|} = \frac{2x\bar{i}+2y\bar{j}}{2|x\bar{i}+y\bar{j}|}$$

$$= \frac{x\bar{i}+y\bar{j}}{\sqrt{x^2+y^2}} = \frac{x\bar{i}+y\bar{j}}{\sqrt{4}} = \frac{x\bar{i}+y\bar{j}}{2}.$$

$$\bar{A}\cdot\hat{n} = (4x\bar{i}-2y^2\bar{j}+z^2\bar{k})\cdot\frac{x\bar{i}+y\bar{j}}{2} = 2x^2-y^3.$$

The cylindrical polar coordinates for S_3 are

$$x = r\cos\theta,\ y = r\sin\theta,\ z = z;\ dS = (r\,d\theta)\,dz,$$

where $r = 2,\ 0\le\theta\le 2\pi,\ 0\le z\le 3$ so that now

$$x = 2\cos\theta,\ y = 2\sin\theta,\ dS = 2\,d\theta\,dz.$$

$$\iint_{S_3} \bar{A}\cdot\hat{n}\,dS = \int_{z=0}^{3}\int_{\theta=0}^{2\pi} [2(2\cos\theta)^2-(2\sin\theta)^3](2d\theta)\,dz$$

$$= 3\int_{\theta=0}^{2\pi} (16\cos^2\theta - 16\sin^3\theta)\,d\theta$$

$$= 48\int_0^{2\pi}(\cos^2\theta - \sin^3\theta)\,d\theta$$

$$= 48\left[\int_0^{2\pi}\frac{1+\cos 2\theta}{2}d\theta\right] - 48\left[\frac{\cos^3\theta}{3} - \cos\theta\right]_0^{2\pi}$$

$$= 24\left[\theta + \frac{\sin 2\theta}{2}\right]_0^{2\pi} + 0 = 48\pi. \quad (3)$$

$$\therefore \quad \iint_S \bar{A}\cdot\hat{n}\,dS = 0 + 36\pi + 48\pi = 84\pi. \quad (4)$$

Method 1.

$$\iiint_V \nabla\cdot\mathbf{A}\,dV = \iiint_V \left(\mathbf{i}\frac{\partial}{\partial x} + \mathbf{j}\frac{\partial}{\partial y} + \mathbf{k}\frac{\partial}{\partial z}\right).(4x\mathbf{i} - 2y^2\mathbf{j} + z^2\mathbf{k})\,dV$$

$$= 2\iiint_V (2 - 2y + z)\,dV$$

$$= 2\int_{x=-2}^{x=2}\int_{y=-\sqrt{4-x^2}}^{y=\sqrt{4-x^2}}\int_{z=0}^{z=3}(2 - 2y + z)\,dz\,dy\,dx = 84\pi.$$

Method 2.

$$\iiint_V \nabla\cdot\mathbf{A}\,dV = \iiint_V \nabla\cdot(4x\mathbf{i} - 2y^2\mathbf{j} + z^2\mathbf{k})\,dV$$

$$= \iiint_V (4 - 4y + 2z)\,dV$$

$$= 4V - 4V\bar{y} + 2V\bar{z},$$

where $\bar{y}$, $\bar{z}$ are the y, z coordinates of the mass centre of the cylinder

$$= 4V - 4V(0) + 2V\left(\frac{3}{2}\right)$$

since the mass centre ($\bar{x}$, $\bar{y}$, $\bar{z}$) of the cylinder here is (0, 0, 3/2)

$$= 84\pi$$

$(\because\ V = \pi.\,2^2.3)$.

Method 3.

Alternatively, using the cylindrical coordinates, we get

$$x = \rho\cos\phi,\ y = \rho\sin\phi,\ z = z,$$

where $\quad 0 \le \rho \le 2,\ 0 \le \phi \le 2\pi,\ 0 \le z \le 3.$

Also $\quad dV = dx\,dy\,dz = \mathrm{J}\,d\rho\,d\phi\,dz,$

where J, the Jacobian of the transformation, is

$$\begin{vmatrix} x_\rho & x_\phi & x_z \\ y_\rho & y_\phi & y_z \\ z_\rho & z_\phi & z_z \end{vmatrix} = \begin{vmatrix} \cos\phi & -\rho\sin\phi & 0 \\ \sin\phi & \rho\cos\phi & 0 \\ 0 & 0 & 1 \end{vmatrix} = \rho.$$

So
$$\iiint_V \nabla\cdot\mathbf{A}\,dV = 2\iiint_V (2-2y+z)\,dV$$

$$= 2\int_{\rho=0}^{\rho=2}\int_{\phi=0}^{\phi=2\pi}\int_{z=0}^{z=3}(2-2\rho\sin\phi+z)\rho\,dz\,d\phi\,d\rho = 84\pi.$$

Thus the divergence theorem stands verified.

30. Verify the divergence theorem for $\mathbf{A} = yz\,\mathbf{i}+2y^2\,\mathbf{j}+xz^2\,\mathbf{k}$ taken over the region bounded by $x=0, y=0, z=0, z=2$ and $x^2+y^2=9$. **(1981)**

[**Hint:** The surface integral $= 81-9+0+0+36=108$, wherein the figures correspond to the curved surface and the plane faces lying in $x=0, y=0, z=0, z=2$ (sum 21, Ch. 3), Volume integral

$$= \int_0^3\int_0^{\pi/2}\int_0^2 (4r^2\sin\theta+2r^2\cos\theta\, z)\,dz\,d\theta\,dr.]$$

31. *Solved sum.* Verify the divergence theorem for $\mathbf{A} = a(x+y)\,\mathbf{i}+a(y-x)\,\mathbf{j}+z^2\mathbf{k}$ taken over the region V bounded by the upper hemisphere $x^2+y^2+z^2=a^2$ and the plane $z=0$. **(1978)**

Let S_1 be the curved surface and S_2, the circular plane surface of the hemisphere. Then the outward unit normal of S_2 is $-\bar{k}$ and that of S_1 is

$$\hat{n} = \frac{\nabla(x^2+y^2+z^2-a^2)}{|\nabla(x^2+y^2+z^2-a^2)|}$$

$$= \frac{2(x\bar{i}+y\bar{j}+z\bar{k})}{2\sqrt{x^2+y^2+z^2}} = \frac{x}{a}\bar{i}+\frac{y}{a}\bar{j}+\frac{z}{a}\bar{k}.$$

$$\therefore \quad \bar{A}\cdot\hat{n} = (x^2+xy)+(y^2-xy)+\frac{z^3}{a}.$$

$$= x^2+y^2+\frac{z^3}{a} \text{ and } \hat{n}\cdot\bar{k} = \frac{z}{a}.$$

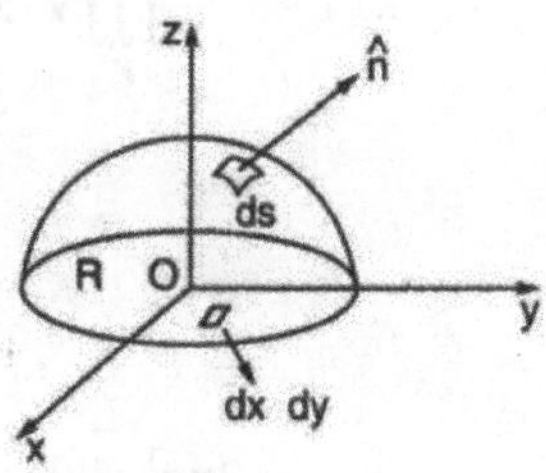

If R is the projection of S_1 on the xOy plane, then

$$dS = \frac{dx\,dy}{|\hat{n}\cdot\bar{k}|}.$$

$$\therefore \quad \iint_{S_1}\bar{A}\cdot\hat{n}\,dS = \iint_R \frac{\mathbf{A}\cdot\mathbf{n}}{\mathbf{n}\cdot\mathbf{k}}\,dx\,dy = \iint_R\left(x^2+y^2+\frac{z^3}{a}\right)\frac{a}{z}\,dx\,dy$$

$$= \iint_R\left(\frac{a(x^2+y^2)}{z}+z^2\right)dx\,dy$$

$$= \iint_R\left[\frac{a(x^2+y^2)}{\sqrt{a^2-x^2-y^2}}+a^2-x^2-y^2\right]dx\,dy.$$

Using polar coordinates,

$$\iint_{S_1}\bar{A}\cdot\hat{n}\,dS = \int_0^a\int_0^{2\pi}\left(\frac{ar^2}{\sqrt{a^2-r^2}}+a^2-r^2\right)r\,d\theta\,dr$$

$$= 2\pi\int_0^a\left(\frac{ar^2}{\sqrt{a^2-r^2}}+a^2-r^2\right)r\,dr.$$

Setting $\quad a^2-r^2=t^2, -r\,dr = t\,dt.$

When $\quad r=0, t=a,$

When $\quad r=a, t=0.$

$$\therefore \quad \iint_{S_1}\overline{A}\cdot\hat{n}\,dS = 2\pi\int_a^0\left(\frac{a(a^2-t^2)}{t}+t^2\right)(-t\,dt)$$

$$= 2\pi\int_0^a(a^3-at^2+t^3)\,dt$$

$$= 2\pi a^4\left[1-\frac{1}{3}+\frac{1}{4}\right]=\frac{11\pi a^4}{6}. \quad \text{... (1)}$$

$$\iint_{S_2}\overline{A}\cdot\hat{n}\,dS = \iint_{S_2}\overline{A}\cdot(-\overline{k})\,dxdy = \iint_{S_2}(-z^2)\,dxdy$$

$$= \iint_{S_2} 0\,dx\,dy = 0 \text{ since } z=0 \text{ on } S_2. \quad \text{... (2)}$$

Adding (1) and (2), we get

$$\iint_A \overline{A}\cdot\hat{n}\,dS = \frac{11\pi a^4}{6}+0=\frac{11\pi a^4}{6}.$$

$$\iiint_V \nabla\cdot\overline{A}\,dV = \iiint_V\left(\frac{\partial a(x+y)}{\partial x}+\frac{\partial a(y-x)}{\partial y}+\frac{\partial z^2}{\partial z}\right)dV$$

$$= \iiint_V (a+a+2z)\,dV = 2aV+2\overline{z}V$$

$$= V(2a+2\overline{z}) = \frac{2\pi a^3}{3}\left(2a+\frac{6a}{8}\right)=\frac{11\pi a^4}{6} \quad \text{... (4)}$$

Since $V=\dfrac{2\pi a^3}{3}$ and $\overline{z}=\dfrac{3a}{8}$. Equality of (3) and (4) verifies the divergence theorem.

32. Verify divergence theorem for

$\overline{A}=2xz\overline{i}+yz\overline{j}+z^2\overline{k}$ and the region bounded by the upper half of the sphere

$x^2+y^2+z^2=a^2$ and the plane $z=0$.

[Hint: As in the example sum,

$$\iint_{S_1}\overline{A}\cdot\hat{n}\,dS = \iint_R[2x^2+y^2+(a^2-x^2-y^2)]\,dx\,dy$$

$$= \int_0^a\int_0^{2\pi}(r^2\cos^2\theta+a^2)r\,d\theta\,dr = \frac{5}{4}\pi a^4.$$

$$\iint_{S_2}\overline{A}\cdot(-\overline{k})\,dS = \iint_{S_2}(-z^2)\,ds = \iint_{S_2}0\,dS = 0.$$

$$\iiint_V \nabla\cdot\bar{A}\,dV = \iiint_V 5z\,dV = 5\bar{z}\,V = \frac{5}{4}\pi a^4]$$

33. Verify the divergence theorem for the vector $\mathbf{A} = 2xz\,\mathbf{i} + yz\,\mathbf{j} + z^2\,\mathbf{k}$ for the region bounded by the upper half of the sphere $x^2 + y^2 + z^2 = a^2$ and the plane $z = 0$. (1984)

[Hint: If S' is the plane surface, the integral on S' is zero since $z = 0$ on S' and if S is the curved surface,

$$\iint_S \mathbf{A}\cdot\mathbf{n}\,dS = \iint_R (2x^2 + y^2 + z^2)\,dx\,dy$$

$$= \iint_R [2x^2 + y^2 + (a^2 - x^2 - y^2)]\,dx\,dy$$

$$= \int_0^a\int_0^{2\pi} (r^2\cos^2\theta + a^2)\,r\,d\theta\,dr = \frac{5}{4}\pi a^4.$$

$$\iiint_V \nabla\cdot\mathbf{A}\,dV = \iiint_V 5z\,dV = 5\bar{z}\,V$$

$$= 5\left(\frac{3a}{8}\right)\left(\frac{2}{3}\pi a^3\right) = \frac{5}{4}\pi a^4]$$

34. *Solved sum.* Apply divergence theorem to evaluate $\iint_S (x+z)\,dy\,dz + (y+z)\,dz\,dx + (x+y)\,dx\,dy$, where S is the surface of the sphere $x^2 + y^2 + z^2 = 1$. (1985)

Now $\mathbf{i}.\,d\mathbf{S} = dy\,dz$, $\mathbf{j}.\,d\mathbf{S} = dz\,dx$, $\mathbf{k}.\,d\mathbf{S} = dx\,dy$. Thus the given surface integral is

$$\iint_S [(x+z)\mathbf{i} + (y+z)\mathbf{j} + (x+y)\mathbf{k}]\cdot d\mathbf{S}$$

$$= \iiint_V \nabla\cdot[(x+z)\mathbf{i} + (y+z)\mathbf{j} + (x+y)\mathbf{k}]\,dV$$

$$= \iiint_V 2\,dV = 2V = 2\left(\frac{4}{2}\pi\right) = \frac{8\pi}{3}.$$

35. *Solved sum.* Verify divergence theorem for $\mathbf{A} = 2x^2y\,\mathbf{i} - y^2\,\mathbf{j} + 4xz^2\,\mathbf{k}$ and for the region in the first quadrant bounded by the cylinder $y^2 + z^2 = 9$ and the planes $x = 0$ and $x = 2$. (1980, 83)

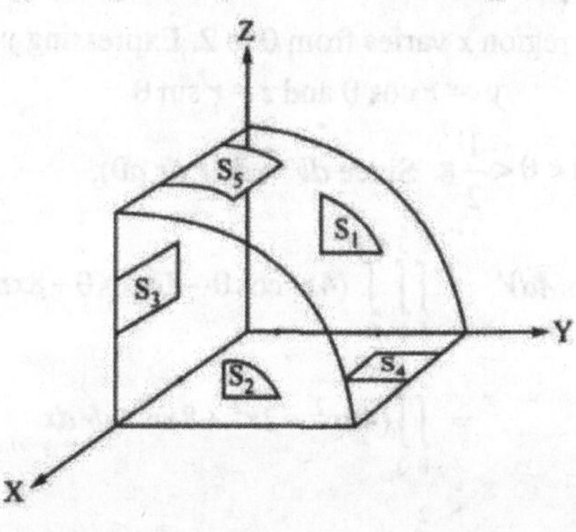

Let us make the following assumptions for the bounding surfaces and their respective unit vectors.

S_1 : Quadrant of the circle in $x = 0$, $\mathbf{n}_1 = -\mathbf{i}$
S_2 : Quadrant of the circle in $x = 2$, $\mathbf{n}_2 = \mathbf{i}$
S_3 : Rectangular surface in $y = 0$, $\mathbf{n}_3 = -\mathbf{j}$
S_4 : Rectangular surface in $z = 0$, $\mathbf{n}_4 = -\mathbf{k}$
S_5 : Curved surface $y^2 + z^2 = 9$, $\mathbf{n}_5 = (y\mathbf{j} + z\mathbf{k})/3$
I_1, I_2, I_3, I_4, I_5: Integral of **A.n** on S_1, S_2, S_3, S_4, S_5.

$$I_1 = \iint_{S_1} \mathbf{A}\cdot(-\mathbf{i})\,dy\,dz = -2\iint_{S_1} x^2 y\,dy\,dz = 0 \text{ since } x = 0 \text{ on } S_1.$$

$$I_2 = \iint_{S_2} \mathbf{A}\cdot\mathbf{i}\,dy\,dz = 2\iint_{S_2} 2^2 y\,dy\,dz \text{ since } x = 2 \text{ on } S_2$$

$$= 8\int_0^3\int_0^{\pi/2} (r\cos\theta)\,r\,d\theta\,dr = 8\int_0^3 r^2 dr = 72.$$

$$I_3 = \iint_{S_3} \mathbf{A}\cdot(-\mathbf{j})\,dx\,dz = \iint_{S_3} y^2 dx\,dz = 0 \text{ since } y = 0 \text{ on } S_3$$

$$I_4 = \iint_{S_4} \mathbf{A}\cdot(-\mathbf{k})dx\,dy = -4\iint_{S_4} xz^2\,dx\,dy = 0 \text{ since } z = 0 \text{ on } S_4.$$

$$I_5 = \iint_{S_5} \mathbf{A}\cdot\mathbf{n}_5\,dS = \frac{1}{3}\iint_{S_5} (-y^3 + 4xz^3)\frac{dx\,dy}{\mathbf{n}\cdot\mathbf{k}}$$

$$= \iint_{S_5}\left(-\frac{y^3}{z} + 4xz^2\right)dx\,dy$$

$$= \int_0^2\int_0^3\left[\frac{-y^3}{\sqrt{9-y^2}} + 4x(9-y^2)\right]dy\,dx$$

$$= \int_0^2 (-18 + 72x)dx = 108.$$

So $$\iint_S \mathbf{A}\cdot d\mathbf{S} = I_1 + I_2 + I_3 + I_4 + I_5 = 0 + 72 + 0 + 0 + 108 = 180. \quad \text{... (1)}$$

For the volume of the region x varies from 0 to 2. Expressing y and z in r and θ,

$$y = r\cos\theta \text{ and } z = r\sin\theta$$

where $0 < r < 3$ and $0 < \theta < \frac{1}{2}\pi$. Since $dV = dx(r\,dr\,d\theta)$,

$$\iiint_V \nabla\cdot\bar{A}dV = \int_0^2\int_0^3\int_0^{\pi/2} (4xr\cos\theta - 2r\cos\theta + 8xr\sin\theta)\,r\,d\theta\,dr\,dx$$

$$= \int_0^2\int_0^3 (4xr^2 - 2r^2 + 8xr^2)\,dr\,dx$$

$$= \int_0^2 (36x - 18 + 72x)dx = 180 \quad \text{... (2)}$$

Equality of (1) and (2) completes the verification.

36. *Solved sum.* If S is a closed surface and $\mathbf{r}$, the position vector of a general point with O as the point of reference, show that $\iint_S \frac{\mathbf{r}}{r^3}\cdot d\mathbf{S}$ is equal to (*i*) zero when O lies outside S and (*ii*) 4π, when O lies inside S. **(1983)**

(*i*) When O lies outside S, $r \neq 0$ and so $\frac{\mathbf{r}}{r^3}$ is defined at all points inside S.

But, $$\nabla\cdot\left(\frac{\mathbf{r}}{r^3}\right) = 0.$$

$$\therefore \quad \iint_S \frac{\mathbf{r}}{r^3}\cdot d\mathbf{S} = \iiint_V \nabla\cdot\left(\frac{\mathbf{r}}{r^3}\right)dV = 0. \quad \text{... (1)}$$

(*ii*) When O lies inside S, $\frac{\mathbf{r}}{r^3}$ is well defined at all points inside S except at O. However, introducing a small sphere S' in V with centre O and radius ε, we get $\frac{\mathbf{r}}{r^3}$ well defined in the region between S and S'. So, applying the divergence theorem for the region between S and S', we get

$$\iint_S \frac{\mathbf{r}}{r^3}\cdot d\mathbf{S} + \iint_{S'} \frac{\mathbf{r}}{r^3}\cdot d\mathbf{S} = 0 \text{ as in (1).}$$

$$\text{Thus} \quad \iint_S \frac{\mathbf{r}}{r^3}\cdot d\mathbf{S} = -\iint_{S'} \frac{\mathbf{r}}{r^3}\cdot d\mathbf{S}$$

$$= \iint_{S'} \frac{1}{\epsilon^2}dS = \frac{1}{\epsilon^2}4\pi\epsilon^2 = 4\pi$$

since $d\mathbf{S}$ on $S' = \mathbf{n}\, dS = -\frac{\mathbf{r}}{\epsilon}dS$ and

$$\frac{\mathbf{r}}{r^3}\cdot d\mathbf{S} = \frac{\mathbf{r}}{\epsilon^3}\cdot\left(-\frac{\mathbf{r}}{\epsilon}\right)dS = -\frac{\epsilon^2}{\epsilon^4}dS = -\frac{1}{\epsilon^2}dS.$$

37. Show that for any closed surface S,

(*i*) $$\iint_S (\phi\nabla\Psi)\cdot d\mathbf{S} = \iiint_V (\phi\nabla^2\Psi + \nabla\phi\cdot\nabla\Psi)\,dV.$$

(*ii*) $$\iint_S (\phi\nabla\Psi - \Psi\nabla\phi)\cdot d\mathbf{S} = \iiint_V (\phi\nabla^2\Psi - \Psi\nabla^2\phi)\,dV.$$

(These results are called Green's first and second identities respectively.)

Setting $\mathbf{A} = \phi(\nabla\psi)$ in (4.2.1), and using (1) of sum 92, Chapter 2,

$$\iint_S (\phi\nabla\Psi)d\mathbf{S} = \iiint_V \nabla\cdot(\phi\nabla\Psi)\,dV$$

$$= \iiint_V [(\nabla\phi)\cdot(\nabla\Psi) + \phi(\nabla\cdot\nabla\Psi)]\,dV$$

$$= \iiint_V [(\nabla\phi)\cdot(\nabla\Psi) + \phi\nabla^2\Psi]\,dV. \quad \text{... (1)}$$

Interchanging ϕ and Ψ in (1), we get

$$\iint_S \Psi\nabla\phi\cdot d\bar{S} = \iiint_V [(\nabla\Psi)\cdot(\nabla\phi)+\psi\nabla^2\phi]dV. \quad ..(2)$$

Now (1) – (2) gives the result as stated.

4.3 Derived results from divergence theorem

38. If S is a closed surface, prove that

(*i*) Surface area of $S = \iiint_V (\nabla\cdot n)\,dV$, (*ii*) $\iint_S d\mathbf{S} = 0.$

(*iii*) $\iint_S r\times d\mathbf{S} = 0$ (*iv*) $\iiint_V (\nabla\times\mathbf{n})\,dV = 0.$

[Hint: (*i*) in (4.2.1) set $\mathbf{A} = \mathbf{n}$, (*ii*) in (4.3.1) set $\phi = 1$, (*iii*) in (4.3.2) set $\mathbf{A} = \mathbf{r}$ and (*iv*) in (4.3.2) set $\mathbf{A} = \mathbf{n}$.**]**

39. For a closed surface S, show that

(*i*) $\iint_S \frac{\mathbf{r}}{r^3}\cdot d\mathbf{S} = 0.$ (*ii*) $\iint_S \frac{\mathbf{r}}{r^2}\cdot d\mathbf{S} = \iiint_V \frac{1}{r^2}dV.$

(*iii*) $\iint_S r^4 d\mathbf{S} = 4\iiint_V r^2\mathbf{r}\,dV.$ (*iv*) $\iint_S f(r)\mathbf{r}\times d\mathbf{S} = 0.$

(*v*) $\iint_S (\nabla r^2)\cdot d\mathbf{S} = 6V.$ (*vi*) $\iint_S (\nabla\times\mathbf{f})\cdot d\mathbf{S} = 0.$ **(1975)**

(*vii*) $\iint_S (r\times c)\times d\mathbf{S} = 2eV$, (c = constant)

[Hint : We know Result 1. $\nabla.\,[f(r)\bar{r}\,] = rf'(r) + 3f(r).$

Result 2. $\nabla f(r) = f'(r)\hat{r} = \frac{f'(r)}{r}\bar{r}$

Divergence theorem: $\iint_S \bar{A}\cdot d\bar{S} = \iiint_V \nabla\cdot\bar{A}\,dV.$

(*i*)

$$\iint_S \frac{1}{r^3}\bar{r}\cdot d\bar{S} = \iiint_V \nabla\cdot\left(\frac{1}{r^3}\bar{r}\right)dV$$

$$= \iiint_V \nabla\cdot[f(r)\bar{r}]dV, \text{ where } f(r) = \frac{1}{r^3}$$

$$= \iiint_V \left[r\left(-\frac{3}{r^4}\right)+3\left(\frac{1}{r^3}\right)\right]dV = 0.$$

(*ii*)

$$\iint_S \frac{1}{r^2}\bar{r}\cdot d\bar{S} = \iiint_V \nabla\cdot\left(\frac{1}{r^2}\bar{r}\right)dV$$

$$= \iiint_V \nabla\cdot[f(r)\bar{r}]dV, \text{ where } f(r) = \frac{1}{r^2}$$

$$= \iiint_V [rf'(r)+3f(r)]dV \text{ by Result 1}$$

$$= \iiint_V \left[r\left(-\frac{2}{r^3}\right)+3\left(\frac{1}{r^2}\right)\right]dV = \iiint_V \frac{1}{r^2}dV.$$

(*iii*) Use (4.3.1).

(*iv*) Use (4.3.2).

(*v*) By result 2, $\nabla r^2 = \frac{2r}{r}\bar{r} = 2\bar{r}$.

$$\therefore \quad \iint_S (\nabla r^2)\cdot d\bar{S} = \iint_S (2\bar{r})\cdot d\bar{S} = 2\iiint_V \nabla\cdot\bar{r}\,dV$$

$$= 2\iiint_V \left(\frac{\partial x}{\partial x}+\frac{\partial y}{\partial y}+\frac{\partial z}{\partial z}\right)dV = 2\iiint_V 3dV = 6V.$$

(*vi*) Set $\mathbf{A} = \nabla \times \mathbf{f}$ in (4.2.1)

(*vii*) Set $\mathbf{A} = \mathbf{r} \times \mathbf{c}$ in (4.3.2)]

4.4 Green's theorem in plane

40. *Solved sum.* Evaluate by Green's theorem

$$\int_C (xy+x^2)dx+(x^2+y^2)dy,$$

where C is the square formed by the lines $x = -1, x = 1, y = -1, y = 1$ in the xOy plane.

Let R be the region enclosed by C. Then, by Green's theorem,

$$\int_C M\,dx+N\,dy = \iint_R (N_x - M_y)\,dx\,dy.$$

But we have

$$\left.\begin{aligned} M &= xy+x^2 \\ M_y &= x \end{aligned}\right\} \qquad \left.\begin{aligned} N &= x^2+y^2 \\ N_x &= 2x \end{aligned}\right\}$$

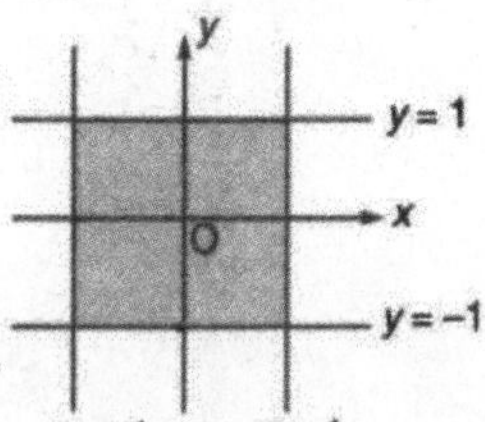

$$\therefore \quad N_x - M_y = x.$$

$$\therefore \quad \int_C M\,dx+N\,dy = \int_{-1}^{1}\int_{-1}^{1} x\,dy\,dx = \int_{-1}^{1} x[y]_{-1}^{1}\,dx$$

$$= \int_{-1}^{1} 2x\,dx = \left[x^2\right]_{-1}^{1}$$

$$= 1-1 = 0.$$

41. *Solved sum.* Using Green's theorem, show that

$$\int_C (3x^2-8y^2)dx+(4y-6xy)dy = 20,$$

where C is the boundary of the rectangular area enclosed by the lines $x = 0, x = 1, y = 0, y = 2$ in the xOy plane.

Let R be the region enclosed by C. Then, by Green's theorem,

$$\int_C M\,dx+N\,dy = \iint_R (N_x - M_y)\,dx\,dy.$$

But we have

$$\left.\begin{aligned} M &= 3x^2-8y^2 \\ M_y &= -16y \end{aligned}\right] \quad \left.\begin{aligned} N &= 4y-6xy \\ N_x &= -6y \end{aligned}\right\} \qquad \therefore N_x - M_y = 10y.$$

$$\therefore \quad \int_C M\,dx + N\,dy = \iint_R 10y\,dx\,dy = 10\int_0^1\left[\int_0^2 y\,dy\right]dx$$

$$= 10\int_0^1\left[\frac{y^2}{2}\right]_0^2 dx = 10\int_0^1\left[\frac{4}{2}\right]dx = 20.$$

42. *Solved sum.* Evaluate by Green's theorem

$\int_C e^{-x}(\sin y\,dx + \cos y\,dy)$, where C is the rectangle with vertices $(0, 0)$, $(\pi, 0)$, $(\pi, \pi/2)$, $(0, \pi/2)$.

Let R be the region enclosed by C. Then, by Green's theorem,

$$\int_C M\,dx + N\,dy = \iint_R (N_x - M_y)\,dx\,dy.$$

But we have

$$\left.\begin{aligned} M &= e^{-x}\sin y \\ M_y &= e^{-x}\cos y \end{aligned}\right] \quad \left.\begin{aligned} N &= e^{-x}\cos y \\ N_x &= -e^{-x}\cos y \end{aligned}\right\} \; N_x - M_y = -2e^{-x}\cos y.$$

$$\therefore \quad \int_C Mdx + Ndy = -\int_0^{\pi}\int_0^{\pi/2} 2e^{-x}\cos y\,dy\,dx$$

$$= -\int_0^{\pi} 2e^{-x}[\sin y]_0^{\pi/2}dx = -\int_0^{\pi} 2e^{-x}dx$$

$$= 2\left[e^{-x}\right]_0^{\pi} = 2(e^{-\pi}-1).$$

43. *Solved sum.* Evaluate $\int_C xy\,dx - x^2\,dy$ by converting it into a double integral. It is given that C is the boundary of the region bounded by the line $y = x$ and the parabola $x^2 = y$.

Let R be the the region bounded by C. To convert the given integral into a double integral, we shall use Green's theorem. Now

$$M = xy. \text{ So } M_y = x.$$
$$N = -x^2. \text{ So } N_x = -2x.$$
$$\therefore \quad N_x - M_y = -3x.$$

The parabola $x^2 = y$ and the line $y = x$ intersect at $(0, 0)$ and $(1, 1)$. Thus R is the region specified by

$$0 \le x \le 1, x^2 \le y \le x.$$

Therefore, by Green's theorem,

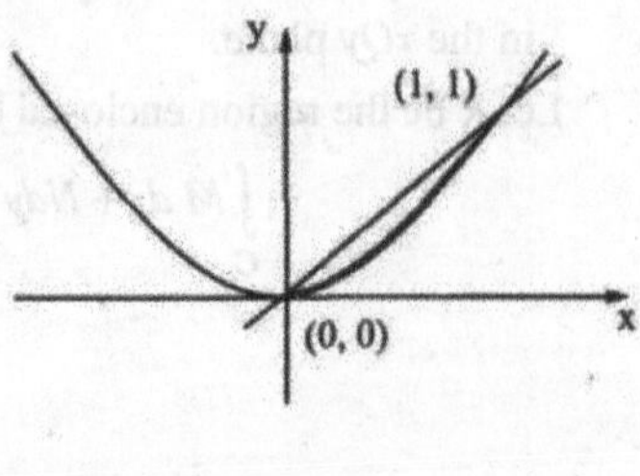

$$\int_C xy\,dx - x^2dy = \iint_R (-3x)\,dx\,dy$$

$$= \int_0^1\int_{x^2}^{x} (-3x)\,dy\,dx$$

$$= \int_0^1 (-3x)[y]_{x^2}^{x}\, dx = -3\int_0^1 x[x-x^2]\,dx$$

$$= -3\int_0^1 (x^2-x^3)\,dx = -3\left[\frac{x^3}{3}-\frac{x^4}{4}\right]_0^1$$

$$= -3\left[\frac{1}{12}\right] = -\frac{1}{4}.$$

44. *Solved sum.* Using Green's theorem, show that

$\int_C (3x+4y)\,dx + (2x-3y)\,dy = -8\pi$, where C is the circle $x^2+y^2=4$.

Now $M = 3x+4y$, $M_y = 4$; $N = 2x-3y$, $Nx = 2$. So

$$N_x - M_y = -2.$$

If R is the region enclosed by C, then

$$\int_C M\,dx + N\,dy = \iint_R (N_x - M_y)\,dx\,dy = -2\iint 1\,dx\,dy$$

$$= -2[\text{Area of the circle}] = -2[\pi.2^2] = -8\pi.$$

45. *Solved sum.* (*i*) Prove that the area enclosed by a simple closed curve C in xOy plane is $\frac{1}{2}\int_C x\,dy - y\,dx$.

Deduce the area of a ellipse $\frac{x^2}{a^2}+\frac{y^2}{b^2}=1$.

(*ii*) Evaluate $\int_C x\,dy - y\,dx$, where C is the circle

$$x^2+y^2 = 4.$$

The Green's theorem is

$$\int_C M\,dx + N\,dy = \iint_R (N_x - M_y)\,dx\,dy.$$

Taking $M = -y$, $N = x$, we have $M_y = -1$, $N_x = 1$. So

$$M\,dx + N\,dy = -y\,dx + x\,dy = x\,dy - y\,dx$$

$$N_x - M_y = 2.$$

$$\therefore \quad \int_C x\,dy - y\,dx = \iint_R 2\,dx\,dy = 2\iint_R 1\,dx\,dy$$

$$= 2(\text{area enclosed by } R)$$

$$\therefore \text{ Area enclosed by } R = \frac{1}{2}\int_C x\,dy - y\,dx.$$

(*i*) **Deduction.** The parametric equations of the ellipse are

$$x = a\cos t,\ y = b\sin t,\ 0 \le t \le 2\pi.$$

$$\therefore \quad dx = -a\sin t\,dt,\ dy = b\cos t\,dt.$$

$$\therefore \quad \text{Area} = \frac{1}{2}\int_C x\,dy - y\,dx = \frac{1}{2}\int_0^{2\pi} (ab\cos^2 t + ab\sin^2 t)\,dt$$

$$= \frac{1}{2}\int_0^{2\pi} ab\,dt = \frac{1}{2}(ab)(2\pi) = \pi ab.$$

(*ii*) Now C is a circle. Its area is 4π. So

$$\int_C x\,dy - y\,dx = 2 \times \text{Area} = 2 \times 4\pi = 8\pi.$$

46. *Solved sum*. Evaluate $\int_C xy^2 dy - x^2 y\,dx$, where C is the cardioid $r = a(1 + \cos\theta)$.

$$\left.\begin{aligned} M &= -x^2 y \\ M_y &= -x^2 \end{aligned}\right\} \left.\begin{aligned} N &= xy^2 \\ N_x &= y^2 \end{aligned}\right\} \quad \therefore N_x - M_y = x^2 + y^2.$$

If R is the region enclosed by C, by Green's theorem,

$$\int_C xy^2 dy - x^2 y\,dx = \iint_R (x^2 + y^2)\,dx\,dy.$$

Changing the right-side integral to polar coordinates,

$$\int_C (xy^2 dy - x^2 y\,dx) = \iint_R r^2 (r\,dr\,d\theta) = \int_0^{2\pi}\int_0^{a(1+\cos\theta)} r^3 dr\,d\theta$$

$$= \int_0^{2\pi}\left[\frac{r^4}{4}\right]_0^{a(1+\cos\theta)} d\theta = \frac{a^4}{4}\int_0^{2\pi}(1+\cos\theta)^4\,d\theta.$$

But we know that, if $f(2a - x) = f(x)$, then

$$\int_0^{2a} f(x)\,dx = 2\int_0^a f(x)\,dx.$$

$$\therefore \quad \int_C (\ldots) = \frac{a^4}{4}\cdot 2\int_0^{\pi}(1+\cos\theta)^4\,d\theta = \frac{a^4}{2}\int_0^{\pi}(2\cos^2(\theta/2))^4\,d\theta$$

$$= 8a^4\int_0^{\pi}\cos^8(\theta/2)\,d\theta$$

Put $\theta/2 = t$
$d\theta = 2dt$
When $\theta = 0$, $t = 0$
When $\theta = \pi$, $t = \pi/2$

$$= 16a^4\int_0^{\pi/2}\cos^8 t\,dt$$

$$= 16a^4 I_8 = 16a^4\cdot\frac{7\cdot 5\cdot 3\cdot 1}{8\cdot 6\cdot 4\cdot 2}\frac{\pi}{2} = \frac{35}{16}\pi a^4.$$

47. Using Green's theorem, evaluate

$$\int_C x^2(1+y)\,dx + (y^3 + x^3)\,dy,$$

where C is the square formed by the lines $y = \pm 1$, $x = \pm 1$.

[**Hint:** $M = x^2(1+y)$, $N = y^3 + x^3$; $M_y = x^2$, $N_x = 3x^2$, $N_x - M_y = 2x^2$.

$$\int_C (\ldots) = \iint_R 2x^2\,dx\,dy = 2\int_{-1}^{1} x^2 dx\int_{-1}^{1} dy = \frac{8}{3}]$$

48. Evaluate $\int_C (2xy - x^2)\,dx + (x + y^2)\,dy$, where C is the closed curve formed by $x^2 = y$ and $y^2 = x$.

[**Hint:** $M = 2xy - x^2, N = x + y^2, N_x - M_y = 1 - 2x$. Hence

$$\int_C (...) = \iint_R (1-2x)\,dx\,dy = \int_0^1 (1-2x)\,dx \int_{x^2}^{\sqrt{x}} dy$$

$$= \int_0^1 (1-2x)(\sqrt{x} - x^2)\,dx = \frac{1}{30}]$$

49. Evaluate, by Green's theorem,

$$\int_C (\cos x \sin y - xy)\,dx + \sin x \cos y\,dy,$$

where C is the circle $x^2 + y^2 = 1$.

[**Hint:** $N_x - M_y = x$.

$$\int_C (...) = \iint_R x\,dy\,dx = \int_{-1}^{1} \int_{-\sqrt{1-x^2}}^{\sqrt{1-x^2}} x\,dy\,dx$$

$$= 2\int_{-1}^{1} x\sqrt{1-x^2}\;dx = 0 \text{ since the integrand is odd}]$$

50. *Solved Sum*. Verity Green's theorem in the plane for the integral $\oint_C xy\;dx + x^2\;dy$, where C is the curve enclosing the region bounded by the parabola $y = x^2$ and the line $y = x$.

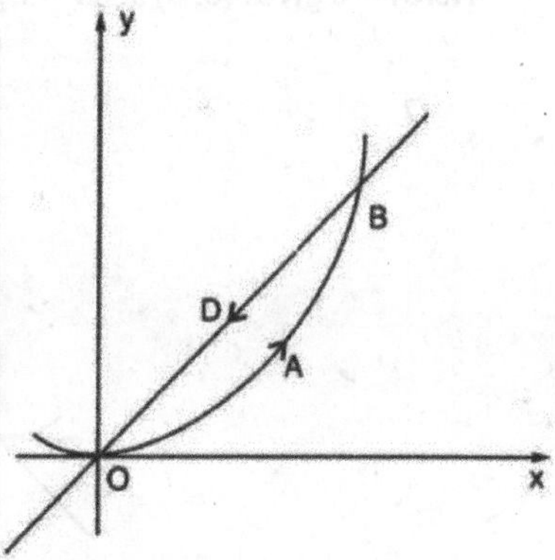

The parabola and the line intersect at the origin and at $B(1, 1)$. In Fig. 9.5. the region in question is shown by the figure OABDO. The parametric equations of the arc OAB are $x = t, y = t^2$, where t runs from 0 to 1 and the parametric equations of the line BDO are $x = t, y = t$, where t runs from 1 to 0. So

$$\oint_C xy\,dx + x^2 dy = \int_{OAB} xy\,dy + x^2 dy + \int_{BDO} xy\,dx + x^2 dy$$

$$= \int_0^1 (t)(t^2)dt + t^2(2t\;dt) + \int_1^0 (t)\,(t)dt + t^2 dt$$

$$= \int_0^1 3t^3 dt + \int_1^0 2t^2\;dt = \frac{1}{12}.$$

Now

$$\iint_R \left\{\frac{\partial(x^2)}{\partial x} - \frac{\partial(xy)}{\partial y}\right\} dx\,dy = \iint_R x\;dxdy$$

$$= \int_0^1 \int_{y=x^2}^{y=x} x\,dy\,dx = \int_0^1 \left(xy\Big|_{y=x^2}^{y=x} \right) dx$$

$$= \int_0^1 (x^2 - x^3)\,dx = \frac{1}{12}.$$

Thus the theorem is verified.

51. *Solved Sum.* Verify Green's theorem in plane for the integral $\int_C (xy+y^2)\,dx + x^2 dy$, where C is the closed curve enclosing the region R bounded by the parabola $y = x^2$ and the line $y = x$. **(1977, 82)**

The Green's Theorem is

$$\int_C M\,dx + N\,dy = \iint_R \left(\frac{\partial N}{\partial x} - \frac{\partial M}{\partial y}\right) dx\,dy. \qquad \text{... (1)}$$

Now $M = xy + y^2$ and $N = x^2$ and R is the region enclosed by C.

Solving $y = x^2$ and $y = x$, we get the common points of the parabola $y = x^2$ and the line $y = x$, as $O(0, 0)$ and $(1, 1)$ say B.

The parabola and the line intersect at the origin and at $B(1, 1)$. The region in question is shown by the figure $OABDO$, *i.e.*, the curve C consists of the parabolic arc OAB, say Γ and the line segment BDO, say Γ'. Now $C = \Gamma + \Gamma'$.

The parametric equations of Γ are $x = t, y = t^2$, where t runs from 0 to 1.

Here $t = 0$ gives $(0, 0)$ and $t = 1$ gives $(1, 1)$. Also $dx = dt$, $dy = 2t\,dt$.

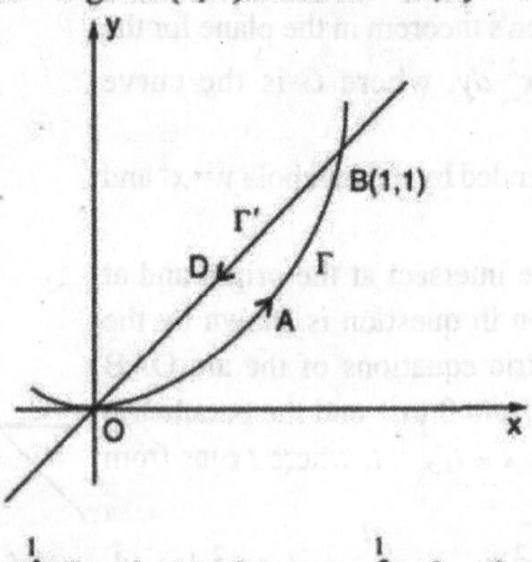

$$\therefore \int_\Gamma (xy+y^2)\,dx + x^2 dy = \int_0^1 (t^3 + t^4)\,dt + t^2(2t\,dt) = \int_0^1 (3t^3 + t^4)\,dt = \left[\frac{3t^4}{4} + \frac{t^5}{5}\right]_0^1 = \frac{3}{4} + \frac{1}{5} = \frac{19}{20}.$$

The parametric equations of Γ' are $x = t, y = t$, where t runs from 1 to 0.

Here $t = 1$ gives $(1, 1)$ and $t = 0$ gives $(0, 0)$. Also $dx = dt$, $dy = dt$

$$\therefore \quad \int_{\Gamma'} (xy+y^2)\,dx + x^2 dy = \int_1^0 (t^2 + t^2)\,dt + t^2 dt = \int_1^0 3t^2 dt = [t^3]_1^0 = -1.$$

Hence the integral along C is $\frac{19}{20} + (-1) = -\frac{1}{20}$. Next,

$$\iint_R \left\{\frac{\partial x^2}{\partial x} - \frac{\partial}{\partial y}(xy + y^2)\right\} dx\,dy = \iint_R (x - 2y)\,dx\,dy$$

$$= \int_{x=0}^{1} dx \int_{y=x^2}^{x} (x-2y)\,dy = \int_0^1 \left[xy - y^2\right]_{y=x^2}^{y=x} dx$$

$$= \int_0^1 \{(x^2-x^2)-(x^3-x^4)\}\,dx = \int_0^1 (-x^3+x^4)\,dx = \left[-\frac{x^4}{4}+\frac{x^5}{5}\right]_0^1 = \frac{-1}{4}+\frac{1}{5} = -\frac{1}{20}.$$

Thus the Green's theorem in plane stands verified.

52. Verify Green's theorem in plane for the integral $\int_C x^2 y\,dx + y\,dy$, where C is the curve enclosing the region R bounded by the line $y = x$ and the parabola $y^2 = x$. **(1973)**
[**Hint:** Common points: (0, 0), (1, 1); the line segment ($x = t, y = t$, where t runs from 0 to 1) lies below the parabolic arc ($x = t^2, y = t$, where t runs from 1 to 0). The integrals are

$$\int_0^1 (t^3+t)\,dt + \int_1^0 (2t^6+t)\,dt = -\frac{1}{28};$$

$$\int_{x=0}^{1} \int_{y=x}^{\sqrt{x}} (-x^2)\,dy\,dx = -\frac{1}{28};$$

since $M = x^2y$, $N = y$ and $N_x - M_y = -x^2$].

53. Verify Green's theorem for $\int_C x^2 dx + xy\,dy$, where C is the curve enclosing the region R bounded by the line $y = 2x$ and parabola $y^2 = 8x$.
[**Hint:** Common points: (0, 0), (2, 4); the line segment ($x = t, y = 2t$, where t runs from 0 to 2) lies below the parabolic arc ($x = 2t^2, y = 4t$, where t runs from 1 to 0). The integrals are

$$\int_0^2 5t^2 dt + \int_1^0 (16t^5+32t^3)\,dt = \frac{8}{3};\quad \int_{x=0}^{2}\int_{y=2x}^{\sqrt{8x}} y\,dy\,dx = \frac{8}{3}$$

since $M = x^2$, $N = xy$ and $N_x - M_y = y$]

54. *Solved sum.* Verify Green's theorem for

$$\int_C (3x^2-8y^2)\,dx + (4y-6xy)\,dy,$$

where C is the boundary of the region R enclosed by the parabolas $y = x^2$ and $y^2 = x$. **(1977)**
The Green's Theorem is

$$\int_C M\,dx + N\,dy = \iint_R \left(\frac{\partial N}{\partial x} - \frac{\partial M}{\partial y}\right) dx\,dy. \qquad (1)$$

Now $M = 3x^2 - 8y^2$ and $N = 4y - 6xy$ and R is the region enclosed by C.
Solving the equations

$$x^2 = y,\ y^2 = x,$$

we get the common points

$$O(0, 0),\ A(1, 1).$$

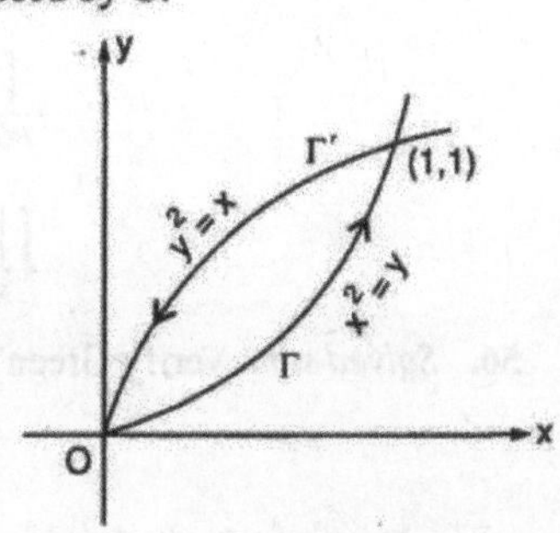

The parabolas intersect at (1, 1) besides (0, 0). Now C is composed of the arc Γ' of the parabola $y = x^2$ and the arc Γ' of the parabola $y^2 = x$. Now $C = \Gamma + \Gamma'$. The parametric equations of Γ: $x^2 = y$ are $x = t, y = t^2$, where t runs from 0 to 1. Here t = 0 gives (0, 0) and t = 1 gives (1, 1). Also $dx = dt$, $dy = 2t\,dt$.

$$\int_{\Gamma} = \int_0^1 (3t^2-8t^4)\,dt + (4t^2-6t^3)(2t\,dt)$$

$$= \int_0^1 (3t^2+8t^3-20t^4)\,dt = 1+2-4=-1.$$

The parametric equations of $\Gamma' : y^2 = x$ are $x = t^2, y = t$, where t runs from 1 to 0. Here $t = 1$ gives (1, 1) and $t = 0$ gives (0,0). Also $dx = 2t\,dt$, $dy = dt$.

$$\int_{\Gamma'} = \int_1^0 (3t^4-8t^2)(2t\,dt) + (4t-6t^3)\,dt$$

$$= \int_1^0 (6t^5-22t^3+4t)\,dt = -1+\frac{22}{4}-2 = \frac{5}{2} = 2\frac{1}{2}.$$

So the sum of the line integrals is $(-1)+2\frac{1}{2} = 1\frac{1}{2}$.

Since $M_y = -16y$, $N_x = -6y$, we get $N_x - M_y = 10y$. So

$$\iint_R \left(\frac{\partial N}{\partial x} - \frac{\partial M}{\partial y}\right) dx\,dy = \int_{x=0}^{1} \int_{y=x^2}^{y=\sqrt{x}} 10y\,dx\,dy$$

$$= \int_0^1 \left[5y^2\right]_{x^2}^{\sqrt{x}} dx = 5\int_0^1 (x-x^4)\,dx = 5\cdot\frac{3}{10} = \frac{3}{2}.$$

Thus Green's theorem in plane stands verified.

55. Verify Green's theorem for

$$\int_C (3x^2-8y^2)\,dx + (4y-6xy)\,dy,$$

where C is the boundary of the region R enclosed by

$$x = 0, y = 0, x + y = 1.$$

[**Hint:** If $OABO$ is the boundary,

OA: $x = t, y = 0$

AB: $x = 1-t, y = t$

BO: $x = 0, y = 1-t$,

where t runs from 0 to 1.

$$\int_{OA} = \int_0^1 3t^2\,dt = 1; \int_{AB} = \int_0^1 (-3+4t+11t^2)\,dt = \frac{8}{3};$$

$$\int_{BO} = \int_0^1 4(t-1)\,dt = -2. \text{ So their sum is } \frac{5}{3};$$

$$\iint_R = \int_{x=0}^{1} \int_{y=0}^{1-x} (-6y+16y)\,dy\,dx = \int_0^1 5(1-x)^2\,dx = \frac{5}{3}.]$$

56. *Solved sum*. Verify Green's theorem for

$$\int_C (x-2y)\,dx + x\,dy,$$

where C is the circle $x^2+y^2=1$ **(1986)**

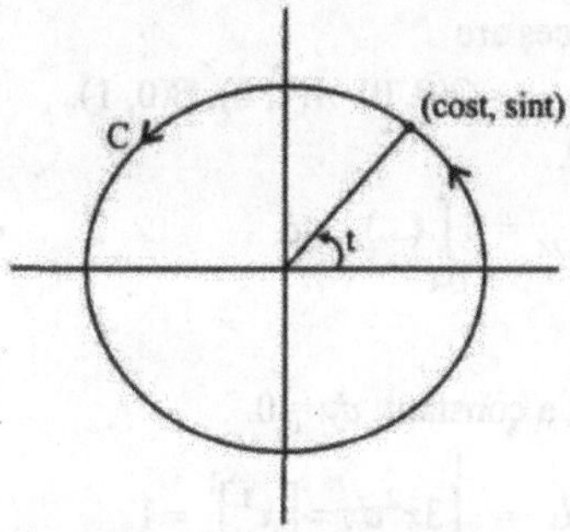

Green's theorem is

$$\int_C M\,dx + N\,dy = \iint_R (N_x - M_y)\,dx\,dy. \quad \text{... (1)}$$

In this, R is the region enclosed by C. Now $M = x - 2y$, $N = x$.

The parametric equations of the circle C are

$x \cos t$, $y = \sin t$; t runs from 0 to 2π.

$$\therefore \quad dx = -\sin t\,dt,\ dy = \cos t\,dt.$$

$$\text{LHS of (1)} = \int_0^{2\pi} (\cos t - 2\sin t)(-\sin t\,dt) + \cos t(\cos t\,dt)$$

$$= \int_0^{2\pi} (-\sin t \cos t + 2\sin^2 t + \cos^2 t)\,dt$$

$$= \int_0^{2\pi} \left[-\frac{\sin 2t}{2} + (1 - \cos 2t) + \frac{1 + \cos 2t}{2} \right] dt$$

$$= \int_0^{2\pi} \left[-\frac{\sin 2t}{2} + \frac{3}{2} - \frac{\cos 2t}{2} \right] dt$$

$$= 0 + \int_0^{2\pi} \frac{3}{2}\,dt = \frac{3}{2}(2\pi) = 3\pi. \quad \text{... (2)}$$

$$\left.\begin{aligned} M &= x - 2y \\ M_y &= -2 \end{aligned}\right\} \quad \left.\begin{aligned} N &= x \\ N_x &= 1 \end{aligned}\right\} \quad \therefore\ N_x - M_y = 3.$$

$$\text{RHS of (1)} = \iint_R 3\,dx\,dy = 3\iint_R 1\,dx\,dy$$

$$= 3\ (\text{area of the circular region } R)\ 3\,(\pi \cdot 1^2) = 3\pi. \quad \text{... (3)}$$

Equality of (2) and (3) verifies Green's theorem.

57. *Solved sum.* Verify Green's thoerem for

$$\int_C (3x^2 - 8y^2)\,dx + (4y - 6xy)\,dy,$$

where C is the boundary of the triangular area region R enclosed by the straight lines

$$y = 0,\ x + y = 1,\ x = 0.$$

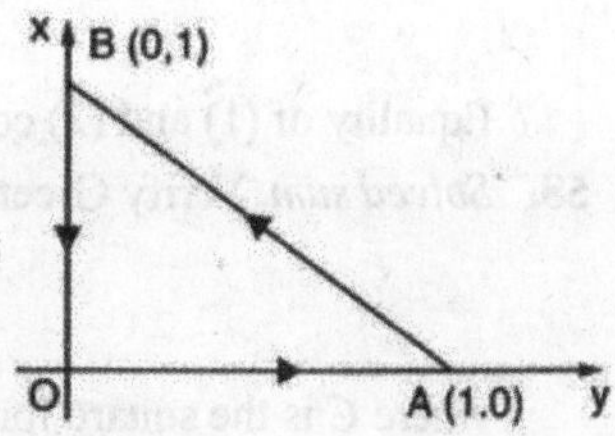

R is a triangle whose vertices are

$$O(0, 0), A(1, 0), B(0, 1).$$

Hence $C = OA + AB + BO$.

Let $$I_{OA} = \int_{OA} (\ldots), \text{ etc.}$$

On OA

x varies from 0 to 1, $y = 0$, a constant; $dy = 0$.

$$\therefore \quad \int_{OA} (3x^2 - 8y^2)\,dx = \int_0^1 3x^2\,dx = \left[x^3\right]_0^1 = 1.$$

On AB

Setting $x = t$ in $x + y = 1$, we get $y = 1 - t$. So, for AB,

$$\left.\begin{array}{rcl} x &=& t \\ y &=& 1-t \end{array}\right\} \left.\begin{array}{rcl} dx &=& dt \\ dy &=& -dt \end{array}\right] \left.\begin{array}{l} t=0 \text{ gives } (1,0) \\ t=0 \text{ gives } (0,1) \end{array}\right\}$$

$$\therefore \quad \int_{AB} M\,dx + N\,dy = \int_1^0 [3t^2 - 8(1-t)^2]\,dt + [4(1-t) - 6t(1-t)](-dt)$$

$$= \int_1^0 (-11t^2 + 26t - 12)\,dt$$

$$= -\left[-\frac{11}{3} + 13 - 12\right] = -\left[-\frac{8}{3}\right] = \frac{8}{3}.$$

On BO

$x = 0$, a constant; $dx = 0$ and y varies from 1 to 0.

$$\therefore \quad \int_{BO} (4y - 6xy)\,dy = \int_1^0 4y\,dy = \left[2y^2\right]_1^0 = -2.$$

$$\therefore \quad \int_C M\,dx + N\,dy = 1 + \frac{8}{3} - 2 = \frac{5}{3}. \qquad \ldots (1)$$

But $$M_y = -16y, N_x = -6y. \text{ Therefore } N_x - M_y = 10y. \text{ So}$$

$$\iint_R (N_x - M_y)\,dx\,dy = \int_0^1 \left[\int_0^{1-x} 10y\,dy\right] dx$$

$$= 10\int_0^1 \left[\frac{y^2}{2}\right]_0^{1-x} dx = 5\int_0^1 (1-x)^2\,dx$$

$$= -5\left[\frac{(1-x)^3}{3}\right]_0^1 = \frac{5}{3}. \qquad \ldots (2)$$

Equality of (1) and (2) completes the verification.

58. *Solved sum.* Verify Green's theorem for

$$\int_C x^2(1+y)\,dx + (y^3 + x^3)\,dy,$$

where C is the square formed by $x = \pm a, y = \pm a$.

Let C be the square $ABCD$ as in the figure and R be the region enclosed by C. Now Green's theorem is

$$\int_C M\,dx + N\,dy = \iint_R (N_x - M_y)\,dx\,dy.$$

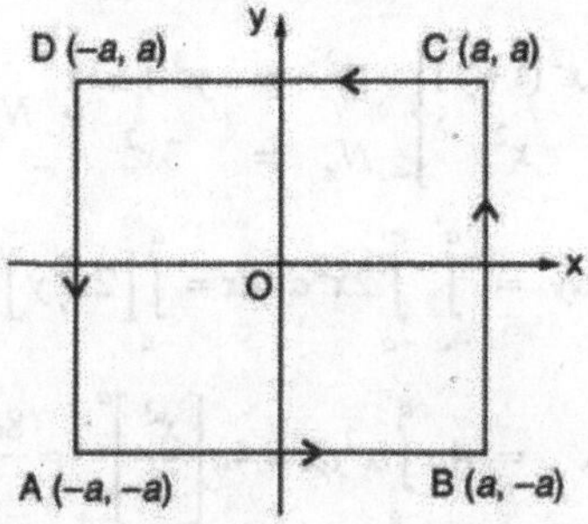

Let $\int_{AB} M\,dx + N\,dy = I_{AB}$, etc.

(*i*) **On AB:** x varies from $-a$ to a; $y = -a$, a constant ; $dy = 0$.

$$I_{AB} = \int_{-a}^{a} x^2(1+y)\,dx = \int_{-a}^{a} x^2(1-a)\,dx = 2(1-a)\int_0^a x^2\,dx$$

$$= 2(1-a)\left[\frac{x^3}{3}\right]_0^a = (1-a)\frac{2a^3}{3}.$$

(*ii*) **On BC:** $x = a$, a constant; $dx = 0$,
y varies from $-a$ to a.

$$I_{BC} = \int_{-a}^{a} (y^3 + x^3)\,dy = \int_{-a}^{a} (y^3 + a^3)\,dy$$

$$= \left[\frac{y^4}{4} + a^3 y\right]_{-a}^{a} = 2a^4.$$

(*iii*) **On CD:** x varies from a to $-a$.
$y = a$, a constant; $dy = 0$.

$$I_{CD} = \int_{a}^{-a} x^2(1+y)\,dx = \int_{a}^{-a} x^2(1+a)\,dx$$

$$= 2(1+a)\left[\frac{x^3}{3}\right]_0^{-a} = -(1+a)\frac{2a^3}{3}.$$

(*iv*) **On DA:** $x = -a$, a constant; $dx = 0$,
y varies from a to – a

$$I_{DA} = \int_{a}^{-a} (y^3 + x^3)\,dy = \int_{a}^{-a} (y^3 - a^3)\,dy$$

$$= \left[\frac{y^4}{4} - a^3 y\right]_{a}^{-a} = 2a^4.$$

$$\therefore \quad I_C = I_{AB}+I_{BC}+I_{CD}+I_{DA}$$

$$= (1-a)\frac{2a^3}{3}+2a^4-(1+a)\frac{2a^3}{3}+2a^4=\frac{8a^4}{3}. \qquad ...(1)$$

Next, we have

$$\left.\begin{aligned} M &= x^2(1+y) \\ M_y &= x^2 \end{aligned}\right\} \left.\begin{aligned} N &= y^3+x^3 \\ N_x &= 3x^2 \end{aligned}\right\} N_x - M_y = 2x^2.$$

$$\therefore \quad \iint_R 2x^2\,dx\,dy = \int_{-a}^{a}\int_{-a}^{a} 2x^2\,dy\,dx = \int_{-a}^{a}\left[2x^2y\right]_{-a}^{a}dx$$

$$= 4a\int_{-a}^{a} x^2 dx = 4a\left[\frac{x^3}{3}\right]_{-a}^{a} = \frac{8a^4}{3}. \qquad (2)$$

Equality of (1) and (2) completes the verification.

59. Verify Green's theorem for the integral

$$\int_C (x^2+y)\,dx - xy^2\,dy$$

taken around the boundary of the square with vertices $O(0, 0)$, $A(1, 0)$, $B(1, 1)$, $C(0,1)$.

[**Hint:** $\int M\,dx+N\,dy$ along OA, AB, BC, CO are $\frac{1}{3}, -\frac{1}{3}, -\frac{4}{3}, 0$ and $\iint_R (N_x-M_y)dx\,dy = -\frac{4}{3}$]

60. Verify Green's theorem for the integral

$$\int_C x^2y\,dx + y\,dy,$$

where C is the curve formed by $y = x$ and $y^2 = x$.

[**Hint:** Let C_1 be the straight line joining $O(0, 0)$ and $A(1, 1)$ and C_2 be the parabolic are from A to O. Then

$C_1 : x = t, y = t$, where t varies from 0 to 1.

$C_2 : x = t^2, y = t$, where t varies from 1 to 0.

$$\int_{C_1+C_2} = \int_0^1 (t^3+t)\,dt + \int_1^0 (2t^6+t)\,dt = -\frac{1}{28}.$$

$$\iint_R = \int_0^1 (-x^2)\int_x^{\sqrt{x}} dy\,dx = \int_0^1 (-x^{5/2}+x^3)\,dx = -\frac{1}{28}]$$

61. Show that Green's theorem in plane can be expressed in the vector form as $\int_C \mathbf{A}\cdot d\mathbf{r} = \iint_R (\nabla\times\mathbf{A})\cdot\mathbf{k}\,dx\,dy$, **where** $\mathbf{A} = M\mathbf{i}+N\mathbf{j}$.

4.5 Stokes' theorem

62. *Solved sum.* Verify stokes' theorem for $\mathbf{A} = (2x-y)\mathbf{i} - yz^2\mathbf{j} - y^2z\mathbf{k}$ taken over the upper half surface of the sphere $x^2+y^2+z^2 = 1$. $z \geq 0$ and the boundary curve C, the circle $x^2+y^2 = 1, z = 0$. **(1980, 81, 84, 86)**

Now we have to verify that

$$\int_C \bar{A}\cdot d\bar{r} = \iint_S (\nabla\times\bar{A})\cdot\hat{n}\,dS.$$

Method 1.

C lies in the xOy plane whose equation is $z = 0$. Since C is a circle of radius 1 its parametric equations are $x = \cos t$, $y = \sin t$, $z = 0$, where t varies from 0 to 2π.

$$\therefore \quad dx = -\sin t\, dt$$
$$dy = \cos t\, dt$$
$$dz = 0.$$

$$\therefore \quad \bar{A}\cdot d\bar{r} = A_1 dx + A_2 dy + A_3 dz$$
$$= (2x - y)\, dx - yz^2\, dy - y^2 z\, dz$$
$$= (2\cos t - \sin t)(-\sin t\, dt) + 0 + 0.$$

Method 2.

Vector equation of the circle is

$$r = \cos t\mathbf{i},\ y = \sin t\mathbf{j},\ \le 0 \le t \le 2\pi$$

because, on C, $x = \cos t$, $y = \sin t$ and $z = 0$. Hence on C $d\mathbf{r} = (-\sin t\mathbf{i} + \cos t\mathbf{j})\,dt$. Again, since $z = 0$ on C,

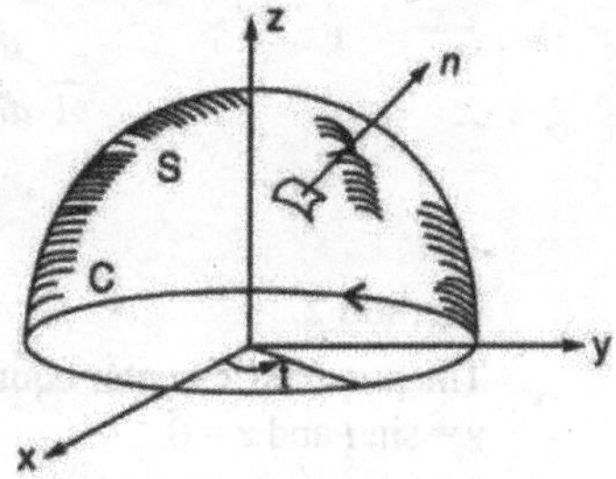

$$\mathbf{A} = (2x - y)\mathbf{i} = (2\cos t - \sin t)\,\mathbf{i}.$$

$$\therefore \quad \int_C \mathbf{A}\cdot d\mathbf{r} = \int_0^{2\pi} (2\cos t - \sin t)(-\sin t)\,dt$$

$$= \int_0^{2\pi} (-\sin 2t + \sin^2 t)\,dt$$

$$= \int_0^{2\pi} \left(-\sin 2t + \frac{1-\cos 2t}{2}\right)dt$$

$$= \left[\frac{\cos 2t}{2} + \frac{t}{2} - \frac{\sin 2t}{4}\right]_0^{2\pi} = \left[\frac{t}{2}\right]_0^{2\pi} = \pi.$$

But

$$\nabla \times \mathbf{A} = \begin{vmatrix} \mathbf{i} & \mathbf{j} & \mathbf{k} \\ \dfrac{\partial}{\partial x} & \dfrac{\partial}{\partial y} & \dfrac{\partial}{\partial z} \\ 2x - y & -yz^2 & -y^2 z \end{vmatrix} = \mathbf{i}(-2yz + 2yz) - \mathbf{j}(0) + \mathbf{k}(0 + 1) = \mathbf{k}.$$

On S,

$$\hat{n} = \frac{\nabla\phi}{|\nabla\phi|} = \frac{\nabla(x^2 + y^2 + z^2 - 1)}{|\nabla(x^2 + y^2 + z^2 - 1)|}$$

$$= \frac{2x\mathbf{i} + 2y\mathbf{j} + 2z\mathbf{k}}{\sqrt{4x^2 + 4y^2 + 4z^2}} = \frac{x\mathbf{i} + y\mathbf{j} + z\mathbf{k}}{\sqrt{x^2 + y^2 + z^2}}$$

$$= \frac{x\mathbf{i} + y\mathbf{i} + z\mathbf{k}}{\sqrt{1}} \quad \text{since the sphere is } x^2 + z^2 + z^2 = 1$$

$$= x\mathbf{i} + y\mathbf{j} + z\mathbf{k}.$$

$$\therefore \quad \iint_S (\nabla \times \mathbf{A})\cdot d\mathbf{S} = \iint_S \mathbf{k}\cdot\mathbf{n}\, dS = \iint_R \mathbf{k}\cdot\mathbf{n}\frac{dx\,dy}{\mathbf{k}\cdot\mathbf{n}}$$

where R is the projection of S on the xOy plane which is the circular area enclosed by C.

$$= \iint_R dx\,dy = \text{area of } R = \pi.1^2 = \pi.$$

Thus Stokes' theorem stands verified.

63. *Solved sum.* **Verify stoke's theorem for $\mathbf{A} = y\mathbf{i} + 2yz\mathbf{j} + y^2\mathbf{k}$ taken over the upper half surface of the sphere $x^2 + y^2 + z^2 = 1, z \geq 0$ and the bounding circle $C, x^2 + y^2 = 1, z = 0$.**

Now we have to verify that

$$\int_C \bar{A} \cdot d\bar{r} = \iint_S (\nabla \times \bar{A}) \cdot \hat{n}\, dS.$$

Method 1.

C lies in the xOy plane whose equation is $z = 0$. Since C is a circle of radius 1 its parametric equations are

$$x = \cos t,\ \ y = \sin t,\ \ z = 0,$$

where t varies from 0 to 2π.

$$\therefore \qquad dx = -\sin t\,dt,\ dy = \cos t\,dt,\ dz = 0$$

$$\therefore \qquad \bar{A} \cdot d\bar{r} = A_1\,dx + A_2\,dy + A_3\,dz$$

$$= y\,dx + 2yz\,dy + y^2\,dz$$

$$= (\sin t)(-\sin t\,dt) + 0 + 0.$$

Method 2.

The parametric vector equation of C is $\mathbf{r} = \cos t\mathbf{i} + \sin t\mathbf{j}, 0 \leq t \leq 2\pi$ because, on $C, x = \cos t$, $y = \sin t$ and $z = 0$.

$$\int_C \mathbf{A} \cdot d\mathbf{r} = \int_C (\sin t\mathbf{i} + \sin^2 t\mathbf{k}) \cdot (-\sin t\mathbf{i} + \cos t\mathbf{j})\,dt$$

$$= -\int_0^{2\pi} \sin^2 t\,dt = -\int \frac{1-\cos 2t}{2}\,dt = -\left[\frac{t}{2} - \frac{\sin 2t}{4}\right]_0^{2\pi} = -\pi.$$

It can be seen that
$$\nabla \times \bar{A} = \begin{vmatrix} \bar{i} & \bar{j} & \bar{k} \\ \frac{\partial}{\partial x} & \frac{\partial}{\partial y} & \frac{\partial}{\partial z} \\ y & 2yz & y^2 \end{vmatrix}$$

$$= \bar{i}(2y - 2y) - \bar{j}(0) + \bar{k}(0 - 1) = -\bar{k}$$

Let R be the circular area enclosed by C.Then R is the projection of S on xOy plane. But, by formula,

$$dS = \frac{dx\,dy}{|\bar{k} \cdot \hat{n}|}.$$

Therefore
$$\iint_S (\nabla \times \mathbf{A}).d\mathbf{S}\ \iint_S (\nabla \times \bar{A}) \cdot \hat{n}\,dS = \iint_S (-\bar{k}) \cdot \hat{n}\,dS = -\iint_R (\bar{k} \cdot \hat{n}) \frac{dx\,dy}{|\bar{k}.\hat{n}|}$$

$$= \iint_R dx\,dy = -(\text{area of } R)$$

$$= -(\pi \times 1^2) = -\pi.$$

since $\mathbf{k}.\ \mathbf{n}\ dS = dx\,dy$ and R, the projection of S on the xOy plane, is a circular area of radius 1. Thus the theorem is verified.

64. Verify Stokes' theorem for $\mathbf{A} = y\mathbf{i} + z\mathbf{j} + x\mathbf{k}$ for the surface S of the upper half of the sphere $x^2 + y^2 + z^2 = 1$ and for the boundary circle C. **(1982)**

[**Hint:** Equation of $C : \mathbf{r} = \cos t\,\mathbf{i} + \sin t\,\mathbf{j},\ 0 \le t \le 2\pi;\ x = \cos t,\ y = \sin t,\ z = 0,\ 0 \le t \le 2\pi$.

On C, $\mathbf{A}\cdot d\mathbf{r} = -\sin^2 t\,dt$ and $\int_C \mathbf{A}\cdot d\mathbf{r} = -\int_0^{2\pi} \sin^2 t\,dt = -\int_0^{2\pi} \frac{1-\cos 2t}{2}\,dt = -\pi.$

On S, $\mathbf{n} = x\mathbf{i} + y\mathbf{j} + z\mathbf{k}.\ dS = \frac{dxdy}{\mathbf{k}\cdot\hat{n}} = \frac{1}{z}\,dx\,dy.$ But $\nabla\times\mathbf{A} = -(\mathbf{i}+\mathbf{j}+\mathbf{k})$.

$$\iint_S (\nabla\times A)\cdot dS = \iint_S (\nabla\times\bar{A})\cdot\hat{n}\,dS = -\iint_R \frac{x+y+z}{z}\,dxdy$$

$$= -\iint_R \left(\frac{x+y}{z}+1\right)dx\,dy$$

$$= -\iint_R \left(\frac{x+y}{\sqrt{1-x^2-y^2}}+1\right)dx\,dy \qquad \text{since } x^2+y^2+z^2=1$$

$$= -\iint_R \frac{x+y}{\sqrt{1-x^2-y^2}}\,dx\,dy - \iint_R dx\,dy$$

$$= -\int_0^1\int_0^{2\pi} \frac{r}{\sqrt{1-r^2}}(\cos\theta+\sin\theta)r\,d\theta\,dr - \pi(1^2) = -\pi.]$$

65. Show that $\iint_S (\nabla\times\mathbf{A})\cdot d\mathbf{S} = 0$, where $\mathbf{A} = y^2\,\mathbf{i} + xz\,\mathbf{j} + yz\,\mathbf{k}$ and S is the curved surface enclosing the region specified by $z = 9 - (x^2 + y^2),\ z \ge 0$.

66. Show that, in the previous sum, the integral vanishes even if S is the curved surface enclosing the region specified by $x^2 + y^2 + z^2 - 4 = 0, z \ge 0$.

67. *Solved sum.* Verify Stokes' theorem for $\mathbf{A} = x^2\mathbf{i} + xy\mathbf{j}$ taken over the square surface S in the xOy plane whose vertices are $O(0, 0, 0)$, $A(a, 0, 0)$, $B(a, a, 0)$, $C(0, a, 0)$ and over its boundary. **(1981)**

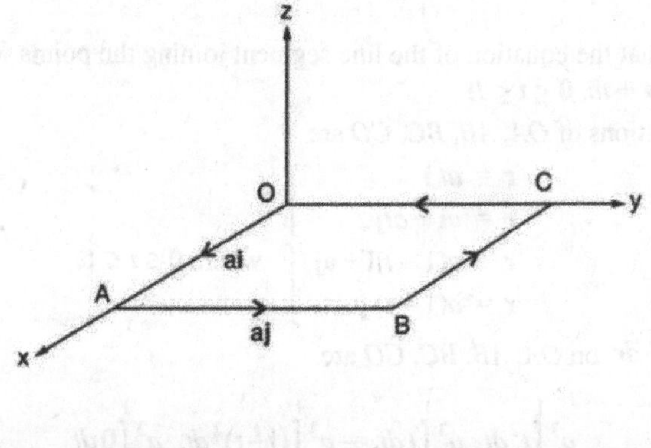

Method 1

(*i*) $(x, 0, 0)$ is a point on the x axis. As x varies from 0 to a, this point describes the line segment OA, from O to A. Thus we have

On OA

$$\bar{r} = x\bar{i} + 0\bar{j} + 0\bar{k}. \qquad \therefore \qquad d\bar{r} = dx\bar{i}.$$

$$\bar{A} = x^2\bar{i} + x(0)\bar{j} = x^2\bar{i}. \qquad \therefore \qquad \bar{A}\cdot d\bar{r} = x^2 dx.$$

$$\int_{OA} \bar{A}\cdot d\bar{r} = \int_0^a x^2 dx = \frac{a^3}{3}. \qquad \text{... (1)}$$

(*ii*) (*a*, *y*, 0) is a point on *AB*. As *y* varies from 0 to *a*, the point describes *AB*. Therefore

On AB

$$\bar{r} = a\bar{i} + y\bar{j} + 0\bar{k}. \qquad \therefore \qquad d\bar{r} = dy\bar{j}.$$

$$\bar{A} = a^2\bar{i} + ay\bar{j}. \qquad \therefore \qquad \bar{A}\cdot d\bar{r} = ay\,dy.$$

$$\int_{AB} \bar{A}\cdot d\bar{r} = \int_0^a ay\,dy = \frac{a^3}{2}. \qquad \text{... (2)}$$

(*iii*) (*x*, *a*, 0) is a point on *BC*. As *x* varies from a to 0, the point describes *BC*. Therefore

On BC

$$\bar{r} = x\bar{i} + a\bar{j} + 0\bar{k}. \qquad \therefore \qquad d\bar{r} = dx\bar{i}.$$

$$\bar{A} = x^2\bar{i} + xa\bar{j}. \qquad \therefore \qquad \bar{A}\cdot d\bar{r} = x^2 dx.$$

$$\int_{BC} \bar{A}\cdot d\bar{r} = \int_a^0 x^2 dx = \frac{-a^3}{3} \qquad \text{... (3)}$$

(*iv*) (0, *y*, 0) is a point on *CO*. As *y* varies from *a* to 0, the point describes *CO*. Therefore

On CO

$$\bar{r} = 0\bar{i} + y\bar{j} + 0\bar{k}. \qquad \therefore \qquad d\bar{r} = dy\bar{j}$$

$$\bar{A} = 0^2\bar{i} + (0)y\bar{j}. \qquad \therefore \qquad \bar{A}\cdot d\bar{r} = 0.$$

$$\therefore \int_{CD} \bar{A}\cdot d\bar{r} = 0. \qquad \text{... (4)}$$

Method 2.

We know that the equation of the line segment joining the points whose P.V.'s are **a**, **b** is $\mathbf{r} = (1-t)\,\mathbf{a} + t\mathbf{b}$, $0 \le t \le 1$.

So the equations of *OA*, *AB*, *BC*, *CO* are

$$\left.\begin{array}{ll} OA: & \mathbf{r} = at\,\mathbf{i} \\ AB: & \mathbf{r} = a\mathbf{i} + at\mathbf{j} \\ BC: & \mathbf{r} = a(1-t)\mathbf{i} + a\mathbf{j} \\ CO: & \mathbf{r} = a(1-t)\,\mathbf{j} \end{array}\right\} \text{ where } 0 \le t \le 1.$$

Hence $\int \mathbf{A}\cdot d\mathbf{r}$ on *OA*, *AB*, *BC*, *CO* are

$$a^3\int_0^1 t^2 dt,\ a^3\int_0^1 t\,dt,\ -a^3\int_0^1 (1-t)^2 dt,\ a^3\int_0^1 0\,dt$$

i.e.,
$$\frac{a^3}{3}, \frac{a^3}{2}, -\frac{a^3}{3}, 0.$$

$$\therefore \quad \int_{OABCO} \bar{A}\cdot d\bar{r} = \frac{a^3}{3}+\frac{a^3}{2}-\frac{a^3}{3}+0=\frac{a^3}{2}. \quad \text{... (5)}$$

Next, $$\nabla\times\bar{A} = \begin{vmatrix} \bar{i} & \bar{j} & \bar{k} \\ \frac{\partial}{\partial x} & \frac{\partial}{\partial y} & \frac{\partial}{\partial z} \\ x^2 & xy & 0 \end{vmatrix} = \bar{i}(0)-\bar{j}(0)+\bar{k}(y) = y\bar{k}.$$

Now S is the area of the square in xOy plane. So $dS = dx\,dy$ and $\hat{n}=\bar{k}$. Therefore

$$\iint_S (\nabla\times\bar{A})\cdot\hat{n}\,dS = \int_0^a\int_0^a (y\bar{k})\cdot\bar{k}\,dy\,dx$$

$$= \int_0^a\int_0^a y\,dy\,dx = \int_0^a\left[\frac{y^2}{2}\right]_0^a dx = \frac{a^2}{2}\int_0^a 1\,dx = \frac{a^3}{2}. \quad \text{... (6)}$$

Equality of (5) and (6) completes the verification.

Thus the Stokes' theorem stands verified.

68. Solved sum. Verify Stokes's theorem for $\mathbf{A} = xy\mathbf{i} + yz\mathbf{j} + zx\mathbf{k}$ taken over the triangular surface S in the plane $x+y+z=1$ bounded by the planes $x=0, y=0, z=0$ and over its boundary.

(1977, 79)

The vertices of the given triangular surface are $A(1, 0, 0)$, $B(0, 1, 0)$, $C(0, 0, 1)$. The equations of the sides AB, BC, CA are

$$\left.\begin{aligned} \mathbf{r} &= (1-t)\mathbf{i}+t\mathbf{j} \\ \mathbf{r} &= (1-t)\mathbf{j}+t\mathbf{k} \\ \mathbf{r} &= (1-t)\mathbf{k}+t\mathbf{i} \end{aligned}\right\}, 0\le t\le 1.$$

Integrals $\int \mathbf{A}\cdot d\mathbf{r}$ on AB, BC, CA are equal and due to symmetry they are $\int_0^1 (1-t)\,t\,(-dt)$.

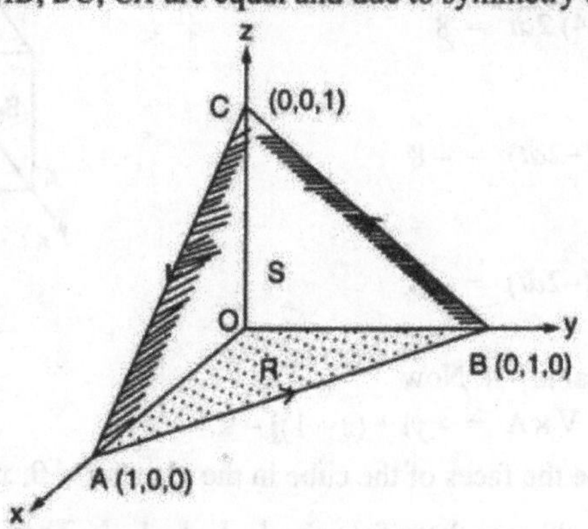

$\therefore$ the integral along the triangular perimeter is

$$3\int_0^1 (-t+t^2)dt = 3\left(-\frac{1}{2}+\frac{1}{3}\right) = -\frac{1}{2}.$$

$$\iint_S (\nabla\times\mathbf{A})\cdot d\mathbf{S} = \iint_S (\nabla\times\mathbf{A})\cdot\mathbf{n}\,dS = \iint_R (\nabla\times A)\cdot\mathbf{n}\frac{dxdy}{|\mathbf{n}\cdot\mathbf{k}|}$$

$$= -\iint_R (y+z+x)\,dx\,dy \qquad \because \mathbf{n} = (\mathbf{i}+\mathbf{j}+\mathbf{k})/\sqrt{3}$$

$$= \iint_R \{y+(1-x-y)+x\}\,dx\,dy \qquad (\because z = 1-x-y)$$

$$= -\iint_R dx\,dy = -\text{Area of } R = -\frac{1}{2}.$$

Thus the Stokes' theorem stands verified.

69. *Solved sum.* Verify Stokes' theorem for

$$\mathbf{A} = (y-z+2)\,\mathbf{i} + (yz+4)\,\mathbf{j} - xz\mathbf{k},$$

where S is the surface of the cube $x = 0, x = 2, y = 0, y = 2, z = 0, z = 2$ above the xOy plane (*i.e.*, excluding the face on the xOy plane). **(1983)**

Let the face on $z = 0$ be OABC (Fig 3.8.3). Then the equations of the edges OA, AB, BC, CO are

$$\left.\begin{aligned} OA:\ \mathbf{r} &= 2t\mathbf{i} \\ AB:\ \mathbf{r} &= 2\mathbf{i}+2t\mathbf{j} \\ BC:\ \mathbf{r} &= 2(1-t)\mathbf{i}+2\mathbf{j} \\ CO:\ \mathbf{r} &= 2(1-t)\mathbf{j} \end{aligned}\right\},\ 0 \le t \le 1.$$

The integrals $\int A\cdot d\mathbf{r}$ on OA, AB, BC, CO are

$$\int_0^1 2\,(2)\,dt = 4$$

$$\int_0^1 (4)\,2dt = 8$$

$$\int_0^1 (2+2)(-2dt) = -8$$

$$\int_0^1 4(-2dt) = -8.$$

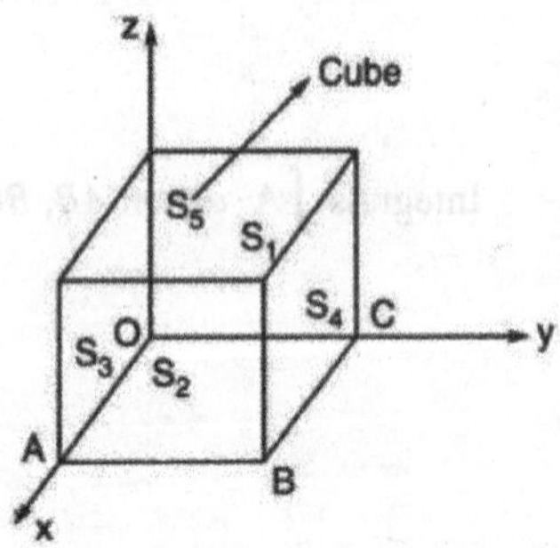

The sum of the integral is -4. Now

$$\nabla\times\mathbf{A} = -y\mathbf{i} + (z-1)\mathbf{j} - \mathbf{k}.$$

Let S_1, S_2, S_3, S_4, S_5 be the faces of the cube in the planes $x = 0, x = 2, y = 0, y = 2, z = 2$. Let the integrals $(\nabla\times\mathbf{A}).\,d\mathbf{S}$ over these faces be I_1, I_2, I_3, I_4, I_5. Then

$$I_1 = \int_0^2\int_0^2 \{-y\mathbf{i}+(z-1)\mathbf{j}-\mathbf{k}\}\cdot(-\mathbf{i})\,dy\,dz = \int_0^2\int_0^2 y\,dy\,dz = 4.$$

$$I_2 = \int_0^2\int_0^2 \{-y\mathbf{i}+(z-1)\mathbf{j}-\mathbf{k}\}\cdot(\mathbf{i})\,dy\,dz = -4.$$

$$I_3 = \int_0^2\int_0^2 \{-y\mathbf{i}+(z-1)\mathbf{j}-\mathbf{k}\}\cdot(-\mathbf{j})\,dz\,dx = \int_0^2\int_0^2 (-z+1)\,dz\,dx = 0.$$

$$I_4 = \int_0^2\int_0^2 \{-y\mathbf{i}+(z-1)\mathbf{j}-\mathbf{k}\}\cdot(\mathbf{j})\,dz\,dx = 0.$$

$$I_5 = \int_0^2\int_0^2 \{-y\mathbf{i}+(z-1)\mathbf{j}-\mathbf{k}\}\cdot(\mathbf{k})\,dx\,dy = -4.$$

Hence $I_1 + I_2 + I_3 + I_4 + I_5 = -4$. Thus the Stoke's theorem stands verified.

70. Verify Stoke's theorem for $\mathbf{F} = xz\mathbf{i} - y\mathbf{j} + x^2y\mathbf{k}$, where S is the surface of the region bounded by $x = 0, y = 0, z = 0, 2x + y + 2z = 8$ which is not included by the xOz plane.

71. By using Stokes' theorem evaluate the integral

$$I = \iint_C [(1+y)z\mathbf{i} + (1+z)x\mathbf{j} + (1+x)y\mathbf{k}]\cdot d\mathbf{r}$$

in the following cases:

(*i*) C is the circle $x^2 + y^2 = 1, z = 1$.

(*ii*) C is the triangle ABC, where A, B, C are $(1, 0, 0), (0, 1, 0), (0, 0, 1)$.

(*iii*) C is a closed curve in the plane $x - 2y + z = 1$.

[**Hint:** $I = \iint_S (\nabla\times\mathbf{A})\,.\,\mathbf{n}\,dS$, where $\nabla\times\mathbf{A} = \mathbf{i}+\mathbf{j}+\mathbf{k}$ and S is the plane-face spanning C.

(*i*) $\mathbf{n} = \mathbf{k}; I = \iint_S 1\,dS$ = area of the circle = π.

(*ii*) $\mathbf{n} = (\mathbf{i}+\mathbf{j}+\mathbf{k})/\sqrt{3}\,; I = \iint_S \frac{3}{\sqrt{3}}dS = \frac{3}{\sqrt{3}}(\text{area}) = \frac{3}{\sqrt{3}}\cdot\frac{\sqrt{3}}{2} = \frac{3}{2}.$

(*iii*) $\mathbf{n} = (\mathbf{i}-2\mathbf{j}+\mathbf{k})/\sqrt{6};\ I = \iint_S 0\,dS = 0.$]

72. Evaluate $\iint_S (\nabla\times\mathbf{A})\cdot d\mathbf{S}$, if $\mathbf{A} = y\mathbf{i} + z\mathbf{j} + x\mathbf{k}$, where S is the paraboloidal surface $x^2 + y^2 = 1 - z, z \geq 0$. **(1975, 84)**

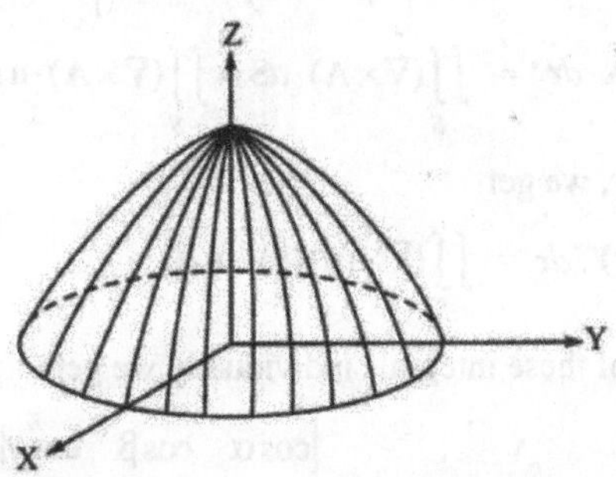

The boundary of the surface S is the circle C on the xOy plane whose equations are $x^2 + y^2 = 1, z = 0$.

Its parametric equations are

$$x = \cos t, y = \sin t, z = 0$$

Its vector equation is

$$\mathbf{r} = \cos t\,\mathbf{i} + \sin t\,\mathbf{j},\ 0 \le t \le 2\pi.$$

$$\therefore \quad \iint_S (\nabla\times\mathbf{A})\cdot d\mathbf{S} = \int_C \mathbf{A}\cdot d\mathbf{r}$$

$$= \int_0^{2\pi} (\sin t\,\mathbf{i} + 0\mathbf{j} + \cos t\,\mathbf{k})\cdot(-\sin t\,\mathbf{i} + \cos t\,\mathbf{j})\,dt$$

$$= -\int_0^{2\pi} \sin^2 t\,dt = -\frac{1}{2}\int_0^{2\pi}(1-\cos 2t)\,dt = -\pi.$$

73. Show that, if A is the area enclosed by a closed curve C in the xOy plane, then

(*i*) $A = \int_C x\,dy$; (*ii*) $-\int_C y\,dx = A$; (*iii*) $\frac{1}{2}\int_C x\,dy - y\,dx = A$.

[Hint:

$$\int_C x\,dy = \int_C (x\mathbf{j})\cdot d\mathbf{r} = \iint_S (\nabla\times x\mathbf{j})\cdot d\mathbf{S} = \iint_S \mathbf{k}\cdot d\mathbf{S} = A;$$

$$-\int_C y\,dx = -\int_C (y\mathbf{i})\cdot d\mathbf{r} = -\iint_S (\nabla\times y\mathbf{i})\cdot d\mathbf{S} = \iint_S \mathbf{k}\cdot d\mathbf{S} = A.]$$

74. By using Stokes' theorem show that $\nabla\times(\nabla\phi) = 0$.

[Hint: Let C be any closed curve and S, any surface spanning it. Then, setting $\mathbf{A} = \nabla\phi$ in Stokes' theorem,

$$\iint_S \nabla\times(\nabla\phi)\cdot d\mathbf{S} = \int_C (\nabla\phi)\cdot d\mathbf{r} = \int_C d\phi = \phi\,\Big|_C = 0.$$

Since S is arbitrary, the integrand $\nabla\times(\nabla\phi) = 0$.]

75. *Solved sum.* If C is the boundary of a plane surface S in space and $\cos\alpha\,\mathbf{i} + \cos\beta\,\mathbf{j} + \cos\gamma\,\mathbf{k}$, the constant unit normal $\mathbf{n}$ to S, then prove that the area of S is given by

$$S = \frac{1}{2}\oint_C \begin{vmatrix} dx & dy & dz \\ \cos\alpha & \cos\beta & \cos\gamma \\ x & y & z \end{vmatrix}$$

Now

$$\oint_C \mathbf{A}\cdot d\mathbf{r} = \iint_S (\nabla\times\mathbf{A})\cdot d\mathbf{S} = \iint_S (\nabla\times\mathbf{A})\cdot\mathbf{n}\,dS.$$

Replacing $\mathbf{A}$ with $\mathbf{n}\times\mathbf{r}$, we get

$$\oint_C (\mathbf{n}\times\mathbf{r})\,.\,d\mathbf{r} = \iint_S \{\nabla\times(\mathbf{n}\times\mathbf{r})\}\cdot\mathbf{n}\,dS$$

Evaluating the values of these integrals individually we get

$$\oint_C (\mathbf{n}\times\mathbf{r})\cdot d\mathbf{r} = \oint_C \begin{vmatrix} \cos\alpha & \cos\beta & \cos\gamma \\ x & y & z \\ dx & dy & dz \end{vmatrix}$$

And

$$\iint_S \{\nabla\times(\mathbf{n}\times\mathbf{r})\}\cdot\mathbf{n}\,dS = \iint_S \{(\mathbf{r}\cdot\nabla)\mathbf{n} - (\mathbf{n}\cdot\nabla)\mathbf{r} + (\nabla\cdot\mathbf{r})\mathbf{n} - (\nabla\cdot\mathbf{n})\mathbf{r}\}\cdot\mathbf{n}dS$$

$$= \iint_S \{0-(\mathbf{n}\cdot\nabla)\mathbf{r}+(\nabla\cdot\mathbf{r})\mathbf{n}-0\}\cdot\mathbf{n}\,dS$$

$$= \iint_S \{-\mathbf{n}+(3)\mathbf{n}\}\cdot\mathbf{n}\,dS = \iint_S 2\,dS = 2S$$

Since n is a constant vector and (n .∇) r equals

$$\left(\cos\alpha\frac{\partial}{\partial x}+\cos\beta\frac{\partial}{\partial y}+\cos\gamma\frac{\partial}{\partial z}\right)\mathbf{r}$$

$$= \cos\alpha\,\mathbf{i}+\cos\beta\,\mathbf{j}+\cos\gamma\,\mathbf{k} = \mathbf{n}.$$

So the result follows as stated.

76. *Solved sum.* Using Stoke's theorem, evaluate

$$\int_C (\sin z\,dx - \cos x\,dy + \sin y\,dz),$$

where C is the boundary of the rectangle

$$0 \le x \le \pi,\ 0 \le y \le 1,\ z = 3.$$

Stoke's theorem: $\int_C \bar{A}\cdot d\bar{r} = \iint_S (\nabla\times\bar{A})\cdot\hat{n}\,dS.$

S is a plane parallel to xOy plane. Since $\hat{n}$ is the unit vector normal to S, it is $\bar{k}$. Now

$$\bar{A} = \sin z\bar{i} - \cos x\bar{j} + \sin y\bar{k}.$$

$$\nabla\times\bar{A} = \begin{vmatrix} \bar{i} & \bar{j} & \bar{k} \\ \dfrac{\partial}{\partial x} & \dfrac{\partial}{\partial y} & \dfrac{\partial}{\partial z} \\ \sin z & -\cos x & \sin y \end{vmatrix}$$

$$= \bar{i}(\cos y - 0) - \bar{j}(0-\cos z) + \bar{k}(\sin x - 0).$$

So the given line integral is equal to

$$\iint_S (\nabla\times\bar{A})\cdot\hat{n}\,dS = \iint_S (\cos y\bar{i} + \cos z\bar{j} + \sin x\bar{k})\cdot\bar{k}\,dS$$

$$= \iint_S \sin x\,dS = \int_0^1\left[\int_0^\pi \sin x\,dx\right]dy = \int_0^1 [-\cos x]_0^\pi\,dy = \int_0^1 2\,dy = 2.$$

77. *Solved sum.* Verify Stoke's theorem for $\bar{F} = (x^2+y^2)\bar{i} - 2xy\bar{j}$ taken round the rectangle bounded by the lines $x = \pm a, y = 0, y = b.$

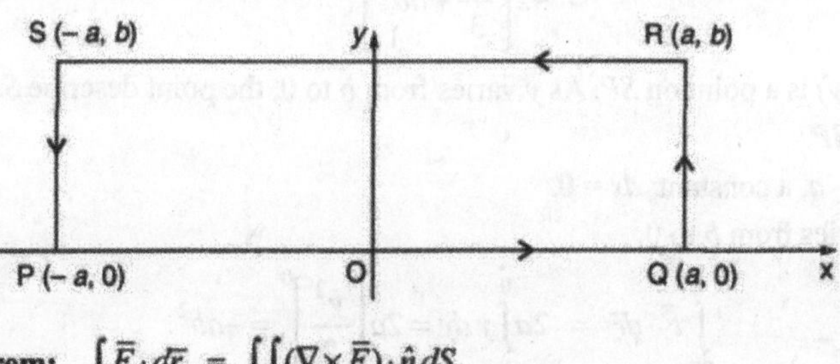

Stoke's theorem: $\int_C \bar{F}\cdot d\bar{r} = \iint_S (\nabla\times\bar{F})\cdot\hat{n}\,dS.$

Now $$\bar{F}\cdot d\bar{r} = [(x^2+y^2)\bar{i}-2xy\bar{j}]\cdot(dx\bar{i}+dy\bar{j})$$
$$= (x^2+y^2)dx - 2xy\,dy.$$

Let the vertices of the rectangle be

$$P(-a, 0), Q(a, 0), R(a, b), S(-a, b).$$

(*i*) Now $(x, 0)$ is a point on the x axis. As x varies from $-a$ to a, this point describes the line segment PQ, from P to Q. Therefore we have

On *PQ*

x varies from $-a$ to a,

$y = 0$, a constant; $dy = 0$.

$$\therefore \int_{PQ} \bar{F}\cdot d\bar{r} = \int_{PQ} (x^2+y^2)dx - 2xy\,dy = \int_{-a}^{a} x^2 dx$$

$$= 2\int_0^a x^2 dx = 2\left[\frac{x^3}{3}\right]_0^a = \frac{2a^3}{3}. \quad \text{... (1)}$$

(a, y) is a point on QR, As y varies from 0 to b,

(*ii*) The point describes QR. Therefore

On *QR*

$x = a$, a constant; $dx = 0$,

y varies from 0 to b.

$$\therefore \int_{QR} \bar{F}\cdot d\bar{r} = -2a\int_0^b y\,dy = -2a\left[\frac{y^2}{2}\right]_0^b = -ab^2. \quad \text{... (2)}$$

(*iii*) (x, b) is a point on RS. As x varies from a to $-a$, the point describes RS. Therefore

On *RS*

x varies from a to $-a$,

$y = b$, a constant; $dy = 0$

$$\therefore \int_{RS} \bar{F}\cdot d\bar{r} = \int_a^{-a} (x^2+b^2)dx = -\int_{-a}^{a} (x^2+b^2)dx$$

$$= -2\int_0^a (x^2+b^2)dx = -2\left[\frac{x^3}{3}+b^2x\right]_0^a$$

$$= -2\left[\frac{a^3}{3}+ab^2\right]. \quad \text{... (3)}$$

(*iv*) $(-a, y)$ is a point on SP. As y varies from b to 0, the point describe SP. Therefore

On *SP*

$x = -a$, a constant; $dx = 0$.

y varies from b to 0.

$$\therefore \int_{SP} \bar{F}\cdot d\bar{r} = 2a\int_b^0 y\,dy = 2a\left[\frac{y^2}{2}\right]_b^0 = -ab^2. \quad \text{... (4)}$$

Thus (1) + (2) + (3) + (4) gives

$$\int_{PQRSP} \bar{F}\cdot d\bar{r} = \frac{2a^3}{3} - ab^2 - 2\left(\frac{a^3}{3} + ab^2\right) - ab^2 = -4ab^2. \quad \text{... (5)}$$

The Stoke's theorem is

$$\int_C \bar{F}\cdot d\bar{r} = \iint_S [(\nabla\times\bar{F})\cdot\hat{n}]dS.$$

Now S is the plane area of the rectangle. So $\hat{n} = \hat{k}$ and $dS = dx\,dy$. But

$$\nabla\times\bar{F} = \begin{vmatrix} \bar{i} & \bar{j} & \bar{k} \\ \frac{\partial}{\partial x} & \frac{\partial}{\partial y} & \frac{\partial}{\partial z} \\ x^2+y^2 & -2xy & 0 \end{vmatrix}$$

$$= \bar{i}(0) - \bar{j}(0) + \bar{k}(-2y-2y) = -4y\bar{k}.$$

$$\therefore \quad \iint_S (\nabla\times\bar{F})\cdot\hat{n}\,dS = \iint_S (-4y\bar{k})\cdot\bar{k}\,dS$$

$$= -4\int_{-a}^{a}\left[\int_0^b y\,dy\right]dx = -4\int_{-a}^{a}\left[\frac{y^2}{2}\right]_0^b dx$$

$$= -4\int_{-a}^{a}\frac{b^2}{2}dx = -2b^2\int_{-a}^{a} 1dx$$

$$= -2b^2[x]_{-a}^{a} = -2b^2[a+a] = -4ab^2. \quad \text{... (6)}$$

Equality of (5) and (6) completes the verification.

78. Solved sum. Verify Stoke's theorem for the vector

$$\bar{F} = (x^2 - y^2)\bar{i} + 2xy\,\bar{j}$$

in the rectangular region in the xOy plane bounded by the lines $x = 0, x = a, y = 0, y = b$.

Stoke's theorem: $\int_C A\cdot d\bar{r} = \iint_S (\nabla\times\bar{A})\cdot\hat{n}dS.$

Now
$$\bar{F}\cdot d\bar{r} = \left[(x^2-y^2)\bar{i} + 2xy\bar{j}\right]\cdot(dx\,\bar{i} + dy\,\bar{j})$$
$$= (x^2-y^2)\,dx + 2xy\,dy.$$

Let the vertices of the rectangle be

$O(0, 0), A(a, 0),$
$B(a, b), C(0, b).$

On OA

x varies from 0 to a.

$y = 0$ a constant; $dy = 0$.

$$\therefore \quad \int_{OA} \bar{F}\cdot d\bar{r} = \int_0^a x^2 dx = \left[\frac{x^3}{3}\right]_0^a = \frac{a^3}{3}.$$

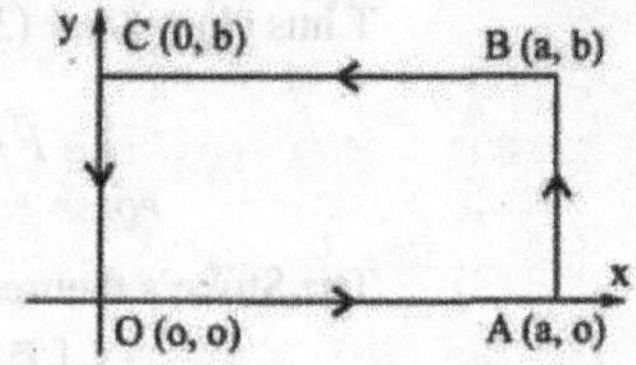

On AB

$x = a$, a constant; $dx = 0$,
y varies from 0 to b.

$$\therefore \int_{AB} \overline{F}\cdot d\overline{r} = \int_0^b 2ay\,dy = \left[ay^2\right]_0^b = ab^2$$

On BC

x varies from a to 0,
$y = b$, a constant; $dy = 0$.

$$\therefore \int_{BC} \overline{F}\cdot d\overline{r} = \int_a^0 (x^2 - b^2)\,dx = \left[\frac{x^3}{3} - b^2 x\right]_a^0 = ab^2 - \frac{a^3}{3}.$$

On CO

$x = 0$, a constant; $dx = 0$,
y varies from b to 0.

$$\therefore \int_{CO} \overline{F}\cdot d\overline{r} = \int_b^0 0\,dy = 0.$$

$$\therefore \int_C \overline{F}\cdot d\overline{r} = \frac{a^3}{3} + ab^2 + \left(ab^2 - \frac{a^3}{3}\right) + 0 = 2ab^2. \quad \text{... (1)}$$

Next we shall evaluate $\iint_S (\nabla\times\overline{F})\cdot\hat{n}\,dS$. Now S is the plane area of the rectangle. So $\hat{n} = \overline{k}$ and $dS = dx\,dy$. But

$$\nabla\times\overline{F} = \begin{vmatrix} \overline{i} & \overline{j} & \overline{k} \\ \frac{\partial}{\partial x} & \frac{\partial}{\partial y} & \frac{\partial}{\partial z} \\ x^2 - y^2 & 2xy & 0 \end{vmatrix} = \overline{k}(2y + 2y) = 4y\overline{k}.$$

$$\therefore \iint_S (\nabla\times\overline{F})\cdot\hat{n}\,dS = \int_0^a\int_0^b (4y\overline{k})\cdot\overline{k}\,dy\,dx$$

$$= \int_0^a\int_0^b 4x\,dy\,dx = \int_0^a \left[2y^2\right]_0^b dx$$

$$= \int_0^a 2b^2\,dx = \left[2b^2 x\right]_0^a = 2ab^2. \quad \text{... (2)}$$

Equality of (1) and (2) completes the verification.

79. Verify Stoke's theorem for the vector
$\overline{F} = (x^2 - y^2)\overline{i} + 2xy\overline{j}$ and the square region bounded by $x = 0, x = 2, y = 0, y = 2$.

80. Verify Stoke's theorem for $\overline{F} = (x^2 + y^2)\overline{i} + 2xy\overline{j}$ taken over the rectangle bounded by $x = 0$, $x = a, y = 0, y = b$.

Stoke's Theorem: $\int_C \overline{A}\cdot d\overline{r} = \iint_S (\nabla\times\overline{A})\cdot\hat{n}\,dS.$

Now $\quad \bar{F}\cdot d\bar{r} = [(x^2+y^2)\bar{i}+2xy\bar{j}]\cdot(dx\bar{i}+dy\bar{j})$

$= (x^2+y^2)\,dx + 2xy\,dy.$

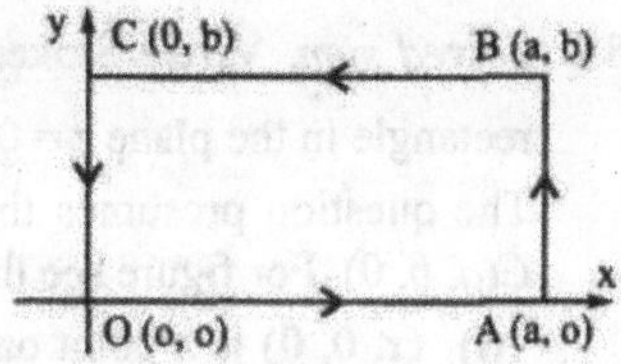

Let the vertices of the rectangle be $O(0, 0)$, $A(a, 0)$, $B(a, b)$, $C(0, b)$.

On OA

x varies from 0 to a,

$y = 0$, a constant; $dy = 0$.

$$\therefore \quad \int_{OA} \bar{F}\cdot d\bar{r} = \int_0^a x^2 dx = \left[\frac{x^3}{3}\right]_0^a = \frac{a^3}{3}.$$

On AB

$x = a$, a constant; $dx = 0$,

y varies from 0 to b.

$$\therefore \quad \int_{AB} \bar{F}\cdot d\bar{r} = \int_0^b 2ay\,dy = \left[ay^2\right]_0^b = ab^2.$$

On BC

x varies from a to 0,

$y = b$, a constant; $dy = 0$.

$$\therefore \quad \int_{BC} \bar{F}\cdot d\bar{r} = \int_a^0 (x^2+b^2)dx = \left[\frac{x^3}{3}+b^2x\right]_a^0 = -\left[\frac{a^3}{3}+ab^2\right].$$

On CO

$x = 0$, a constant; $dx = 0$,

y varies from b to 0.

$$\therefore \quad \int_{CO} \bar{F}\cdot d\bar{r} = \int_b^0 0\,dy = 0.$$

$$\therefore \quad \int_C \bar{F}\cdot d\bar{r} = \frac{a^3}{3}+ab^2-\left[\frac{a^3}{3}+ab^2\right]+0 = 0. \qquad \text{... (1)}$$

Next we shall evaluate $\iint_S (\nabla\times\bar{F})\cdot\hat{n}\,dS$. Now S is the plane area of the rectangle. So $\hat{n} = \bar{k}$ and $dS = dx\,dy$. But

$$\nabla\times\bar{F} = \begin{vmatrix} \bar{i} & \bar{j} & \bar{k} \\ \dfrac{\partial}{\partial x} & \dfrac{\partial}{\partial y} & \dfrac{\partial}{\partial z} \\ x^2+y^2 & 2xy & 0 \end{vmatrix}$$

$$= \bar{i}(0)-\bar{j}(0)+\bar{k}(2y-2y) = \bar{0}$$

$$\therefore \quad \iint_S (\nabla\times\bar{F})\cdot\hat{n}\,dS = \iint_S 0\,dS = 0. \qquad \text{... (2)}$$

Equality of (1) and (2) completes the verification.

Note. In the following two Examples, the vector equations of the line segments are used.

81. *Solved sum.* Verify Stokes theorem for the function $\overline{A}=x^2z\overline{i}+xy\overline{j}$ integrated along the rectangle in the plane $z=0$, where the sides are along the lines $x=0, y=0, x=a, y=b$.

The question presumes the vertices of the rectangle as $O(0, 0, 0)$, $A(a, 0, 0)$, $B(a, b, 0)$, $C(a, b, 0)$. For figure see the previous sum.

(*i*) $(x, 0, 0)$ is a point on the x axis. As x varies from O to a, this point describes the line segment OA, from 0 to A. Thus we have

On OA

$$\overline{r} = x\overline{i}+0\overline{j}+0\overline{k}. \quad \therefore d\overline{r}=dx\overline{i}$$

$$\overline{A} = x^2(0)\overline{i}+x(0)\overline{j} \quad \therefore \overline{A}\cdot d\overline{r}=0.$$

$$\therefore \int_{OA}\overline{A}\cdot d\overline{r}=0. \qquad \text{... (1)}$$

(*ii*) $(a, y, 0)$ is a point on AB. As y varies from 0 to b, the point describes AB. Therefore

On AB

$$\overline{r} = a\overline{i}+y\overline{j}+0\overline{k}. \therefore d\overline{r}=dy\overline{j}.$$

$$A = a^2(0)\overline{i}+ay\overline{j} \quad \therefore \overline{A}\cdot d\overline{r}=ay\,dy.$$

$$\therefore \int_{AB}\overline{A}\cdot d\overline{r} = \int_0^b ay\,dy=\frac{ab^2}{2}. \qquad \text{... (2)}$$

(*iii*) $(x, b, 0)$ is a point on BC. As x varies from a to 0, the point describes BC. Therefore

On BC

$$\overline{r} = x\overline{i}+b\overline{j}+0\overline{k}. \therefore d\overline{r}=dx\overline{i}$$

$$\overline{A} = x^2(0)\overline{i}+xb\overline{j}. \therefore \overline{A}\cdot d\overline{r}=0.$$

$$\int_{BC}\overline{A}\cdot d\overline{r}=0. \qquad \text{... (3)}$$

(*iv*) $(0, y, 0)$ is a point on CO. As y varies from b to 0, the point describes CO. Therefore

On CO

$$\overline{r} = 0\overline{i}+y\overline{j}+0\overline{k}. \quad \therefore \quad d\overline{r}=dy\,\overline{j}$$

$$\overline{A} = 0^2(0)\overline{i}+0(y)\overline{j}. \quad \therefore \quad \overline{A}\cdot d\overline{r}=0.$$

$$\int_{CO}\overline{A}\cdot d\overline{r} = 0. \qquad \text{... (4)}$$

$$\therefore \int_{OABCO}\overline{A}\cdot d\overline{r} = 0+\frac{ab^2}{2}+0+0=\frac{ab^2}{2}. \qquad \text{... (5)}$$

Next, we have

$$\nabla\times\overline{A} = \begin{vmatrix}\overline{i} & \overline{j} & \overline{k}\\ \frac{\partial}{\partial x} & \frac{\partial}{\partial y} & \frac{\partial}{\partial z}\\ x^2z & xy & 0\end{vmatrix} = \overline{i}(0)-\overline{j}(-x^2)+\overline{k}(y) = x^2\overline{j}+y\overline{k}.$$

Now S is the area of the rectangle in xOy plane. So $dS = dx\,dy$ and $\hat{n} = \hat{k}$. Therefore

$$\iint_S (\nabla\times\bar{A})\cdot\hat{n}\,dS = \int_0^a\int_0^b [x^2\bar{j} + y\bar{k}]\cdot\bar{k}\,dy\,dx$$

$$= \int_0^a\int_0^b y\,dy\,dx = \int_0^a\left[\frac{y^2}{2}\right]_0^b dx$$

$$= \frac{b^2}{2}\int_0^a 1\,dx = \frac{ab^2}{2}. \qquad (6)$$

Equality of (5) and (6) completes the verification.

82. Verify Stoke's theorem for $\bar{A} = xy\bar{i} + yz\bar{j} + x^2\bar{k}$ over the surface of unit cube $x = 0, y = 0, z = 0, x = 1, y = 1, z = 1$ above xOy plane.

[Hint: $\int_{OA}\bar{A}\cdot d\bar{r} = \int_{AB}\bar{A}\cdot d\bar{r} = \int_{CO}\bar{A}\cdot d\bar{r} = 0, \int_{BC}\bar{A}\cdot d\bar{r} = -\frac{1}{2}$.

Their sum is $-1/2$. Let S_1, S_2, S_3, S_4, S_5 be the faces lying in the planes $x = 1, x = 0, y = 1, y = 0, z = 1$.

Then the outward unit normals to them are

$$\bar{i}, -\bar{i}, \bar{j}, -\bar{j}, \bar{k}.$$

The integrals of $(\nabla\times\bar{A})\cdot\hat{n}\,dS$ on these faces are respectively $-1/2$, $1/2$, -1, 1, $-1/2$. Their sum $= -1/2$]

83. Verify Stoke's theorem for $\mathbf{A} = (2 - z)\,\mathbf{i} + yz\mathbf{j} - xz\mathbf{k}$, where S is the surface of the cube formed by the planes $x = 0, y = 0, z = 0, x = 1, y = 1, z = 1$ excluding the surface of the face which lies in $z = 0$.

84. *Solved sum.* Verify Stokes' theorem for $\bar{F} = y^2\bar{i} + y\bar{j} - xz\bar{k}$ over the upper half of the surface of the sphere $x^2 + y^2 + z^2 = a^2, z \geq 0$.

Now the Stoke's theorem is

$$\int_C \bar{A}\cdot d\bar{r} = \iint_S (\nabla\times\bar{A})\cdot\hat{n}\,dS.$$

Here C lies in the xOy plane whose equation is $z = 0$. Since C is a circle of radius a, its parametric equations are

$x = a\cos t, y = a\sin t, z = 0,$

where t varies from 0 to 2π.

$$\therefore \quad dx = -a\sin t\,dt,$$

$$dy = a\cos t\,dt,$$

$$dz = 0.$$

$$\therefore \quad \bar{A}\cdot d\bar{r} = y^2\,dx + y\,dy - xz\,dz$$

$$= (a\sin t)^2(-a\sin t\,dt) + (a\sin t)(a\cos t\,dt) - 0$$

$$\therefore \quad \int_C \bar{A}\cdot d\bar{r} = a^3\int_0^{2\pi}\sin^3 t\,dt + a^2\int_0^{2\pi}\sin t\cos t\,dt$$

$$= a^3 \int_0^{2\pi} \frac{3\sin t - \sin 3t}{4} dt + a^2 \int_0^{2\pi} \frac{\sin 2t}{2} dt = 0. \quad \text{... (1)}$$

since $\int_0^{2\pi} \sin nt \, dt = 0$, where n is an integer.

Now the projection R of S on the xOy plane is the circular area enclosed by C. We know that, if $\hat{n}$ is the unit vector normal to S,

$$\iint_S \phi \, dS = \iint_R \phi \frac{dx\,dy}{|\hat{n}\cdot\bar{k}|}.$$

$$\therefore \quad \iint_S [(\nabla\times\bar{A})\cdot\hat{n}]\,dS = \iint_R [(\nabla\times\bar{A})\cdot\hat{n}]\frac{dx\,dy}{|\hat{n}\cdot\bar{k}|}. \quad \text{... (2)}$$

But
$$\hat{n} = \frac{\nabla(x^2+y^2+z^2-a^2)}{|\nabla(x^2+y^2+z^2-a^2)|} = \frac{2x\bar{i}+2y\bar{j}+2z\bar{k}}{2|x\bar{i}+y\bar{j}+z\bar{k}|}$$

$$= \frac{x\bar{i}+y\bar{j}+z\bar{k}}{\sqrt{x^2+y^2+z^2}} = \frac{x\bar{i}+y\bar{j}+z\bar{k}}{a}$$

$$\nabla\times\bar{A} = \begin{vmatrix} \bar{i} & \bar{j} & \bar{k} \\ \frac{\partial}{\partial x} & \frac{\partial}{\partial y} & \frac{\partial}{\partial z} \\ y^2 & y & -xz \end{vmatrix}$$

$$= \bar{i}(0-0) - \bar{j}(-z-0) + \bar{k}(0-2y) = z\bar{j} - 2y\bar{k}.$$

$$\therefore \quad (\nabla\times\bar{A})\cdot\hat{n} = \frac{(z\bar{j}-2y\bar{k})\cdot(x\bar{i}+y\bar{j}+z\bar{k})}{a} = \frac{zy-2yz}{a} = -\frac{yz}{a}.$$

Also
$$\hat{n}\cdot\bar{k} = \frac{(x\bar{i}+y\bar{j}+z\bar{k})}{a} = \frac{z}{a}.$$

Thus (2) becomes

$$\iint_S (\nabla\times\bar{A})\cdot\hat{n}\,dS = -\iint_R \frac{yz}{a}\,\frac{dx\,dy}{z/a} = -\iint y\,dx\,dy = -A\bar{y} = 0, \quad \text{... (3)}$$

where A is the area of R and $\bar{y}$ is the y coordinate of the mass centre of the area R which is 0.

Thus equality of (1) and (3) completes the verification.

85. Solved Sum. Verify Stoke's theorem for $\bar{A} = y\bar{i} + 2yz\bar{j} + y^2\bar{k}$, taken over the upper half surface of the sphere $x^2 + y^2 + z^2 + 1$, $z \geq 0$, and the bounding circle C

$$x^2 + y^2 = 1, z = 0.$$

The parametric vector equation of C is

$$\bar{r} = \cos t\,\bar{i} + \sin t\,\bar{j}, 0 \leq t \leq 2\pi \text{ because on C, } x = \cos t, y = \sin t \text{ and } z = 0.$$

$$\therefore \quad \int_C \bar{A}\cdot d\bar{r} = \int_C (\sin t\,\bar{i} + \sin^2 t\,\bar{k})\cdot(-\sin t\,\bar{i} + \cos t\,\bar{j})\,dt$$

$$= -\int_0^{2\pi} \sin^2 t\, dt = -\pi. \qquad \text{... (1)}$$

It can be seen that $\nabla \times \bar{A} = -\bar{k}$. Therefore

$$\iint_S (\nabla \times \bar{A}) \cdot d\bar{S} = -\iint_S \bar{k} \cdot d\bar{S} = -\iint_S \bar{k} \cdot \hat{n}\, dS$$

$$= -\iint_R dx\, dy = -\pi 1^2 = -\pi \qquad \text{... (2)}$$

since $\bar{k} \cdot \hat{n}\, dS = dx\, dy$ and R, the projection of S on the xOy plane, is a circular area of radius 1. Thus the theorem is verified by (1) and (2).

86. Verify Stoke's theorem for $\bar{A} = y\bar{i} + 2yz\bar{j} + y^2\bar{k}$ taken over the hemisphere $x^2 + y^2 + z^2 = 1$, $z \geq 0$ and the bounding circle $x^2 + y^2 = 1, z = 0$.